Bridge Construction Equipment

Bridge Construction Equipment

Marco Rosignoli

Published by ICE Publishing, One Great George Street,
Westminster, London SW1P 3AA.

Full details of ICE Publishing sales representatives and distributors can
be found at:
www.icebookshop.com/bookshop_contact.asp

Reprinted 2016

Other titles from ICE Publishing:

Prestressed Concrete Bridges, Second edition.
N. Hewson. ISBN 978-0-7277-4113-4
Designers' Guide to Eurocode 1: Actions on bridges.
J.-A. Calgaro, M. Tschumi and H. Gulvanessian.
ISBN 978-0-7277-3158-6
Steel-concrete Composite Bridges, Second edition.
D. Collings. ISBN 978-0-7277-5810-1

www.bookshop.com

A catalogue record for this book is available from the British Library

ISBN 978-0-7277-5808-8

© Thomas Telford Limited 2013

ICE Publishing is a division of Thomas Telford Ltd, a wholly-owned
subsidiary of the Institution of Civil Engineers (ICE).

Commissioning Editor: Rachel Gerlis
Production Editor: Imran Mirza
Market Specialist: Catherine de Gatacre

Typeset by Academic+Technical, Bristol
Printed and bound in Great Britain by TJ International Ltd, Padstow

FSC
www.fsc.org
MIX
Paper from
responsible sources
FSC® C013056

Contents

Foreword

The modern bridge builder is, in many ways, confronted with challenges beyond those experienced by his predecessors. This may be most evident in bridges with a short-to-medium span range, where the complexity of the bridges has been substantially augmented by the geometrical demands of modern expressways and high speed railway lines.

Formerly, bridges were decisive for the layout of an entire infrastructure project, so that the alignment of the approaching roads or railways was chosen to make the bridge geometry simple, for example, by a perpendicular and straight crossing of the underlying obstacle. This situation is still valid for very large bridges with the longest spans, but in the short to medium span range the bridges will, in general, have to submit to the overall alignment chosen for the expressways or high speed railways. As a result, these bridges will often have to be built with a pronounced skewness and a varying horizontal and vertical curvature. Furthermore, the width of the bridge deck and its crossfall might have to be varied along the bridge length, for example, to allow bifurcations of the traffic lanes or change of horizontal curvature.

For many years, it was believed that concrete bridges, with a very complex geometry, could only be constructed if cast in place on individual ground scaffolding, but, through the development of more sophisticated bridge building machines, it has also become possible to mechanise the construction of bridges characterised by a complicated spatial geometry.

Many modern bridges for elevated highways, light rail transit or high speed railways are characterised by a large number of similar spans. In these cases, the traditional use of wooden scaffolding for the formwork of concrete superstructures becomes uneconomical as it is too labour intensive. Instead, the concrete superstructures are cast on-site with movable scaffolding systems allowing a large degree of repetitive use or cast as large precast elements on stationary formwork at a purpose-built work site area (or in a factory) and subsequently transported by sea or on land to the actual building site.

In the case of moderate spans, it might be possible to use precast elements comprising a full span and this will allow a very fast erection but it presents a number of problems in transporting very heavy elements from the casting yard to the erection site. If built across navigable waters the elements can be transported on barges and erected by floating cranes or special lifting equipment. On land, heavy elements can be transported on multi-wheel trailers or along the completed spans and erected with overhead or underslung gantries.

For long span bridges, precast elements cannot comprise a full span but only segments to be joined at the site.

On-site casting of full spans can be mechanised by using movable scaffolding systems and, in the case of balanced cantilever construction, overhead or underslung form travelers can imply repetitive use of the equipment.

Application of gantries or form travelers will, in general, require clearances under or over the bridge superstructure beyond what is required for the completed bridge. To allow a rational mechanised construction it is, therefore, important that additional geometrical demands in the construction phase are taken into account by the conceptual design phase. Here, it is also important to consider that the dimensions of some structural elements might be influenced by loading conditions in the construction phase, for example, when precast elements or erection equipment is moved to a new position along the completed spans. Also, the critical lateral load might occur when an overhead gantry is positioned on top of the superstructure.

In many cases, the early conceptual designs are prepared by architects, with engineers only considering the final structure – possibly due to a lack of knowledge about the demands imposed by mechanised construction methods. Consequently, this publication fulfils the need for a publication describing the construction methods and the equipment used in modern mechanised bridge building.

Niels Jørgen Gimsing
Emeritus Professor and Bridge Design Consultant,
Department of Civil Engineering,
Technical University of Denmark, Denmark

Preface

The technological aspects of construction influence the modern bridge industry from the very first steps of design. Entire families of bridges, such as the launched bridges, the span-by-span bridges and the balanced cantilever bridges, take their names from the construction method. The full-span method so frequently applied in high-speed rail projects is another example.

Bridge industry is moving to mechanised construction because this saves labour, shortens project duration and improves quality. This trend is evident in many countries and affects most construction methods. Mechanised bridge construction is based on the use of specialised bridge erection equipment.

Beam launchers are used to erect precast beams. Self-launching gantries and lifting frames are used to erect precast segmental bridges. Movable scaffolding systems (MSSs) and form travellers are used for in-place casting of spans and segments of prestressed concrete (PC) bridges. Forming carriages are used for segmental casting of the concrete slab of composite bridges. Portal carriers with underbridge and span launchers fed by tyre trolleys are used for ground transportation and placement of precast spans. Lifting platforms are used to hoist macro-segments for suspension bridges. Alternative configurations of machines are also available for most construction methods.

New-generation bridge construction machines are complex and delicate structures. They handle heavy loads on long spans under the same constraints that the obstruction to overpass exerts onto the bridge. Safety of operations and quality of the final product depend on complex interactions between human decisions, structural, mechanical and electro-hydraulic components, control systems, and the bridge being erected.

In spite of their complexity, these machines must be as light as possible. Weight governs the initial investment, the cost of shipping and site assembly, the erection stresses, and sometimes even the cost of the bridge. Weight limitation dictates the use of high-grade steels and designing for high stress levels in different load and support conditions, which makes these machines potentially prone to instability.

Bridge erection equipment is assembled and decommissioned many times, under different conditions and by different crews. It is modified, reconditioned and adapted to new work conditions. Connections and field splices are subjected to hundreds of load reversals. The nature of loading is often highly dynamic, the equipment may be exposed to strong wind, and the full design load is reached multiple times and sometimes exceeded. Impacts are not infrequent, vibrations may be significant, and most machines are actually quite lively (they deflect, vibrate and exhibit highly dynamic behaviour) because of their high structural efficiency.

Movement adds the very important complication of variable geometry. Loads and support reactions are applied eccentrically, the support sections are often devoid of diaphragms, and most machines have flexible support systems. Indeed, such design conditions are almost inconceivable in permanent structures subjected to such loads.

The level of sophistication of new-generation machines requires adequate technical culture. Long subcontracting chains may lead to loss of communication, problems not dealt with during planning and design must be solved on site, the risks of incorrect operation are not always evident in so complex and sophisticated machines, and human error is the prime cause of accidents.

Experimenting with new solutions without due preparation may lead to catastrophic results. Several bridge erection machines have collapsed over the years, leading to a heavy cost in terms of fatalities, wounding, damage to property, delays in the project schedule and legal disputes. Technological improvement alone cannot guarantee a decrease in failures of bridge construction equipment, and may even increase them. Only a deeper consciousness of our human and social responsibilities can lead to a safer work environment. A level of technical culture adequate for the complexity of mechanised bridge construction would save human lives and facilitate the decision-making processes with more appropriate risk evaluations.

In a perfect world, bridge construction equipment would be purchased to meet clear performance requirements, be designed according to international standards and project-specific technical specifications, be subjected to independent design checking, be fabricated and commissioned within quality assurance/quality control procedures, and be operated by experienced supervisors and trained crews according to procedures issued by the manufacturer.

However, we do not live in a perfect world. Bridge construction equipment is often purchased by procurement personnel who have only a vague idea about what they are buying and who tend to recommend decisions to management based on the only aspect they can compare – the cost. The final cost for the contractor is typically higher than the figure written at the end of the offer, and the overall value of two apparently similar machines may also be pretty different. Inspection may clarify whether a machine is in good condition or is a freshly repainted pile of rust; however, other aspects influence the value of a machine. The demands on labour and crane usage of site assembly of the machine, for example, may be a bitter surprise if hundreds of field splices are designed with friction bolts and lap plates instead of through pins or stressed bars.

Because of its weight and dimensions, bridge construction equipment is assembled on site. A rule of thumb is that site assembly may take 7–10 hours of labour per metric ton of steel, and two cranes for the entire period. This means a lot of money – so the contractor frequently takes care of the costs of labour and

cranes for assembly to save the supplier's overheads. Identifying
what should be assembled in the workshop (paid for with the cost
of the machine) and what should be done on the site (paid for by
the contractor) may lead to interesting discussions if not specified
in the contract. Most of us would guess that primary hydraulic
systems should be assembled and tested in the workshop and only
the hoses should be applied on site. However, the author has
witnessed the building of all the hydraulic systems of an 800 ton
MSS on site – from bending cold-drawn steel tubing to painting –
because the contract did not explicitly itemise this work within the
list of workshop tasks. The result was that, in addition to the extra
cost to the contractor and a 2-month delay in the project schedule,
a delay penalty was shifted into an accelerated delivery fee.

The absence of comprehensive design standards further
complicates the situation. Although construction is the most
critical moment in the lifetime of a bridge and poor workmanship
may irremediably affect quality and durability, bridge
construction equipment does not receive any attention or research
funding. Controlling bridge design with state-of-the-art standards
is just the first step: assuring quality and durability of the final
product requires similar levels of attention and control during
construction.

Safety is another hidden problem. In several industrialised
countries the loss of lives during bridge construction is one order
of magnitude higher than the loss of lives due to structural failure
of bridges in operation. As bridge construction takes a few years
and bridge service covers decades, the risks to bridge workers are
two orders of magnitude higher than the risks to the bridge users.
All of this should suggest a need for deep reflection.

In an imperfect world machine operations are also uncertain.
A machine that is overly complex to operate may lead to slow
progress in construction, and thus lead the site supervisor to take
risky shortcuts to keep to the schedule. The rate of progress may
even be so slow as to force the contractor to purchase a second
machine. It is therefore in the best interests of the contractor to
bind the supplier to a stated performance, productivity and labour
demand.

If the machine is brand new, procurement may be even more
complex. Who will identify the performance requirements and
technical specifications for such an expensive piece of equipment?
Who will review and compare the suppliers' proposals? Who will
make an independent check of the design? Where will the machine
be fabricated if there are heavy import duties? Who will audit the
manufacturer's quality control processes during fabrication? Who
will supervise load testing and site commissioning? Who will
inspect the machine during operations?

Mechanised bridge construction is widespread all over the world.
When comparing different proposals, one often notices large cost
differences between apparently similar machines. However, many

aspects should be considered when making such a comparison: the different average quality of steelwork in different countries, the different weight resulting from steel grade and structural efficiency, the degree of mechanisation and access, the durability and energy efficiency, modularity and ease of reconditioning for future reuse, and ease of shipping and site assembly.

Decisions about bridge construction equipment are transdisciplinary in nature. Safety is the first concern, performance and productivity govern planning and investments, structure – equipment interaction affects the design of bridge and special equipment, risk mitigation is a major issue for contractors and insurance carriers, and quality assurance/quality control of design, fabrication and operation is closely related.

Little has been written about these machines, in spite of their fundamental role in the modern bridge industry. The purpose of this book is to convey multidisciplinary information on special equipment and bridge construction technologies to engineering professionals who are involved in mechanised bridge construction. Bridge owners, architects and engineers will find comprehensive information on how the bridge will be built and how it will interact with special equipment during construction. Educators will find a solid technological basis for advanced courses of bridge architecture and engineering. Contractors will find information on procurement, operations, performance and productivity of special equipment and a guide to value engineering, time scheduling, risk analysis, bidding, safety planning and the formation of management and site personnel for the risks (and the advantages) of mechanised bridge construction. Designers and manufacturers of special equipment will find comprehensive information on design loads and load combinations, calibration of load and resistance factors, design for robustness and redundancy, numerical modelling and analysis, out-of-plane buckling and prevention of progressive collapse, human error, failure of materials and systems, repair, reconditioning and industry trends. Erection engineers, resident engineers, inspectors and safety planners will find information on operations, casting cycles, cycle times, loading and structure–equipment interaction. Forensic engineers will find numerous case studies on equipment failure.

The book identifies all the aspects of special equipment: procurement (what you buy), design, fabrication and commissioning (what you get), operations and structure–equipment interaction (what you should do), and things that have gone wrong and forensics (what you can miss).

The book is divided into three sections. The first, comprising Chapters 2 to 8, is concerned with the technological aspects of bridge construction and the technical features of related equipment. Each of these chapters has the same structure, describing: bridge construction technology; details of special equipment; loading, kinematics and typical features; support, launch and lock systems; performance and productivity; and structure–equipment interaction.

Many machines share conceptually similar components, which are discussed throughout the book rather than in specific chapters. This may make searching for specific information more difficult, but should make reading the book more progressive and interesting. Specific information may be retrieved by searching the index.

Chapter 1 introduces the reader to the content of the book and identifies the typical features of bridge construction equipment. Chapter 2 deals with beam launchers and shifters, and Chapter 3 describes the self-launching gantries for span-by-span erection of precast segments. Chapter 4 looks at overhead gantries for macro-segmental construction – a hybrid erection method between segmental precasting and in-place casting for medium-span box girders.

Chapter 5 is devoted to the different types of MSS for span-by-span in-place casting – twin-girder overhead MSSs, single-girder overhead MSSs, modular single-truss overhead MSSs for long spans, and underslung MSSs are discussed in detail. Chapter 6 deals with the self-launching forming carriages used for in-place segmental casting of the concrete slab of steel bridges.

The special equipment required for balanced cantilever construction is discussed in Chapter 7. Precast and in-place segmental bridges are based on the same design principles and the construction equipment is also conceptually similar. Chapter 7 deals with lifting frames, cable cranes, lifting platforms for suspension bridges, self-launching gantries for precast segmental bridges, form travellers and suspension-girder MSSs for in-place casting, and special travellers for arches and cable-stayed bridges. Finally, Chapter 8 deals with self-propelled modular transporters (SPMTs), portal carriers with underbridge and span launchers fed by tyre trolleys for full-span precasting.

The second set of chapters deals with the design of bridge construction equipment. Chapter 9 illustrates design loads and load combinations for MSSs and heavy lifters, and Chapter 10 details specific issues of structural analysis and design such as numerical modelling, analysis of instability, robustness, redundancy, material-related failures, and design of connections and field splices.

The third and final set of chapters discusses contractual aspects. Chapter 11 deals with risk management, procurement and quality assurance/quality control of fabrication and operations, and Chapter 12 discusses the forensic aspects of structural and functional failure of bridge construction equipment as illustrated through a number of case studies.

This book does not deal with the construction equipment required for launched bridges. In this case the deck itself works like an erection machine during construction, and the design issues related to this construction method are therefore very particular. The design and construction of launched bridges are discussed in the author's book *Bridge Launching*, also published by Thomas Telford (2002).

The author would welcome comments and dialogue at marcorosignoli.BCE@gmail.com or through his LinkedIn group: Bridge Construction Equipment.

Acknowledgements

There are many people I would like to thank for their contribution to this work.

My first thoughts go to my family – my wife Carla, my daughter Chiara and my son Luca – who in these last 8 years have accepted as a normal thing that I spent almost all of my weekends working on a book instead of doing something funnier with them. And since, at just 28, my daughter is a brilliant bridge engineer with some great publications in her background and has also prepared the drawings for this book, I guess I must apologise also for this insane influence.

Many other people have helped me throughout my 30-year career as a bridge engineer: employers, clients, and designers of bridges and bridge erection equipment for which I performed the independent design checking. I am thankful for what I learned from them and also for their kind authorisation to publish photographs and information related to those assignments in this book.

This book is not related to my activity as founder and chairman of the International Association for Bridge and Structural Engineering (IABSE) working group WG-6 Bridge Construction Equipment. It is undeniable, however, that several members of WG-6 provided precious information and suggestions for this book. I got good ideas from the WG-6 Seminar State-of-the-art Bridge Deck Erection: Safe and Efficient Use of Special Equipment, which I had the privilege of chairing in November 2010 in Singapore, and I also got good ideas from the publications on bridge construction equipment that were coordinated and peer reviewed by WG-6 for the 4/2011, 2/2012 and 3/2012 issues of *Structural Engineering International*. The book by Robert Ratay on forensic engineering practice in the USA (*Forensic Structural Engineering Handbook*, 2nd edition, McGraw Hill Professional, 2009) also provided precious guidance in reordering Chapter 12.

Many firms provided photographs for this book. I have acknowledged their help in the captions and in the list that follows, but this is far from being enough. I am sincerely thankful for the material and not less honoured for their trust. The photographs without credits derive from my professional activity and show machines of Alpi, Comtec, Deal, De Nicola, Hünnebeck-Röro, MCT, Sercam, Sica, SPIC and Wito, taken within assignments from Bonatti, Codelfa, Cogefar, Cogefar-Impresit, Italstrade-Torno, Modena Scarl and Pizzarotti. The copyright of the other photographs belongs to the following firms:

- Antonio Povoas – Bridge Construction Services (AP-BCS)
- BERD
- Beijing Wowjoint Machinery (BWM)
- Comtec
- DEAL
- De Nicola
- Deutsche Bahn

- Doka
- Fagioli
- Freyssinet
- HNTB (the gantry in most of these photographs was manufactured by Deal)
- NRS (the traveller shown in Figure 7.42 was also manufactured by NRS)
- OVM International (OVM)
- SDI
- Strukturas
- Systra
- ThyssenKrupp
- VSL International (VSL).

The numerical models illustrated in this book have been implemented on ADINA, LUSAS and SAP2000 platforms.

Notation

a_g peak ground acceleration corresponding to a 475-year return period

a_{gc} peak ground acceleration of the seismic design event during construction

a_{mean} mean longitudinal acceleration of the suspension point of payload

$A_{A.eq}$ equivalent cross-sectional area of the assembly for stressed bars

A_C area of the concrete slab

A_S area of the steel girder

b breadth of member

b_R effective breadth of rail

C_{eq} carbon equivalent

d diameter

d_B notional diameter of the bolt

$d_{B.min}$ minimum diameter of the bold at the base of the screw thread

d_{CS} diameter of the contact surface of the bolt nut

$d_{D.min}$ minimum diameter of winch drum and sheaves

d_H diameter of the hole lodging the bar

d_{HR} notional diameter of the hoisting rope

d_W diameter of the wheels of crane bogies

E_C elastic modulus of concrete

$E_{C.c}$ elastic modulus of concrete during the cooling phase

$E_{C.h}$ elastic modulus of concrete during the heating phase

E_S elastic modulus of steel

f_r resonance frequency

f_t tensile strength

$f_{t,eff}$ effective tensile strength

f_y nominal yield strength

F characteristic force or load

F_B tensile force applied to the bar, longitudinal buffer force

F_D design inertial force, longitudinal drive force

F_L longitudinal force

F_ϕ characteristic value of dynamic crane action

G shear modulus

h height from the ground

I_i moment of inertia about the axis of rotation

I_2 moment of inertia about the vertical principal axis

J torsion constant

K_P spectral factor

L suspension length of load, span length of the bracing system

L_e effective length of compression member

L_{tot} total nut-to-nut length of the clamped assembly

m_C mass of crane and hoist load

M bending moment

M_D mass of design payload

M_E equivalent mass of components in movement

M_L mass of elements in longitudinal movement

n	number of rolls or wheels
n_D	number of single wheel drives on each runway
n_i	number of load applications
n_{tot}	total number of work cycles
p	probability of exceeding the seismic design event during construction
p_T	pitch of the thread
P	axial force, payload
P_D	design hoist load
P_i	generic load level
P_L	compression load in the tower column
P_V	vertical load applied to the runway
q	distributed self-weight
q_{cr}	critical distributed load for out-of-plane buckling
Q_i	force effect
r	radius of gyration
R_n	nominal resistance
S_B	spring constant of the buffer
t	thickness of member
t_c	duration of construction of the bridge
t_{rc}	return period of the seismic design event during construction
t'	notional thickness of laminated shapes
T	tension force
T_L	longitudinal oscillation period
T_{mean}	mean duration of acceleration or deceleration
T_{min}	minimum temperature
v_L	longitudinal velocity of the load suspension point
v_V	vertical (lifting or lowering) steady hoisting speed
Z_i	evaluation index for influence i
Z_{tot}	total evaluation index
α	local gradient, coefficient of thermal expansion of concrete
α_{max}	maximum inclination of the hoisting rope
β	relative period, ratio of cross-sectional areas, reliability index
φ_i	dynamic allowance
φ_L	longitudinal dynamic allowance
φ_V	vertical dynamic allowance
ϕ	rotation, friction angle, resistance factor, angular deviation from verticality
γ_i	load factor
γ_{LC}	load factor for a specific load condition
δ	displacement, deflection, logarithmic decrement of damping
ΔF	additional force
ΔL_A	shortening of assembly under the clamping force in the bar
ΔL_B	bar elongation at tensioning
ΔM	additional bending
ΔP	released portion of payload

ΔT	difference in temperature
λ	slenderness ratio
μ	relative mass, friction coefficient
μ_{T}	friction coefficient of the thread
η	lateral displacement
η_{D}	load multiplier related to ductility
η_i	combined load multiplier related to ductility, redundancy and operational classification of the bridge
η_I	load multiplier related to the operational classification of the bridge
η_{M}	load multiplier for the design of mechanical components of a heavy lifter
η_{R}	load multiplier related to redundancy
η_{M}	load multiplier for the design of structural components of a heavy lifter
ω	angular velocity
σ_{all}	allowed axial stress
σ_{B}	axial stress in the bolt generated by tensioning
$\sigma_{\mathrm{B.max}}$	maximum serviceability limit state tensile stress applicable to the bar
σ_{con}	tensile stress resulting from constant load
σ_{LC}	linear contact pressure in a roll
σ_{r}	residual tensile stress in the concrete slab
τ_{B}	tangential stress generated by bolt tensioning
ζ_{B}	buffer characteristic

Bridge Construction Equipment
ISBN 978-0-7277-5808-8

ICE Publishing: All rights reserved
http://dx.doi.org/10.1680/bce.58088.001

Chapter 1
Introduction

1.1.   Construction cost of a bridge deck

Every bridge construction method has its own advantages and weak points. In the absence of requirements that make one solution immediately preferable to the others, the evaluation of the possible alternatives is always a difficult task.

Comparisons based on the quantities of structural materials are poor indicators of the overall efficiency of design, and may mislead. The technological costs of processing raw materials (labour, investments for special equipment, shipping and site assembly of equipment, energy) and the indirect costs related to project duration (administration, depreciation of investments, financial exposure) often govern in industrialised countries. Higher quantities of raw materials due to the use of an efficient and rapid construction method rarely make a solution anti-economical (Rosignoli, 2002).

Steel trusses are light and structurally efficient, but less efficient I-girders permit robotised welding and may be supported at any location. In industrialised countries, the savings in raw materials in a truss are soon offset by the higher labour costs of hand welding. In prestressed concrete (PC) bridges, tendon splicing by overlapping is often less expensive than the use of couplers, even if the tendons are longer and the quantity of strand is therefore higher. Rebars with multiple bents are structurally more efficient than straight bars, but the latter are often preferred because of the lower labour costs. Optimised cross-sections may diminish the quantity of concrete, but they may also involve prohibitive forming costs (Rosignoli, 2002).

Similar considerations apply for bridge construction equipment. A ground falsework may lead to optimum quantities of structural materials due to the absence of design constraints from the construction process; these savings, however, are soon offset by the labour costs of poorly mechanised construction, and crane demand and long project duration add indirect and financial costs. In spite of savings on materials, therefore, ground falsework is used in industrialised countries only for small bridges or specific tasks within more mechanised projects.

PC bridges built by incremental launching are good indicators of the complexity of the problem (Rosignoli, 2001). The quantities of concrete and prestressing are higher than those achievable with a falsework due to the design impacts of temporary launch stresses, but casting long deck segments on the ground is simple and repetitive and requires small crews, the learning curve is short, the segments do not require handling, the risks of construction are minimised, prestressing tendons are long, and construction equipment is inexpensive and easily reusable on other projects. Although the quantities of raw materials are higher, incremental launching is a first-choice solution for medium-span bridges of medium length.

1

Several parameters influence the cost-effectiveness of bridge design and project organisation, the break-even point is different for the same construction methods in different countries, and the choice is, therefore, hardly standardisable.

1 Length, uniformity and sequence of the spans, deck width and plan curvature govern the choice of materials and structural solutions.
2 Bridge context, accessibility of the area beneath the bridge, the height of the piers and weight of the deck all influence the erection method.
3 Labour cost, time schedule, bridge length and logistics influence the optimal level of mechanisation of the construction process.
4 The number of bridges and their geometric features influence the choice of special erection equipment.

The investment in special equipment generates direct costs and financial exposure; however, the service life of these machines is longer than the project duration, and different strategies are available to diminish the impact of the investment on the project. Leasing the machine for the project duration, selling the machine when no longer required, or modifying the machine and reusing it on new projects are typical strategies. Local fiscal rules also play an important role by defining the number of years for full fiscal depreciation of the investment.

The bridge contractor will estimate these costs during the bidding process, monitor the actual costs accrued during construction, and use the data thus acquired to refine future bidding. This information is not in the public domain – it is kept confidential as it directly affects the contractor's ability of carrying out business. However, comparisons of alternatives based on raw material quantities, labour, investments, energy, indirect costs and risks of construction are often a good starting point in defining the cost-effectiveness of bridge design.

1.2.   Introduction to bridge construction equipment

The construction method that comes closest to incremental launching is segmental precasting (ASBI, 2008). Hundreds of PC bridges have been built with precast segments, even though the need to avoid joint decompression increases the cost of prestressing. Compared to incremental launching, however, the investment in special equipment and precasting facilities is also higher, and the break-even point shifts to longer bridges. On shorter bridges, prefabrication is limited to the beams, and the concrete slab is cast in place.

If the piers are not very tall and the area beneath the bridge is accessible, precast beams are erected with ground cranes. Cranes are readily available for rental in many countries, and the spans can be erected according to pier availability. Tall piers, sensitive environments, inaccessible sites, steep slopes and inhabited areas often require assembly with beam launchers, and the technological costs increase.

Span-by-span erection of precast segmental decks requires more expensive equipment because the spans are not self-supporting (Meyer, 2011). A few spans on short piers may be erected on ground falsework, but high labour costs soon suggest the use of an erection gantry as the number of spans increases. Poor soil conditions, tall piers and sensitive environments are other situations that suggest the use of a gantry. If access from the ground is possible, ground cranes may be used to load the segments onto the gantry; in the most delicate cases, the segments are delivered on the deck.

Ground cranes can also erect precast segmental bridges as balanced cantilevers; however, balanced cantilever bridges are typically chosen in response to inaccessible sites or tall piers, which discourages the use of cranes. A crane on the deck may also be used to lift the segments, but self-launching gantries and lifting frames offer several advantages as soon as the number of spans increases.

The use of special equipment is indispensable for full-span precasting (Rosignoli, 2011a). Lifting a span with a floating crane is often possible in long sea bridges. Long ground bridges may be accessible for most of their length with ground cranes; however, cranes beyond 200–300 ton are not readily available, must be assembled on site, are slow to operate and to redeploy, and may require ground improvement on poor soils. In addition, full-span precasting is chosen for large-scale projects that require placement of hundreds of spans, and the project dimensions permit the investment for multiple erection machines (Liu, 2012).

The use of ground cranes for full-span precasting is limited to single-track U-spans of light rapid transit (LRT) bridges, which can be delivered overnight with standard trucks (Vion and Joing, 2011). Two ground cranes can erect two complete spans every night, and the precasting facility is designed to maximise the productivity of the erection lines. The use of self-propelled modular transporters (SPMTs) is limited to single spans as an alternative to monolithic launching (Rosignoli, 2001; UDT, 2008).

Short- and medium-span PC decks may also be cast in place. For bridges with more than two or three spans it is convenient to advance in line by reusing the same formwork several times, and the deck is built span by span. Casting is done in either fixed or movable formwork. The choice of equipment is governed by economic factors, as the labour costs associated with a fixed falsework and the investments required for a movable scaffolding system (MSS) are both considerable (Däbritz, 2011; Pacheco et al., 2007, 2011).

Since the 1940s, wooden falsework has been replaced by modular steel frame systems. In spite of refined proprietary details and good adaptability, this equipment involves long assembly times and high labour costs. Modular trusses may be used for long spans and tall piers; the truss length may exceed 30 m (see Figure 1.1), but as the length increases the forming systems become so flexible that from 15–18 m onwards it is often preferable to introduce intermediate supports, which mean additional costs. Assembly duration and labour demand increase with the height above the ground.

The construction cost of the spans of the bridge shown in Figure 1.1, constructed in Italy in the 1990s, was 62% labour, 29% raw structural materials and 9% depreciation of investments. With such labour burdens, span-by-span casting on ground falsework is a viable solution only for small bridges in industrialised countries; however, it is often the most competitive solution in emerging countries with cheap labour. Obstruction of the area beneath the bridge and high crane demand are additional limitations.

An MSS comprises a casting cell assembled on a self-launching frame. MSSs are used for span-by-span casting of long bridges with spans of 30–70 m. If the piers are not tall and the area under the bridge is accessible, 90–120 m spans can be cast with 45–60 m MSSs supported by a temporary pier in every span. Construction is a compromise between in-place casting and prefabrication: mechanisation and repetitive operations diminish the labour demand, the

Figure 1.1. Modular trusses on temporary towers

quantities of materials are unaffected, and the quality is higher than that achievable with a ground falsework.

PC bridges with longer spans are often cast in place as balanced cantilevers. Two casting cells are assembled on self-launching travellers supported by the deck, which is designed to sustain their weight during construction. Two casting cells may also be suspended from a self-launching girder supported on the bridge piers. The girder accelerates construction, diminishes deck prestressing, lengthens the deck segments and facilitates access to the working areas. The girder also stabilises the hammer during construction; however, the investment is higher, and deck construction must proceed from abutment to abutment. The use of a suspension girder MSS is suitable to spans of 100–120 m and for long parallel bridges.

Cable-stayed bridges are also built as balanced cantilevers. The casting cells may be supported on underslung travellers that do not interfere with the fabrication of the stay cables; overhead travellers may also be used, especially in bad weather or when thermal curing is applied to the segments. Precast PC deck segments are erected with lifting frames or cranes located on the on deck. Composite deck grillages are more commonly used in modern cable-stayed bridges due to the lighter deck; steel grillage and precast deck panels are both erected with light deck-supported derrick platforms.

Steel bridges are lighter during construction, and the use of special equipment is therefore less frequent. Although ground cranes solve most of the erection needs, lifting frames are often used for strand jacking of heavy loads. Special lifting platforms supported on the main cables of suspension bridges are used to lift macro-segments of a stabilising truss or a streamlined box girder. Special equipment is also necessary for the incremental launching construction of U-girders and steel grillages, and the concrete slab may be cast in place with forming carriages that shuttle back and forth along the girders. Construction of most types of bridge, in other words, requires the use of special construction equipment.

New-generation bridge erection machines are sophisticated and complex. They resist huge loads (often the weight of an entire span) on long spans. Operations and repositioning occur under the same constraints that the obstruction to overpass exerts on the permanent structure, and with the most advanced machines it is possible to reposition their support systems to avoid the need for ground cranes. Deck erection and launching must be compatible with plan and vertical curves in the bridge, which requires adjustable supports and generates load eccentricity and torsion.

In spite of such complexity, bridge construction equipment must also be light. Weight governs the initial investment, the shipping and assembly costs, and indeed the feasibility of self-launching. Weight limitations dictate the use of high-grade steels and designing for high stress levels from different load conditions and support configurations.

These machines are reused several times, in different conditions and by different crews, who often are not aware of previous problems or anomalous work conditions. Equipment is modified and adapted to new work conditions and is repositioned many times during operations. Members, connections and field splices are subjected to hundreds of load reversals, the nature of loading is often highly dynamic, and the equipment may be exposed to impacts and strong wind. The support reactions are applied eccentrically, the support sections are often devoid of diaphragms, and the support systems are far from rigid.

Mechanical components and hydraulic systems interact with the structural components and govern load distribution (Rosignoli, 2011b). The safety itself of some machines depends on complex interactions among mechanical, hydraulic, electronic and structural systems. Sensor-controlled operations are becoming more common as the bridge erection equipment becomes 'smarter'. Indeed, such design and service conditions are almost inconceivable in a permanent structure subjected to such loads.

1.3. Typical configurations of bridge construction equipment

The manufacture of bridge erection equipment is a highly specialised niche industry. Every machine is initially conceived for a range of functions, every manufacturer has its own techno-logical habits, and every contractor has preferences and reuse expectations. Bridge length, time schedule and labour cost govern the mechanisation level, and the country of fabrication also affects the design of erection equipment. Machines fabricated in developing countries have lower automation levels, oversized welds and simpler connections. In emerging-economy countries light trusses are preferred to save materials, while heavier I-girders are preferred in industrialised countries to save labour via robotised welding. Nevertheless, the conceptual schemes are few.

Many beam launchers comprise two triangular trusses made of long, transportable, welded modules. The diagonals may be pinned to long chord modules for easy shipping, although site assembly is more expensive. Pins or stressed bars are used for the field splices in the chords.

Two winch trolleys bridge between the trusses, with the beam suspended underneath, and no braces are installed between the trusses to permit full-length operation. The span is short (50 m spans are rarely exceeded in precast beam bridges), the payload is low (a precast beam is only a small portion of the weight of the span) and the winch trolleys operate far from each other as the beam is suspended at the ends. The beam launchers are, therefore, light and lively during operations.

A self-launching gantry for span-by-span erection of precast segments also operates on 30–50 m spans, but the payload is higher as the gantry supports an entire span of segments during erection. These machines are versatile and adaptable to different deck geometries; however, the light gantries used for LRT bridges cannot carry the weight of road bridges, and gantries used for road bridges are often too heavy and cumbersome for use on LRT bridges. The trusses of some gantries can be reused in MSS configurations; the design load increases as it also includes the casting cell, but the winch trolleys are lighter and the nature of the loading is less dynamic.

The most versatile machines for span-by-span construction comprise two overhead girders that suspend deck segments or the casting cell and carry runways for winch trolleys or portal cranes. The main girders are supported on tower–crossbeam assemblies, and are equipped with light truss extensions to prevent overturning during launching. These heavy workhorses become expensive when used on long bridges, due to their weight, the labour demand, the complexity of operations, and the frequent need for repositioning of the supports with ground cranes.

Lighter and more automated single-girder overhead machines are fabricated around a central three-dimensional (3D) truss or two braced I-girders. A light front nose prevents overturning during launching in combination with a rear C-frame that rolls on the completed span. Telescopic machines comprising a rear main girder and a front underbridge are also available for use on bridges with tight plan radii. Single-girder overhead gantries and MSSs are lighter than the twin-girder overhead machines and do not require crossbeams, but they are also less versatile and adaptable. These machines are stable, compact and less labour intensive, and require ground cranes only for site assembly and decommissioning. When used in MSS configuration, the mechanisation level achievable for in-place casting is also higher (Rosignoli, 2010).

Underslung gantries and MSSs comprise two three-dimensional trusses or pairs of braced I-girders supported on pier brackets on either side of the piers. Props from foundations may be used to increase load capacity or the design span when the piers are short. A rear C-frame rolling on the completed span may be used to shorten the machine and to avoid the need for rear pier brackets. The underslung machines offer a lower level of automation than the single-girder overhead machines, and are more affected by ground constraints and clearance requirements.

Trussed extensions are typically applied to the main girders to control overturning. The front end of the main girders may also be connected by a crossbeam that slides on a central underbridge during launching. When equipped with a rear C-frame, these telescopic machines are able to erect spans with tight plan and vertical radii but require a special design of the pier caps to operate the underbridge. The most advanced underslung machines are able to reposition the pier brackets without the need for ground cranes.

The girders project beneath the deck, and may cause clearance issues at the abutments and when overpassing highways or railroads. The length of the girders is often a problem in curved bridges. Compared to a twin-girder overhead MSS, the main advantage of an underslung MSS is the lower cost of forming operations. Also the underslung gantries for span-by-span erection of precast segments are less expensive and labour intensive than the twin-girder overhead units (Rosignoli, 2007).

Span-by-span macro-segmental construction requires heavy twin-truss overhead gantries with a rear pendular leg that takes support on the deck prior to lifting the segment. Transverse wet joints at the span quarters and a full-length longitudinal joint at the bridge centreline divide 80–100 m continuous spans into four segments. The segments are cast on the deck with casting cells that shuttle back and forth and are rotated and fed with the prefabricated cage behind the abutment.

The overhead gantries for balanced cantilever erection of precast segments reach spans of 100–120 m (Harridge, 2011). Compared to span-by-span erection (Meyer, 2011), the design load is lighter, as the segments are handled individually or in pairs and no entire span is suspended from the gantry. The negative moment from the long, front cantilever and the launch stresses on such long spans govern design. Short 1.5-span varying-depth trusses have been used extensively in the past, while the current trend is towards full two-span constant-depth trusses. Stay cables are rarely used in new-generation machines.

The suspension-girder MSS for in-place macro-segmental casting of balanced cantilever bridges operate in a similar way (Matyassy and Palossy, 2006). Two long casting cells suspended from one or two self-launching girders shift symmetrically from the pier toward midspan to cast the two cantilevers. After midspan closure, the girders are launched to the next pier and the casting cells are moved forward to the new pier table. These machines can be easily modified for strand jacking of macro-segments cast on the ground.

The deck itself can support lifting frames for balanced cantilever erection of precast segments or form travellers for in-place casting. These light machines are used for short or curved bridges, PC spans up to 300 m, and cable-stayed bridges. Lifting frames and form travellers permit the erection of several hammers at once and different erection sequences than from abutment to abutment, but they require more deck prestressing and increase the demand for labour and ground cranes (ASBI, 2008).

The forming carriages are motorised casting cells that roll along the steel girders of composite bridges for in-place casting of long, concrete slab segments. They shuttle along the bridge to cast the span segments first and the pier segments second, in order to avoid permanent tension from negative bending in the pier segments.

Portal carriers with underbridge and span launchers fed by tyre trolleys are used for trans-portation and placement of full-length precast spans (Rosignoli, 2011a; Liu, 2012). The spans are rarely longer than 40 m in LRT and HSR bridges and 50 m in highway bridges, due to the prohibitive loads applied to the carrier and the deck during transportation. Longer precast spans have been handled with floating cranes when the bridge length permitted amortisation of such investments.

REFERENCES

ASBI (American Segmental Bridge Institute) (2008) *Construction Practices Handbook for Concrete Segmental and Cable-supported Bridges*. ASBI, Buda, TX, USA.

Däbritz M (2011) Movable scaffolding systems. *Structural Engineering International* **21(4)**: 413–418.

Harridge S (2011) Launching gantries for building precast segmental balanced cantilever bridges. *Structural Engineering International* **21(4)**: 406–412.

Liu Y (2012) Erecting prefabricated beam bridges in a mountain area and the technology of launching machines. *Structural Engineering International* **22(3)**: 401–407.

Matyassy L and Palossy M (2006) Koroshegy Viaduct. *Structural Engineering International* **16(1)**: 36–38.

Meyer M (2011) Underslung and overhead gantries for span by span erection of precast segmental bridge decks. *Structural Engineering International* **21(4)**: 399–405.

Pacheco P, Guerra A, Borges P and Coelho H (2007) A scaffolding system strengthened with organic prestressing – the first of a new generation of structures. *Structural Engineering International* **17(4)**: 314–321.

Pacheco P, Coelho H, Borges P and Guerra A (2011) Technical challenges of large movable scaffolding systems. *Structural Engineering International* **21(4)**: 450–455.

Rosignoli M (2001) Deck segmentation and yard organization for launched bridges. *ACI Structural Journal* **23(2)**: 2–11.

Rosignoli M (2002) *Bridge Launching*. Thomas Telford, London.

Rosignoli M (2007) Robustness and stability of launching gantries and movable shuttering systems – lessons learned. *Structural Engineering International* **17(2)**: 133–140.

Rosignoli M (2010) Self-launching erection machines for precast concrete bridges. *PCI Journal* **2010(Winter)**: 36–57.

Rosignoli M (2011a) Industrialized construction of large scale HSR projects: the Modena bridges in Italy. *Structural Engineering International* **21(4)**: 392–398.

Rosignoli M (2011b) Bridge erection machines. In *Encyclopedia of Life Support Systems*, Chapter 6.37.40, 1–56. United Nations Educational, Scientific and Cultural Organization (UNESCO), Paris, France.

UDT (Utah Department of Transportation) (2008) *Manual for the Moving of Utah Bridges Using Self Propelled Modular Transporters (SPMTs)*. UDT, Salt Lake City, UT, USA.

Vion P and Joing J (2011) Fabrication and erection of U-through section bridges. *Structural Engineering International* **21(4)**: 426–432.

Bridge Construction Equipment
ISBN 978-0-7277-5808-8

ICE Publishing: All rights reserved
http://dx.doi.org/10.1680/bce.58088.009

publishing

Chapter 2
Beam launchers

2.1.   Technology of beam precasting

Prestressed concrete beams were among the first structural elements to take full advantage of precasting techniques. A great number of bridges and viaducts for highway and railway infrastructure with moderate curvature and constant span length have been constructed with precast beams and cast in-place concrete slab. I- or T-beams are used in highway bridges on spans ranging from 20 m to more than 50 m, while U-beams are frequently used in railway bridges on shorter spans.

Standard precast beams are used in some countries, mostly on short spans. Custom beams are used in other countries and on longer spans, although the investment needed to set up a precasting facility and to provide special transportation and placement equipment limits the competitiveness of this construction method to long bridges (Rosignoli, 2011).

The beams are often placed on elastomeric bearings to simplify erection. They can be joined together with in-place stitches to form a continuous structure, or continuity can be limited to the concrete slab. Structural continuity can be achieved at different levels, ranging from live loads only to full continuity. In the presence of delivery difficulties, long precast beams can be divided into shorter segments match-cast against each other and reassembled on site with epoxy joints before placement.

The optimum number of beams in the deck cross-section depends on the road restrictions, which dictate the top flange width, and on the capacity of the handling devices. The longer beams usually require special equipment for handling and placement, and on large-scale projects it is common to purchase special erection equipment.

Prefabrication avoids cumbersome falsework, and deck erection is much faster. Beam fabrication can start during the construction of foundations and piers when an adequate storage area is available. Concrete poured in factory conditions is of better quality than concrete poured on falsework, and quality assurance/quality control (QA/QC) is also enhanced. Even in those rare cases where the design strength of concrete is not reached, a precast beam can be easily replaced, while a cast in-place span requires demolition or expensive strengthening (Hewson, 2003).

The forms are reused many times for accelerated depreciation of the investment, and can be equipped with steam curing and thermal insulation to shorten the cycle time. The reinforcement cage is fabricated on the bottom forms at multiple locations for parallel tasks and better risk mitigation. A 2- or 3-day cycle time is easy to achieve, but daily cycles are also possible with short beams. Factory work conditions maximise task repetition, minimise the labour demand and shorten the learning curve.

Several forms are used in the precasting facility to shorten construction duration and to accelerate depreciation of the handling systems. The beams are cast in one pour, from one end of a full-length steel form to the opposite one. The form is equipped with an external vibration plant, and the bottom form is raised from the ground to install external vibrators also for the beam soffit. The lateral forms are the most expensive component of the forming system; they are often mobile so that they can work with multiple bottom forms during beam curing.

After casting the beam is cured until it reaches sufficient strength for application of prestressing. Beams shorter than 30 m are typically pre-tensioned with adherent strands, which are released and cut. Hybrid prestressing systems may be necessary for longer beams where rectilinear pre-tensioned strands sized for the full design load are too powerful in the early curing stages. In this case, most of the final prestressing is applied with pre-tensioned strands, and integrative post-tensioning is installed and tensioned before beam transportation. The length of the precast beams may thus reach 60 m, even though such long units pose transportation challenges and require heavy lifters.

Integrative post-tensioning can also be applied after casting the concrete slab, although anchorage pockets in the slab may pose problems of durability. When precast beams are made continuous at the supports, the application of integrative post-tensioning to the continuous beam generates favourable hyperstatic effects. These hybrid solutions can diminish the total cost of prestressing and simplify the strand anchor systems in the precasting facility.

Precast-beam bridges are durable structures, as demonstrated by many statistics of bridge pathologies. Durability problems are rare, even in the case of partial prestressing, provided that the post-tensioning tendons are accurately protected. Even where not readily available, precast beams are economic for long viaducts with constant spans, where the investment in the precasting facility and handling equipment can be depreciated over many uses.

The disadvantages include the cost of the special equipment and the extensive areas needed to cast and store the units. Precast-beam bridges are not suited for highly curved alignments, and require wide and expensive pier caps. The beams have minimal torsional stiffness, and the side wings in the concrete slab must be short if they are not to overload the edge beam.

2.2.   Straddle carriers and heavy lifters

Straddle carriers are off-road lifting vehicles for use in the casting yard (Figure 2.1). They are used to lift precast beams and segments from the casting cell, to move them within the precasting facility, to stack them for storage, and to load them onto the transportation means for final delivery. Straddle carriers are also used during the incremental launching erection of steel girders to handle girder segments in the assembly yard behind the abutment (Figure 2.2).

Custom lifting beams and spreader beams are used to pick up the load with the desired suspension configuration. The straddle carriers for precast beams are not tall, as the beams are rarely stacked. The straddle carriers for precast segmental facilities are taller and can stack up to three segments.

Figure 2.1. Twin straddle carriers for handling of U-beams (Photo: Comtec-Deal)

Figure 2.2. A 50 ton straddle carrier for handling of steel girder segments

The straddle carriers are capable of relatively high speeds, up to 30 km/h, and are available in a variety of configurations, with load capacity ranging from 50 to 150 ton. Good manoeuvrability and the ability to operate on typical heavy vehicle roads provide flexibility to the organisation of the precasting facility. Operation with multiple stockyards with linear or herringbone layouts and complex routings is possible, which gives straddle carrier-based transport systems unrivalled flexibility.

The straddle carriers handle precast beams and segments quickly and effectively. Loading, transporting and unloading cycles can be performed rapidly, and the ability to self-load and unload decouples the carrier operation from the plant being served, thus minimising delays and boosting carrier utilisation.

A single diesel engine drives a hydraulic system to provide power to the hoist, steering, drive, braking and charging pumps. Hydraulic pumps and motors are controlled with solenoid valves. Hydraulic lines, wiring and fluid tanks are often encased internally within the frame for protection from loads and sunlight.

Hydraulic winches or long-stroke double-acting cylinders are used to lift the load. Integrated position sensors provide cylinder position feedback for closed-loop synchronisation of multiple hoist cylinders. Variable displacement hoist pumps supply pressure and flow to operate the hoist and swing cylinders. Lift system pressures and temperatures are monitored by control valves. Counterbalance valves are mounted within the hoist cylinders for load-holding capability.

Full flow-return filtration is typically installed. Pressure filtration is utilised for the drive pump charging circuits and steering circuit. Charging pumps prevent contamination from reaching the main drive pumps. Heat exchangers with hydraulically driven fans are used to keep operating fluids cool. The fan speed is thermostatically controlled to maintain optimum fluid viscosity.

Multi-disc hydraulic brakes provide adequate braking performance. Spring-set hydraulic release parking brakes are installed on the non-driven wheels. All service brakes are cooled by forced lubrication. Hydraulic accumulators are installed to provide hydraulic pressure after an engine shutdown. Accumulator discharge valves eliminate stored energy hazards after the machine is shut down. Braking systems include redundancies to ensure safety against failure. Proportional solenoid valves are used for parking brake release and service brake actuation.

An independently controlled steering cylinder is mounted at each wheel. Position transducers are installed in each cylinder to provide feedback to the steering computer. The straddle carriers typically offer four steering modes: four-wheel coordinated steering, two-wheel (front) automotive steering, two-wheel (rear) forklift steering, and four-wheel parallel (crab) steering. The steering computer synchronises the wheels during manoeuvring and virtually eliminates tyre scrub due to steering geometry errors.

Heavy lifters rolling on rails are also used in the precasting facilities. Three-hinge portal cranes are used to lift heavy loads with hydraulic winches, pull cylinders or hoist bars. Portal cranes are less expensive and easier to operate than the straddle carriers, but do not offer the same manoeuvring capability.

2.3.   Beam launchers

The most common method for erecting precast beams is with ground cranes. Cranes usually give the simplest and most rapid erection procedures with the minimum of investment, and the deck may be cast in several spans at once. Good access is necessary along the entire length of the bridge to position the cranes and deliver the beams. In the presence of tall piers, steep slopes or obstructions such as rivers, railroads or highways, crane erection may be impossible. Ground transportation and crane erection of precast beams are often impossible in urban areas.

Tall straddle carriers and portal cranes rolling on rails have been used to lift the beams into position where level access is available for the entire length of the bridge and the piers are short. The beams are transported under the crane, which lifts them up. Less versatile, but also less expensive, than commercially available cranes, straddle carriers and portal cranes are used for erecting long urban viaducts along existing roads. Twin lifters may be necessary for the longest beams in order to halve the payload and to limit the negative moment generated in the beams by long end cantilevers.

The use of a beam launcher solves any difficulty. A beam launcher is a light self-launching crane comprising two triangular trusses and two winch trolleys spanning between the trusses. The length of the trusses is about 2.3 times the typical span of the bridge, but this is rarely a problem as the gantry operates above the deck (Figure 2.3). Crossbeams support the trusses at the pier caps and permit lateral shifting to erect the edge beams and to traverse the gantry when launching along curves (Rosignoli, 2011).

The winch trolleys roll along the top chords of the trusses and carry two winches each. The hoisting winch lifts the beam, and the haulage winch acts on a capstan that drives the trolley along the gantry. The capstan is anchored at the ends of the gantry and kept in tension by lever counterweights to avoid rope slippage at the haulage winch. One of the winch trolleys carries a power-pack unit that supplies power to gantry operation (Figure 2.4). A third trolley carrying only the PPU may also be used to lighten the winch trolleys and to diminish the wheel load applied to the runways.

When the beams are delivered on the deck, the hoisting winches may be replaced with long-stroke double-acting cylinders (Figure 2.5). The vertical movements are better controlled and the hoists are lighter but also unable to lift the beams from the ground if necessary. Both types of hoist are assembled on crabs that roll transversely along the crane bridge to increase the load eccentricity achievable with the gantry; capstans, hydraulic motors or double-acting cylinders are used to drive the crab along the crane bridge. The pull cylinders are gimbal-mounted to avoid bending due to load oscillations. The precast beam is lowered beneath the bottom chord of the truss and shifted laterally. Shifting the crab within the winch trolley shortens the pier crossbeams but overloads the outer truss due to the transverse load eccentricity.

The twin-truss launchers are relatively inexpensive and easily adaptable in both the length and the spacing of the trusses. The trusses are flexible, and alignment wedges are used at both ends to recover the deflection of the cantilever trusses when landing at the new pier and to release the rear support reaction. These gantries are able to cope with variations in span length and deck geometry, and, as they are located above the deck, they are unaffected by ground-level constraints and the plan curvature of the bridge.

Figure 2.3. A 102 m, 90 ton launcher for 45 m, 120 ton beams (Photo: Comtec)

New-generation single-girder launchers are based on two braced I-girders (Figure 2.6). The girder is less expensive than two trusses, the winch trolleys are smaller and lighter, and the number of articulated support saddles halved. Windowed webs lighten launch noses and tails, and are compatible with full-length robotised welding. A C-frame supports the rear end of the gantry for the beams to pass through when delivered on the deck.

The C-frame is not necessary when the beams are delivered on the ground as the launcher lifts them up and shifts them into position within the same span (Figure 2.7). Single-girder launchers and shifters handle the precast beams beneath the main girder, and are, therefore, tall and potentially less stable than the conventional twin-truss gantries.

The hoist picks up the beam on both sides, and lifting accessories are therefore necessary. Since the beams are delivered under the span to be erected, the launch saddles of the gantry do not interfere with lifting and placement. The PPU can be placed at the rear end of the shifter to act as a counterweight during launching.

Figure 2.4. Winch-trolley carrying the PPU (Photo: Deal)

2.3.1 Loading, kinematics, typical features

The twin-truss beam launchers operate in one of two ways, depending on how the beams are delivered. If the beams are delivered on the ground, the gantry lifts them up to the deck level, shifts them laterally and releases them onto the bearings. If the beams are delivered at the

Figure 2.5. Pull cylinder (Photo: Deal)

Figure 2.6. Single-girder launcher (Photo: Strukturas)

abutment, the gantry is moved back to the abutment and the winch trolleys are moved to the rear end of the trusses. The front trolley picks up the front end of the beam and moves it forward, with the rear end suspended from a straddle carrier. When the rear end of the beam reaches the rear winch trolley, the latter picks it up to release the carrier.

The trusses are braced to each other at both ends and are moved as a whole. The longitudinal movement of the gantry is a two-step process. Automatic clamps anchor the trusses to the launch saddles (Figure 2.8), and the winch trolleys carry the beam one span ahead; the winch trolleys are then anchored to the pier crossbeams with crossed ropes, the anchor clamps are released, and the haulage winches of the winch trolleys are reversed to push the trusses to the next span. Redundant anchorages are necessary in both phases of repositioning for safe operations along inclined planes.

The sequence can be repeated many times so that, when the beams are delivered at the abutment, the gantry can place them several spans ahead. When the bridge is long, shuttling the gantry over many spans slows the cycle time, and it is often faster to cast the concrete slab as soon as the beams are placed and to deliver the next beams on the deck.

Truss deflection at landing is recovered with alignment wedges. The alignment force is small but very inclined, and the launch saddles are anchored to the crossbeams to avoid displacements or overturning. Realignment may also be achieved with long-stroke double-acting cylinders that rotate articulated arms pinned to the tip of the truss (Rosignoli, 1999). Alignment systems are

Figure 2.7. A 74 m, 98 ton single-girder shifter for 28 m, 60 ton beams (Photo: Deal)

Figure 2.8. Automatic anchor clamp (Photo: Deal)

Figure 2.9. Placement of the edge beam (Photo: Comtec)

also applied to the rear end of the trusses to release the support reaction when launching forward and to recover the deflection when launching backward.

2.3.2 Support, launch and lock systems

Crossbeams anchored to the pier caps by stressed bars carry runways for lateral shifting of the gantry. The crossbeams have side overhangs for placement of the edge beams (Figure 2.9), and to traverse the gantry for launching along curves. Adjustable support legs, located so as not to

Figure 2.10. Service cranes for repositioning of pier crossbeams (Photo: Deal)

Figure 2.11. Support saddle (Photo: Deal)

interfere with the precast beams, are used to set the crossbeams horizontal on stepped pier caps. Some beam launchers have light service cranes at the ends to reposition the crossbeams during launching without the need for ground cranes (Figure 2.10).

The support saddles comprise two bottom bogies that roll transversely along the crossbeam and two top bogies that support the bottom chords of the truss during operations and launching (Figure 2.11). Cast-iron rolls are assembled on equalising beams that balance the load in the rolls; four-roll assemblies or twin two-roll assemblies are typically used for the bogies of these machines to diminish the diameter of the rolls and to lengthen the patch load applied to the chords of the truss. The top bogies are articulated to cope with flexural rotations in the truss and the launch gradient. A vertical pivot connects the two assemblies of bogies to provide rotation in the horizontal plane for traversing during launching along curves.

Capstans, hydraulic motors or long-stroke double-acting cylinders drive the support saddles along the crossbeams. Rectangular rails welded to the top flanges of the crossbeam facilitate load dispersal within the webs and take contrast against the flanges of the rolls for transfer of lateral forces due to wind and operations. Rectangular rails are also welded under the bottom flanges of the truss for guidance and lateral load transfer. Launch and transverse drive systems provide lateral constraint only during normal operations, and stronger anchorages are often used in out-of-service conditions.

Figure 2.12. Single capstan

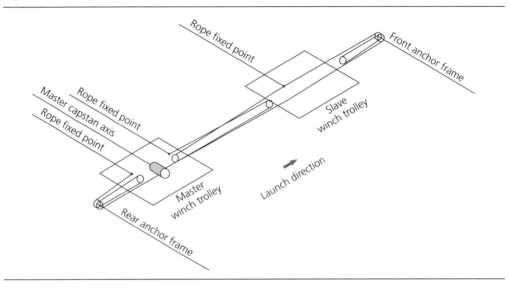

The two types of capstan used to drive the winch trolleys along the trusses and to launch the gantry are illustrated in Figures 2.12 and 2.13. One line or both lines of support saddles are equipped with block systems to avoid involuntary movements of the gantry. Most bridges have a gradient and the beam launchers shuttle along low-friction inclined planes, so the block systems are critical for safe operation. Automatic clamps prevent involuntary movements of the gantry if the winch trolleys are not anchored to the pier crossbeams.

2.3.3 Performance and productivity

Beam launchers have been used successfully on spans ranging from less than 20 m to about 60 m. The lifting speed may vary in the range 0.1–0.5 m/min with slow machines to 1.5–4.0 m/min with

Figure 2.13. Double capstan

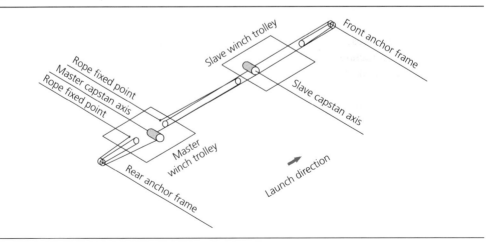

fast machines. The beams can be delivered on the ground or on the deck. The design of the winch trolleys is different in the two cases: long reeved hoisting ropes are used in the first case, while long-stroke double-acting cylinders provide faster operations and better control of movements and lighten the hoist in the second case.

The longitudinal launch speed varies in the range 0.1–0.5 m/min with hydraulic launchers and 4–12 m/min with capstans. The transverse shift speed varies between 0.1 and 0.5 m/min with hydraulic cylinders and 0.5 and 1.0 m/min with motorised wheels or capstans.

Erecting the precast beams for a span takes a few days, with one shift per day in relation to the number of beams. A few heavy beams in every span require a powerful launcher, but accelerate bridge construction with fewer placement cycles. During placement of the beams of span N, the activities on span $N-1$ include forming the concrete slab, assembly of reinforcement, casting and finishing. The concrete slab of span $N-2$ is curing, and span $N-3$ may often be used for beam delivery.

Accelerated erection sequences (e.g. two 10-hour shifts) permit casting of two spans per week and curing of another two spans per week, so the beams are delivered on span $N-5$ and the launch speed of the gantry influences the cycle time.

2.3.4 Structure-equipment interaction
The interaction between a beam launcher and the bridge being erected is minimal. The gantry is light and lively, but this does not cause secondary stresses as the beam is suspended at the ends and the gantry does not take support on the spans during operation. The beam is subjected to dynamic effects during handling, and the dynamic allowance depends on the equipment and procedures adopted. In normal circumstances it is sufficient to check the gantry for a $\pm20\%$ difference between the weight of the precast beam and the lifted accessories to cater for the dynamic effects.

The design of the pier caps must account for the presence of the crossbeams, but this is rarely a problem. When the pier cap has an inverted T-section, the crossbeams are typically supported on the stem of the section.

REFERENCES
Hewson NR (2003) *Prestressed Concrete Bridges: Design and Construction*. Thomas Telford, London.
Rosignoli M (1999) Nose optimization in launched bridges. *Proceedings of the Institution of Civil Engineers, Structures and Buildings* **134**: 373–375.
Rosignoli M (2011) Bridge erection machines. In *Encyclopedia of Life Support Systems*, Chapter 6.37.40, 1–56. United Nations Educational, Scientific and Cultural Organization (UNESCO), Paris, France.

Bridge Construction Equipment
ISBN 978-0-7277-5808-8

ICE Publishing: All rights reserved
http://dx.doi.org/10.1680/bce.58088.023

Chapter 3
Self-launching gantries for span-by-span erection of precast segments

3.1. Technology of precast segmental construction

Segmental precasting of PC bridges is a well-established construction method that offers many benefits on suitable projects (ASBI, 2008). The advantages include economies from mechanised, repetitive construction procedures that reduce costs and construction time. Factory production enhances quality and facilitates QA/QC. Rapid erection, minor site disruption and easy mainten-ance of highway and railway traffic at the erection site are additional advantages. Segmental bridges are easily adaptable to curved alignments and may have pleasant aesthetics.

The disadvantages include the high cost of precasting facilities and special equipment necessary to handle, transport and erect the segments, and the great number of joints in the structure. Although the segmental construction concept is relatively simple, the level of technology involved in the design and construction is more demanding than that required for other types of bridge construction. Deep understanding of the technological requirements is necessary to facilitate construction, to avoid problems encountered in the past, and to reduce delays and costs related to concerns about non-critical issues or the lack of understanding of critical issues (ASBI, 2008).

Several studies have been made in relation to the bridge length necessary to amortise the investments of segmental precasting. Deck surfaces of around 20 000 m^2 may warrant feasibility analyses, but the cost of labour and the presence of local precasting facilities are the real discriminating factors (Harridge, 2011). In some countries labour is simply so cheap as to make in-place casting on ground falsework hardly beatable when the area under the bridge can be disrupted. Other factors that may affect the decision are as follows (Homsi, 2012).

1   *Height of the bridge* to the ground and accessibility with ground cranes.
2   *Ground constraints.* Bridges over water are more likely to be precast than are bridges over land. Bridges with short piers in accessible areas are likely to be erected using ground cranes rather than a gantry. Bridges with tight plan radii or in congested urban areas can be difficult for both gantries and ground cranes that obstruct traffic.
3   *Delivery of segments on the deck, on the ground or both.* Deck delivery requires erecting the spans in sequence, which involves higher risk profiles, lengthens the cycle time and prevents finishing works on the bridge. Ground delivery provides more flexibility, and the area under the bridge may be used for segment storage; ground delivery, however, is affected by ground constraints and requires more expensive hoists in gantries and lifting frames.
4   *Variability of the cross-section.* Varying width or bifurcations add complexity, lengthen the cycle time and mostly prevent the use of underslung gantries. Span variations in

23

varying-depth decks also affect precasting operations, especially when the webs are inclined and the width of the bottom slab varies.

5   *Plan curvature*. The casting cells for tight plan radii are more expensive and complex to operate. Segment erection requires telescopic overhead gantries or underslung gantries with articulated trusses and so specialised machines add costs and complexity. Varying deck crossfall also complicates the use of overhead gantries and lifting frames.

6   *Vertical curvature* affects the operation of both types of erection gantry, particularly during launching.

Precast segmental bridges are used for spans ranging from 25–45 m to 110–130 m. Below 40 m the use of precast beams and a cast-in-place top slab is often more cost-effective, although the overall quality is lower. Where site constraints prevent crane erection, a beam launcher is also lighter and less expensive than a self-launching gantry for span-by-span erection. Balanced cantilever precast segmental spans of 160 m have been reached with lifting frames but they represent the upper bound of the technology due to the depth and weight of the hammerhead segments and the cost of top-slab prestressing.

Most box girders for highway bridges have a single-cell arrangement with side wings. The top slab width varies between 6 m and 18 m with a bottom slab width between 3.5 m and 9 m. Bi-cellular box girders may be wider but the two inner forms complicate the design and operation of the precasting cells. Wider single-cell segments have been achieved by supporting the top slab with diagonal struts or transverse ribs. Struts and ribs are frequently combined in cable-stayed PC box girders with one central plane of cables.

The length of the segments is governed by handling and transportation requirements. Segments up to 3.6 m long are often transported on public roads without major restrictions. When the precasting facility is close to the bridge and the segments can be transported within the right-of-way of the project or on barges, the segments are made as long as is practicable in order to accelerate construction and diminish the number of joints (ASBI, 2008). Segments weighting up to 70–80 ton are usually within the lifting capacity of cranes available locally.

Three different techniques can be used to erect a precast segmental span: span-by-span assembly, balanced cantilever assembly, and progressive placement with temporary stays or props. With the span-by-span method, all the segments for a span are positioned prior to installing and tensioning the prestressing tendons, and the span is released onto the bearings. The balanced cantilever method involves erecting the segments as a pair of cantilevers around each pier, and the pairs of opposite segments are prestressed with tendons that cross the entire hammer. In progressive placement, a lifting frame or a crane lifts and places the segments in one direction from the starting point, passing over the piers in the process.

Span-by-span erection is the most common, simplest and often most cost-effective construction method for precast segmental bridges (Meyer, 2011). It is typically used on long bridges having a great number of 25–45 m spans of large plan radius, and it is compatible with simple and continuous spans. Although the span-by-span method has been used for spans up to 60 m, for spans longer than 45–50 m this method loses much of its appeal due to the cost of the erection gantry. The segment weight ranges between 30 ton and 150 ton and the span weight ranges between 200 ton and more than 2000 ton. The segments are erected using heavy gantries that carry an entire span of segments. External prestressing facilitates segment production and enhances strand protection in case of leaking joints.

The balanced cantilever method can also be used on spans of 40–60 m but is best suited to longer spans. The segments are erected individually or in pairs by means of cranes on the ground or on barges, lifting frames or cranes on deck, and self-launching gantries (Harridge, 2011). Convertible gantries are able to erect long balanced cantilever spans and shorter span-by-span units. Spans longer than 110–120 m are cast in-place with form travellers because of the height, weight and poor stability of precast hammerhead segments. Lifting frames are the typical solution for long precast segmental cable-stayed spans.

Progressive placement with cranes or lifting frames is the most time-consuming erection technique because of the single work location (ASBI, 2008). Special equipment can be particularly inexpensive, especially when ground cranes can erect the segments along the entire length of the bridge and props from foundations support the segments prior to application of pre-stressing. Progressive placement has been applied, or considered, for 30–90 m spans in environmentally sensitive locations where construction access was restricted to one or both ends of the bridge.

3.1.1 Fabrication and delivery of precast segments

The setup cost for the precasting facility is a significant portion of the financial exposure on a project and of the construction cost of the bridge. The cost depends on the project time schedule and the ground conditions. The time schedule dictates the number of casting cells and the dimensions of the segment storage area. The foundations of the casting cells are particularly rigid to ensure geometry control, pile foundations are often necessary when there are poor soil conditions, and this increases the setup cost (ASBI, 2008).

The segments can be cast using either the short- or the long-line method. With either method, a cycle time of one segment per day per casting cell is typically achieved for the standard segments. Pier segments, deviators and other complex segments may require 2–3 days. The casting area of the precasting facility may be sheltered so that production can be continuous through bad weather, and concrete is often batched on-site to ensure good QC and uninterrupted supply.

The casting cells for short-line match-casting have a 2–6 month lead time, depending on the number of forms and the complexity of segments. During this time, the casting yard is prepared to receive the casting cells for assembly. Site preparation includes clearing, levelling, grading, remediation from old foundations, tanks and contaminated materials, casting of foundation slabs, installation of drainage and storm-water retention basins, installation of utilities, and setting up a docking facility if the segments will be delivered using barges. The best soil locations on the site are typically chosen for casting cells and batching plant (ASBI, 2008).

The site size depends on the number of casting cells and rebar jigs, access and the size of the storage areas for rebars and post-tensioning materials, segment length and width, use of portal cranes on rail or straddle carriers for handling of segments, hoist width, whether the segments will be double stacked, space between stored segments for work crews, layout of the shipping area, and dimensions of the batching plant. Because of the temporary nature of a precasting facility, the working area can often be zoned to maximise the use of the site area, as this can bring jobs to the local economy. Setting up a batching plant often requires a transformer or upgrading of existing power lines, which can also be a long lead-time item. The water demand of the batching plant, truck washout area and steam generator may require the drilling of wells or provision of new utility lines.

The storage area is designed to maximise efficiency. The efficient operation of yard crews and equipment operators avoids wasting time looking for segments at the time of delivery or waiting for a crane that is being used on different tasks. The segments are handled with crawler cranes, portal cranes on rails or straddle carriers. Safety details such as wheel guards, travel alarms and proper communications between the operator and yard crew are essential to the safe operation of large lifters in the narrow runways of the storage area (ASBI, 2008).

Crawler cranes can be used to load the segments on tractors/trailers or special haulers, or to crawl directly with the segment to the storage platform. This method is not the most efficient, but is sometimes used when the layout of the storage area can accommodate large runways for the travel and swing radii of crane and transporters.

If the precasting facility is rectangular and the segments are stored on long, narrow strips of land, a portal crane on rails is often the best solution. Portal cranes require narrow runways between the rows of segments, and the segments can be double stacked for optimal storage efficiency. Rail-mounted cranes require foundation beams for the runways and a flat terrain. The casting cells are aligned with the storage area to avoid transverse shifting of the crane. Power supply is typically through retractable cables, but long runways may suggest the use of a carried PPU.

If the storage area is square or irregularly shaped, a motorised straddle carrier is typically preferred. These wheeled cranes straddle one or two segments and provide flexible operations but require multiple large runways. Four-wheel steering and four-wheel drive are helpful in negotiating tight turns. Soil preparation, well-maintained runways and well-designed drainage are necessary to avoid premature wear and tear of the steering and drive mechanisms of the crane. Several parameters are evaluated when choosing a straddle carrier (ASBI, 2008).

- *Hoist capacity*. In addition to the maximum weight of the segments and lifting accessories, different uses on the project or other projects should be investigated.
- *Load-handling capability*. Hoist crabs simplify load handling but are more expensive than fixed hoists. Hoist capacity is higher at the centre of the crane than over the wheels.
- *Hook height*. The segments are often handled over stored segments. The lifting beam and accessories, support saddles for varying-depth segments with an inclined soffit, and barrier rebars protruding from the top slab add to the total height of the segment.
- *Wheelbase*. This establishes the lane width of the storage area and the geometry of the twin loading piers for barge loading.
- *Turning radius*. The layout of several components of the precasting facility is designed based on the turning radius.
- *Hoisting and travelling speed*. The cycle time and other coordination factors depend on the operation speeds.
- *Drive train*. Precasting facilities are typically on rough ground, which is the perfect environment for four-wheel drive. Two-wheel drive straddle carriers are also available.

The storage area is prepared to prevent settlement. The segments are stored on hard timber blocks using three-point support configurations to avoid warping. Two support points are used under one web and one point under the other, and an inverted support scheme is used over the top slab when segments are double stacked. Varying-depth segments require more complex supports due to the sloped bottom soffit; double stacking is not recommended. Concrete pads or crane mats may be necessary to spread the load. The storage area is monitored periodically and after heavy rain to check for settlement and to remediate it if it occurs.

The erection sequence governs the casting schedule of the segments. Rows or series of rows in the storage area are designated to store individual spans. Storing entire spans of segments together facilitates finishing a span all at once without needing to relocate equipment and crews, and the loading crews can locate the segments more easily at the time of delivery. Double stacking of segments reduces the storage area required but complicates finishing and requires accurate planning of casting and storage to avoid double handling.

The use of auxiliary equipment increases the productivity of the precasting facility and enhances coordination (ASBI, 2008). The versatility of telescopic boom forklifts is enhanced by the use of interchangeable attachments such as buckets and working platforms. Aerial personnel lifts are often necessary for maintenance of tall casting cells and finishing of double-stacked segments. Front-end loaders are used to fill the aggregate bins of the batching plants, to load and unload trailers, to maintain the storage area and construction roads, and to move air compressors and post-tensioning equipment around the storage area.

3.1.1.1 Short-line match-casting

In the short-line method, a new segment is match-cast against the preceding segment. When the new segment has hardened, it is moved to the match-cast position to serve as a front bulkhead for the next segment, after which it is moved to the pickup position for transportation to storage. The casting cells are stationary and designed to facilitate the six operations of the casting cycle: insertion of the reinforcement cage into the casting cell, forward insertion of the core form into the cage, pouring of the concrete, backward removal of the core form, transfer of the new segment to the match-casting position, and transfer to the pickup position for transportation to storage.

The short-line method requires less space than the long-line method. The manufacturing process is centralised, repetitive and easily adaptable to variations in geometry such as plan and vertical curves and crossfall transitions, and the casting cells are reusable in other projects (ASBI, 2008). The main disadvantages are the complexity of geometry control and the need for accuracy and precision. Large and repetitive casting errors and poor workmanship are critical, because problems that will affect the structure during erection will not be discovered until many of the segments have been cast.

Geometry control is different for short- and long-line casting. In both cases, the segments for balanced cantilever spans are cast from the pier to the cantilever tip, and the casting direction therefore reverses for up- and down-station segments. This requires care in determining the casting coordinates and their nomenclature, and in setting up the geometry-control program. Three-dimensional CAD programs are often used to plot the casting coordinates of the entire bridge to check for misalignment.

Each segment is equipped with four level bolts for the control of elevations and two notched centreline markers for the control of alignment. Segment geometry is determined for the six markers and is set based on the casting curve of the span. The casting curve varies from span to span and is attained by superimposing three geometry components.

27

1 The three-dimensional design geometry shown on the plans, which includes plan and vertical alignment and crossfall transitions. The design data typically result from a three-dimensional drawing of the bridge that includes anchorages and deviators of post-tensioning and entry and exit points of deviator pipes from concrete surfaces.

2 Compensation for the deflections during erection in relation to age and elastic modulus of concrete, weight of segments and auxiliary erection equipment, application sequences of post-tensioning, and loading from main construction equipment. The data result from a time-dependent analysis of the erection, including the structure–equipment interaction.

3 Compensation for the long-term deflections due to the time-dependent behaviours of the concrete and prestressing steel. Testing of concrete may be necessary to develop accurate creep and shrinkage parameters, which may require 6 months of curing or more. The data result from a time-dependent structural analysis of the bridge in service.

The casting cells for short-line match-casting are designed to produce a segment per day. Rebar cage and post-tensioning ducts are prefabricated in rebar jigs and templates, while tendon anchorages are often installed in the casting cell due to their weight. Two or three rebar jigs serve each casting cell to allow the cages to be assembled over a 2–3 day period. Partial jigs for the bottom slab and webs, top slab and pier diaphragms are often used to assemble the cage into a final template prior to its transfer to the casting cell.

The rebar jigs meet the versatility of the casting cell they serve. The jigs include a stiff framing system that supports an adjustable bottom floor, side walls, two bulkhead templates marked to the proper section size and with locators for longitudinal post-tensioning ducts, and side wings and end templates with locators for transverse post-tensioning ducts. Spacers are provided at the bottom floor, walls and side wings to ensure proper concrete cover. Positioning frames supported on the end bulkheads are used to keep in place vertical ducts for hanger bars and rail fasteners for plinthless LRT elevated guideways. String lines are used to check rebar, cover and duct positions from open surfaces (ASBI, 2008).

Upon completion of the cage, a lifting truss is connected to the cage with slings and cross-ties. Padded slings are used to handle epoxy-coated rebars to avoid damaging the coating. Forklifts, straddle carriers, portal cranes or tower cranes are used to pick up the lifting truss and transfer the cage into the casting cell. Lifting trusses with scalable anchor points may be necessary to lift the cage of expansion joint segments. Cage prefabrication of the end segments of the span accelerates the casting process, especially when the end segments are the starter segments for the span. When pier and expansion joint segments are cast in a short casting cell to compensate for the additional weight of the pier diaphragm, the cage is often fabricated in the casting cell.

The casting cell comprises outer steel forms for rear bulkhead, bottom form table, webs and wings, and a movable core form for the box cell that rolls along a runway behind the bulkhead. The bulkhead is fixed and the geometry of the new segment is set by rotating the match-cast segment. Multiple casting cells are used in the precasting facility according to the project time schedule. The casting cells are aligned to simplify feeding of prefabricated cages and the removal of completed segments, and to minimise the number of survey towers needed for geometry control. A small number of survey towers reduce the amount of surveying equipment required and accelerate the casting cycle.

Segments for span-by-span erection have constant depth, while segments for balanced cantilever spans longer than 60–70 m have varying depth. Constant-depth casting cells are easier to operate as the soffit width and web depth do not change; when the segment length varies, the match-cast segment is shifted longitudinally on the bottom form table. Because of the large number of segments to be cast, the steel forms for soffit, webs and wings are at least 6–7 mm thick. External vibrators may be used to ensure an aesthetic surface finish in complex segments or congested cages.

The core form cantilevers out from a counterweighted tower frame that rolls longitudinally on runway rails and carries hydraulic systems for form opening and closure. The core form is inserted into the casting cell through the fixed bulkhead after cage insertion, and is extracted backward prior to transferring the new segment to the match-casting position. The core form is a modular assembly of components that facilitate setting the form geometry for casting of bar blisters, anchor blisters and deviation saddles on a regular basis (ASBI, 2008). Light interchangeable panels allow hand installation to rapidly reconfigure the casting cell within the daily cycle time. Special casting cells are often used for the expansion-joint segments in long bridges.

Casting cells for varying-depth segments are designed to accommodate changes in depth without disrupting the daily casting cycle. Two-piece bulkheads are shortened or lengthened by bolting together different pieces in combination with adjustable soffit tables and core forms. Even if a casting cell may accommodate a wide range of depths, it may be cost-effective to use one cell for the deepest segments and another cell for the shallowest ones. In this case, the runways of segment manipulators are designed with orthogonal crossovers to transfer the match-cast segments from cell to cell.

The fixed bulkhead is an important part of the casting cell. In a balanced cantilever bridge, the bulkhead also serves as a template for the ducts of internal longitudinal post-tensioning. The ducts are secured to holes in the bulkhead, and the bulkhead must therefore include a hole at every location where a tendon crosses a joint. The post-tensioning layout in the top and bottom slab is designed to use the same tendon locations at the joints in order to avoid unnecessary or overlapping holes in the bulkhead.

The forms for the shear keys can be fabricated from different materials. Machined solid steel bars are durable but expensive; they can withstand impacts from the rebar cages and may be cost-effective if reused many times. Bent steel plates and solid polyethylene bars are more susceptible to damage. Recess pockets for tendon anchorages are made from welded steel plates for repeated use, and from plywood if they are to be used only once.

Steel mandrels or inflatable hoses are inserted into the ducts throughout the segment to stiffen the ducts and to prevent demolition in case of grout seepage. The ducts are supported and tied at close intervals to maintain alignment during concrete pouring. Tight duct support is particularly necessary in the bottom slab when concrete is placed with transverse flow. Supplemental stirrups are used in the bottom slab to secure the top and bottom mats of rebars to control delamination due to angle breaks in the ducts at the joints of varying-depth segments.

The match-cast segment is set as close as possible to the design position prior to inserting the rebar cage. After completing the cage, applying tendon anchorages, inserting the mandrels and closing the casting cell around the match-cast segment, the segment is checked and its position fine-tuned.

Figure 3.1. Thermal bowing of segments during short-line match-casting

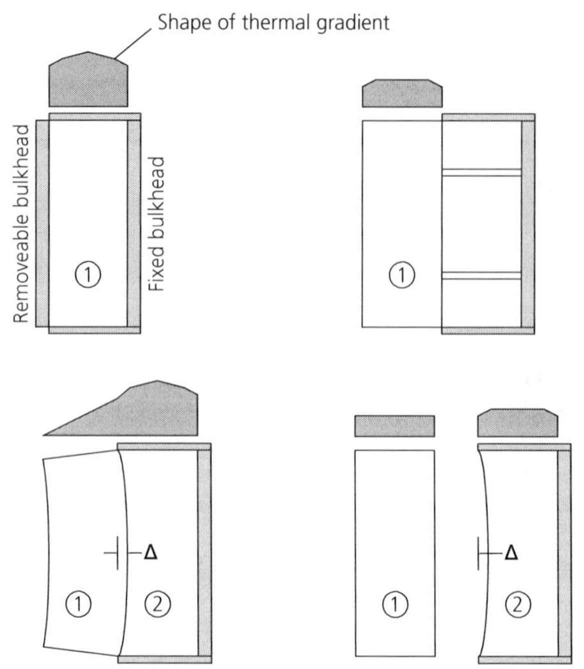

The first concrete is poured into the central portion of the bottom slab via a chute through the bulkhead or a window in the top slab form. The bottom corners of the webs are filled and compacted to complete the bottom slab, the webs are filled symmetrically in 0.50–0.60 m lifts, and the top slab is poured, starting from the centre and side edges and proceeding toward the webs. The use of retarders in the concrete mix simplifies pouring and screed finishing. Mechanical screeding is followed by a final check with a straight aluminium edge from the bulkhead to the match-cast segment. Screeds are rarely used for the bottom slab.

Thermal warping of segments is one of the few weak points of short-line match-casting (Rombach and Abendeh, 2004). The segments are cast using high-strength concrete for early form stripping and to achieve daily casting cycles. The heat of hydration in the new segment causes a thermal gradient in the match-cast segment. This gradient may cause a temporary backward bowing of the match-cast segment. The bowing that occurs before the new concrete has achieved its initial set becomes a permanent curvature in the front joint of the new segment (Figure 3.1). Thus the new segment has one straight and one curved side, whereas a match-cast segment returns to its original shape after cooling to ambient temperature.

Warped segments may cause several problems during erection. The size of the gap progressively increases as each joint is closed, and may reach several millimetres in long cantilevers comprising many segments (Figure 3.2). Tensioning the cantilever tendons causes irregular stress distributions in the cross-section, with increased longitudinal compressive stress in the side wings and reduced axial stress in the central top slab panel between the webs. Most design standards require zero tension in the epoxy joints between precast segments, and compression deficiencies

Figure 3.2. Progressive increase in gap size

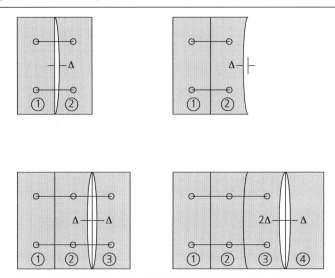

in regions of the top slab directly exposed to traffic may cause shear and flexural failure of the epoxy joints, and serviceability and durability issues. Time-dependent shortening of overloaded side wings eventually decompresses the closure pours at midspan, and may actively pull match-cast joints apart.

Thermal warping of segments may also cause excessive joint thickness, which is frequently the cause of voids, insufficient filling and stress concentration (Kamaitis, 2008). Joint thickness should be kept as small as possible to minimise the influence of the joints on the behaviour of segmental structures. Stress concentration and localised creep have often been associated with horizontal cracks in the webs radiating from thick epoxy joints.

Thermal warping may be noticeable in segments with a large width-to-length ratio and a wide bottom slab (Podolny, 1985). It can be controlled by enclosing the new segment and the match-cast segment into an isothermal enclosure to minimise thermal gradients; mild steam curing may help in maintaining a controlled temperature. Curing blankets and plastic sheeting may be sufficient in warm climates. With a production rate of one segment per day, the curing process in the casting cell cannot be more than a few hours, and a controlled environment is therefore essential.

The required concrete strength for stripping is often specified at 15–20 MPa. The geometry of the new and match-cast segment is surveyed before stripping the forms. The outer form is stripped by lowering and rotating web and wing shutters away from the segment, and the core form is collapsed and extracted backward. Both forms may be moved manually or with double-acting hydraulic cylinders. Partial application of transverse post-tensioning may be required if the top slab reinforcement is unable to carry the cantilever weight of the side wings; in this case the concrete strength is often specified as 25–30 MPa. Transverse pre-tensioning is rarely used in the top slab of precast segmental box girders.

Figure 3.3. Hydraulic manipulator for short-line match-casting (Photo: HNTB)

Hydraulic manipulators rolling on rails are used to move the new segment from the casting cell to the match-cast position, to rotate it to the casting geometry for the new segment, to support it during casting of the new segment to strip it from the new segment, and to move it to the pickup position for transportation to storage. Four double-acting cylinders at the corners of the segment soffit control rotation in the longitudinal and transverse planes (Figure 3.3). Rotation in the horizontal plane is achieved with sliding supports for the manipulator and lateral double-acting cylinders acting against foundation blocks.

Some casting cells are equipped with two form tables and two manipulators, while others have two form tables and one manipulator that shuttles back and forth from soffit to soffit. Modular manipulators are used for major adjustment of the elevation of the bottom form table in varying-depth decks (Figure 3.4). Capstans or motorised wheels are used to drive the manipulators along the runways.

The match-cast segment is separated from the new segment by retracting the rear support cylinders of the manipulator (the ones on the side far from the new segment) and by pushing the manipulator away from the new segment. The match-cast segment tilts and the bond breaks. A bond breaker is applied to the concrete surface of the match-cast segment prior to casting the new segment in order to facilitate clean separation. After moving the match-cast segment to the pickup position, the same operations are repeated on the new segment to separate it from the fixed bulkhead and to move it into the match-cast position. Stripping and pulling back the segments require care to avoid damaging the shear keys.

Figure 3.4. Modular manipulators (Photo: HNTB)

Short-line match-casting requires accurate geometry control. Size, thickness and geometry toler-ances of the segment affect the overall quality of the bridge, and are important QC requirements; geometry control of match-casting, however, is only based on the relative as-cast position of the new segment in relation to the position of the match-cast segment.

Six survey markers are used in every segment; three markers are placed at a short and constant distance from the match-cast joint, and three markers are placed at the same distance from the fixed bulkhead. Four domed-head level control bolts are placed over the webs of the box girder, where the elevation is not affected by transverse bending or top-slab post-tensioning. Two alignment hairpin markers are placed close to the centreline, and the centreline of the casting cell is punch-marked into the hairpins to obtain the surveying alignment.

A theodolite mounted on a shaded instrument tower behind the fixed bulkhead is used to survey plan alignments and elevations relative to the sight to a reference target placed on a permanent tower at the opposite end of the foundation plate of the casting platform, beyond the segment pickup point. Benchmarks are maintained in the precasting facility to frequently check the position of the instrument, target and fixed bulkhead, and to re-establish the geometry control base should the instrument or target be disturbed during segment production.

After finishing the top slab of the new segment, the six survey markers are installed and surveyed. Prior to stripping the forms, the six markers of the match-cast segment are also surveyed to determine whether the segment has moved during casting of the new segment due to settlement of supports, concrete vibration or loads applied during form closure. The critical readings are those taken after casting: although accurate setup before casting is indispensable, movements are very frequent and must be compensated for in the next segment (ASBI, 2008).

The new segment is moved to the match-cast position and reset to follow the casting curve and to compensate for the geometry imperfections that have occurred during casting. The plan offset of the far alignment marker from the instrument–target alignment determines the horizontal curvature, and the vertical offset of the far elevation markers determines vertical curvature and crossfall. The forms of the casting cell are not part of geometry control and are kept plumb and levelled.

Most casting cells are fabricated so that the bulkhead is orthogonal to the cell centreline, the bottom form table is rectangular, and the web and wing forms are parallel to the cell centreline. Horizontal alignment is attained by rotating the match-cast segment in plan. As a result, the bulkhead joint is not radial in curved bridges; it is orthogonal to the chord connecting the centreline end points of the segment.

3.1.1.2   Long-line match-casting

With the long-line method, the segments of a casting run are match-cast over a long support rail by moving the casting cell from one segment to the next without moving any segment. When the segments are erected span-by-span, the casting run includes all the segments of the span. When the segments are erected as balanced cantilevers, the casting run includes all the segments of one cantilever or of both cantilevers. When both cantilevers are cast at once, the hammerhead assembly is also cast in the long-line bed.

The long-line method was the first method used to precast segments (ASBI, 2008). The advantages include simple casting cells and procedures, as all the geometry control is done when setting up the soffit. The main disadvantages are the cost of the foundations for the support rails, the need for a larger precasting facility, and minimal allowance for plan and vertical curvature.

A stiff support rail is necessary to prevent settlement due to the segment weight affecting the geometry of the casting run. Two longitudinal concrete beams located under the deck webs replicate the casting curve. The bottom form table spans the support beams or is supported on shoring towers, and may be as long as the entire rail to save labour. Shorter form tables rolling along the rail have also been used to close the casting cell at multiple locations and to allow some vertical adjustment with shims or screw jacks. The side shutters, core form and front bulkhead are as long as one segment, and roll along the foundation slab of the support rail. The bulkhead and core form are adjustable to the geometry of varying-depth segments. Two sets of forms are used to cast the two cantilevers symmetrically.

After completion of a casting run, the segments are progressively separated and transported to storage, and the vertical profile of the rails is adjusted for the new span. Wide-flange beams are supported on the concrete rails with two-nut bolts for geometry adjustment, as used in the casting cells of incrementally launched bridges. The deck webs are match-cast over the top flange of the support beams.

Long-line match-casting was initially conceived for varying-depth segments of balanced cantilever bridges. Nowadays it is also used to cast multiple short segments of incrementally launched bridges between two subsequent launches (Rosignoli, 2001, 2002) (Figure 3.5) or to cast 35–50 ton U-segments for dual-track LRT bridges (Vion and Joing, 2011) (Figure 3.6). The pier segments for these bridges weigh 50–75 ton, depending on segment length and integration of support ribs, and are match-cast against the adjacent typical segment with the short-line

Figure 3.5. Long-line match-casting of 10 m segments for launched bridges

Figure 3.6. Long-line match-casting of dual-track U-segments for LRT spans (Photo: Systra)

method. One segment is cast per day per casting bed. The shutters are relatively inexpensive and can accommodate different geometries.

3.1.1.3 Post-casting

After casting, there is still much work to do to prepare the segments for delivery and erection (ASBI, 2008). When the segments have reached the required strength, the transverse tendons in the top slab are fabricated, stressed and grouted, and their anchorages are sealed. Some cosmetic work will also be necessary to ensure that the segments meet appearance requirements (AASHTO, 2003).

The final step is light sandblasting of the match-cast joints to remove laitance, dirt and debonding agents prior to delivery for erection. Project specifications and manufacturer's recommendations for the epoxy joint adhesive with regard to surface preparation define the sandblasting equipment, compressor capacity, and composition and grit size of the sandblasting material. Excessive sandblasting (and also repairs of surface defects and epoxy-grouted cracks) may prevent matching of joint surfaces during erection.

The segments are finished either in a special finishing area or at the storage location; some cosmetic work may also be done in the pickup position prior to transportation to storage. The use of a finishing area requires double handling, but finishing is faster and more cost-effective in a stationary setup, and quality, safety and supervision are also enhanced.

In most cases, the majority of the finishing work is done in the storage area. When the segments are finished during storage, enough space is left between the rows of segments to allow aerial personnel lifts or scissor platforms to manoeuvre between the segments and gain access to the transverse tendons with a stressing jack (ASBI, 2008). The equipment used dictates the room required between the rows of segments.

Many solutions have been developed for the lifting beams used to handle the segments in the precasting facility and during erection. Most of these schemes use holes in the top slab of the segment to secure hanger bars. The use of slings is rarely compatible with balanced cantilever erection because the erection equipment holds the segment in place during joint closure. For balanced cantilever segments it is difficult to identify a single hole arrangement that will always avoid the tendon ducts in the top slab, and the lifting beams are therefore designed to accommodate variations in the lifting points.

Securing the hanger bars under the soffit or the side wings requires access outside the segment for bar release after gluing. The lifting holes are therefore located at the top web-slab nodes within the box cell. The holes are formed of oversized corrugated post-tensioning ducts or tapered inserts. Positioning frames are used in the casting cell to keep the ducts in place during casting of the segment. Multiple positioning frames may be needed to accommodate the variations in the lifting hole layout, and proper QC procedures are developed to ensure that the right layout is used. Drilling cores to create new holes at the right location is an expensive remedy to poor QC. After segment erection, the lifting holes are plugged with non-shrink grout, and the surface must therefore ensure a proper bond in order to prevent the plug from falling out.

Depending on segment weight and geometry, four, eight or more hangers are used to handle the segment. High-strength post-tensioning bars with diameters ranging from 32 mm to 76 mm are used for the hangers. Bars and anchor nuts are reused 20–30 times when stressed to 50% of the ultimate tensile strength, and the use of couplers should be avoided. The bars are stressed to generate friction shear capacity that prevents bar bending and shearing due to segment movements during handling. Gradient and crossfall of the top slab of the segment are compensated for by changing the connection points of the lifting beam and shimming at the hangers, by varying the sling length or applying pull cylinders to the slings for synchronous hoisting, or with lifting beams that hydraulically adjust the location of the frame relative to the segment.

Embedded inserts protruding from the top slab have also been used for handling. Embedded inserts include high-strength bars, looped strand bundles or special mechanical devices. High-strength bars are encased in post-tensioning ducts, and the bottom nut and anchor plate are embedded into the segment web. Full engagement of the anchor nut cannot be confirmed visually, and the use of these hangers therefore requires proper QC. After segment erection, the bar portion protruding from the top slab is trimmed to provide adequate concrete cover, and the duct is grouted.

Special inserts and looped strand bundles are also used to handle the segments. The top of the strand loop is inserted in a steel pipe that is bent into a U-shape to ensure that the load is evenly distributed between the strands. The bottom of the loop is secured to web rebars, and a recess form is used at the top. After erection, the strands are flame cut within the recess to provide proper concrete cover, and the recess is filled with patching mortar (ASBI, 2008).

Slings and nylon straps are frequently used to handle the segments for span-by-span erection with underslung gantries. Rubber, plastic or wood shoes are used around the edges of the segment to prevent damage. Combinations of handling devices are often used on the same segment: slings are used in the precasting facility in combination with more specialised means at the erection site.

3.1.1.4 Delivery
Before transporting the segments to the erection site, the QC documents are checked for segment acceptance. Checking includes concrete strength, curing duration, stressing and grouting of transverse tendons, repairs, alignment of inserts and permanent and temporary prestressing ducts, proper identification and orientation of the segment, and preparation of the match-cast faces.

Depending on site conditions and location of the precasting facility relative to the erection site, the segments may be transported via water or land. When the bridge is over deep water, segment delivery via tugs and barges offers many advantages. If the precasting facility does not have access to a navigable waterway, the segments can be transported to a docking facility, although this requires double handling and a staging area. Triple handling may be necessary when a portion of the bridge is over land.

The loaded draft of barges and tugs is an important factor in analysing a potential docking facility. Water depth is analysed in the navigation channel and the working areas outside the channel. Analysis includes tides, seasonal variations, current, prevailing winds and wave actions. Freezing waterways can have a major impact on segment delivery. Circumstances that may cause a fully loaded barge to bottom out or to damage the docking facility must also be evaluated.

An existing docking facility offers the advantage of not requiring permits to be obtained, but the facility must be checked for structural integrity above and below water according to the required operating limits and maximum vessel size. Waterway width and the potential for blocking the channel during segment loading should also be considered.

Twin loading piers protruding in the waterway are often the least expensive solution; they consist of two rows of piles that support runways for a portal crane or a straddle carrier. Bulkheads parallel to the shoreline are easier to permit, but may block the channel during loading of barges and require a loading crane with enough capacity to reach the centreline of the barge. Loading slips dug within the property are expensive due to the size of the barges used for segment delivery.

On arrival at the erection site, match-cast faces and exposed rebars are pressure washed to remove any salt spray. If the post-tensioning ducts were not sealed prior to shipping, they are washed with potable water and dried using oil-free compressed air.

If the segments can be delivered within the reach of the erection equipment, they are unloaded directly from the barge. Wave action complicates securing the lifting beam to the segments, and special attachments may be necessary during application and stressing of the hangers. Dynamic allowance for impact loads is used in the design of all components of the hoist system (AASHTO, 2003, 2012; BSI, 2006; CNR, 1985a; FEM, 1987). If the segments have to be transferred from the barges to land haulers to reach the erection site, a docking facility with a staging area is needed.

When the precasting facility is not at the erection site, the segments are often transported on land by tractors/trailers or special haulers. The segments are loaded with the portal cranes and straddle carriers used for handling the segments within the precasting facility.

Several issues are considered when analysing segment transportation over land (ASBI, 2008). Maximum size and weight of segments in relation to legal loads, allowable axle load and axle layout, over-weight and over-dimension permitting procedures, load rating of bridges, box culverts and underground utilities along the delivery route, height and width restrictions along the route, distance from the erection site, restrictions on the hauling time related to traffic congestion, number of permitted loads anticipated and shipments per day, cycle time and availability of haulers, signage, escorts, convoy allowances, police details, hours of operations, and upcoming construction projects along the route all affect the optimal tractor/trailer combination and their number for delivery.

Depending on segment weight and permit requirements, multi-axle trailers are often used to haul the segments. Heavy segments may require self-levelling hydraulic haulers with steering wheels pulled by special counterweighted tractors. When possible, the segments are delivered directly to the erection equipment in order to avoid double handling and storing onsite.

3.1.2 Segment connection at the erection site

On early bridges the segments were connected with thin in-place concrete or mortar joints to avoid match-casting. Nowadays wet joints are used only to close continuous spans, and the segments are match-cast and glued with epoxy. Some projects have also used match-cast segments erected without any epoxy in the joints, as this 'dry joint' technology in combination with external post-tensioning leads to faster erection rates.

Different two-component epoxy formulations are used depending on the environment temperature. The pot life of the epoxy may range from a few minutes to a few hours depending on the formulation, ambient and storage temperature, and quantity. Span-by-span erection suggests the use of slow-set resins to glue and press most of the span, while balanced cantilever erection suggests the use of normal-set resins (ASBI, 2008). After application of the epoxy, the joints are closed and pressed rapidly. When hardened, the epoxy provides structural continuity between the segments. The epoxy has a minimum compressive strength of 60–70 MPa, which gives a connection stronger than concrete on either side. The elastic modulus of epoxy is lower than that of concrete and the creep coefficient is typically much lower. These factors, combined with the time-dependent rheological behaviour of the epoxy, require the use of thin joints (Kamaitis, 2008).

The match-cast joints to be glued are lightly sandblasted prior to shipping to remove the bond breaker and curing compound and to provide a good bond with the epoxy. The segment is initially offered to the previously erected segment on a 'dry run', with the gluing bars already in place to ensure that everything fits together and matches. The segment is then moved back and the epoxy applied to one or both joint surfaces, after which the segment is pulled into position and the gluing bars stressed.

Epoxy is usually applied by hand. Spread evenly over the surface, the adhesive lubricates the joint during erection, allowing the segment to slide into position. Stressing the gluing bars squeezes the epoxy across the joint to provide a 1.0–1.5 mm water- and grout-tight seal, devoid of voids and irregularities. A plastic sheet is suspended beneath the joint to catch the epoxy as it drops. Plastic pipes are applied to external gluing bars for protection. Epoxy squeezed onto the top deck is swabbed down flush prior to hardening to facilitate deck grinding or the application of waterproof membranes prior to asphalt paving.

Span-by-span precast segmental bridges typically use external tendons for longitudinal post-tensioning, and therefore no ducts cross the joints (Meyer, 2011). Balanced cantilever bridges use internal cantilever tendons in the top slab and often also internal continuity tendons in the bottom slab at midspan (Harridge, 2011). The cantilever tendons stretch from end to end of the hammer, and segments near the pier accommodate one set of tendons for each segment in the cantilever.

Rubber O-rings glued to the segment face around the holes and leaving the epoxy 20–30 mm clear around the holes was the standard solution to prevent the epoxy from being squeezed into the ducts during joint pressing (Hewson, 2003). Duct couplers have been recently introduced to ensure leak-tight connections for internal tendons at the joints and to eliminate the need for swabbing the ducts after stressing the gluing bars.

Two plastic inserts are incorporated in the segment faces during casting, and a central sealing element is screwed into one insert and mechanically engaged into the other insert (ASBI, 2008). Plastic leak-tight connections do not corrode, and prevent ingress of epoxy during segment gluing and of aggressive agents over the service life of the bridge. Leak-tight connections also prevent grout crossover and can be used with high-density poly-ethylene (HDPE) ducts to provide a complete leak-tight plastic encapsulation of the tendons. Some duct couplers do not require increased duct spacing, and the tendon can cross the joint with a slight angle to the bridge axis. Conventional mandrels are used at standard bulkheads to support the ducts.

Commercial post-tensioning bars are used for joint pressing during gluing; they are quick to install and hold the segments in place until the permanent tendons are installed and tensioned. Fast securing of the segment is particularly important for balanced cantilever erection, as early release of the erection hoist shortens the cycle time.

The gluing bars may be either internal or external, and are designed to generate an average compressive stress of 0.2–0.3 MPa during the curing time, which varies in relation to the type of epoxy used and the temperature. The internal bars are left in place and grouted to become part of the permanent prestressing; they are stressed to full design capacity and the subsequent bars are coupled onto the previous set. Combinations of internal and external bars are often used in wide decks, as permanent bars in the top slab are quick to couple, are very effective in controlling shear-lag effects, and do not require forming operations for anchor blisters.

The advantage of using external bars is that they are easy to de-stress and reuse. In some applications the external bars have been positioned above the slabs and anchored to steel brackets stressed down to the segment (ASBI, 2008). Small anchor blisters cast on the interior of the box girder (two at the top web-slab nodes and one at the centre of the bottom slab) have become the standard solution to avoid holes in the slabs, to save labour during segment erection, and to facilitate access to both ends of the bars for fast removal.

The gluing bars may be overlapped so that individual bars or coupled bars extend over a few segments, or they may be coupled throughout the span. The bars are typically stressed to 50% of the guaranteed ultimate tensile strength and reused a few tens of times, and a greater number of bars is therefore required. When it is necessary to exceed 50% of the tensile capacity, bars and anchor nuts are marked (painted) for 'no reuse and disposal' immediately after de-stressing.

When the segments are erected, they join up to the adjacent segments with the same horizontal and vertical angle break as existed when they were match-cast. This is facilitated by the presence of shear keys on the joints that guide the segment into position. Shear keys cover as much of the web surface as is achievable to ensure shear transfer throughout the joint. Alignment keys are also used in the top and bottom slabs to ensure proper segment alignment and to transfer local shear at the joints.

Span-by-span erection is less susceptible to geometry discrepancies than is balanced cantilever erection, because the spans are shorter and the segments are longer. One or two closure pours per span also facilitate geometry adjustment. Shimming the epoxy joints with woven glass-fibre matting to provide wedge-shaped joints for alignment correction is, therefore, rarely necessary.

The joint surfaces at the closure pours must be clean, free of laitance and roughened to expose coarse aggregate. AASHTO (2003) requires 6 mm roughness, which is easy to achieve by bush-hammering or chipping. Application of bond enhancer to the joint surface is often specified to increase adhesion of fresh concrete.

Several precast segmental decks have been built with no epoxy in the joints (Rombach, 2004). Avoiding the epoxy speeds up erection, removes risks and saves costs, although more prestressing may be required to maintain the ultimate shear and moment capacities. Dry joints do not achieve a watertight seal between the segments, and are therefore used only in tropical climates where freeze–thaw cycles do not occur. The need for external prestressing makes dry-joint erection a valid alternative only for span-by-span bridges. In the balanced cantilever bridges, most tendons are designed for negative bending and are therefore located within the top slab.

Wire brushing is used to clean the dry joints, as sandblasting may damage the mating surfaces. Joint pressing is not needed, although a few bars may be useful to pull the segments together during erection and to hold them in place until the permanent tendons are installed. The top of the joint is sealed to prevent water seepage through the top slab. Early techniques included inserting a rubber tube into a groove running across the full width of the top slab. Dry joints have recently been made watertight with epoxy seals poured into narrow recesses along the top slab surface (ASBI, 2008).

3.2. Technology of span-by-span erection of precast segments

Span-by-span construction was initially conceived for in-place casting and was subsequently extended to precast segmental bridges (Meyer, 2011). This method is used for simple and continuous spans. Simple spans of highway bridges may be connected with link slabs at the top-slab level to minimise the number of expansion joints without the complication of full structural continuity. The simple spans of LRT and railway bridges with direct track fixation are connected with inexpensive waterproof strips, as the joints are not affected by train traffic. Special joints are used for ballasted railway bridges.

Four structural solutions are used in combination with span-by-span erection (Meyer, 2011).

1 All spans are simply supported. This is the simplest method with regard to segment production, span erection and prestressing layout. Each pier restrains a span longitudinally and the expansion joints improve stress distribution in the continuous welded rail; simple spans are, therefore, the typical choice for LRT and railway bridges. The disadvantages include a great number of expansion joints, a double number of bearings, and poor control of train-induced span resonance.
2 The spans are erected as simply supported, and made continuous after repositioning the gantry. A first set of tendons makes the span self-supporting and the gantry releases the span onto jacks. Unreinforced wet joints connect the ends of the span to the pier-head segments, which are positioned on the permanent bearings prior to span erection. Two wet joints in every span simplify geometry control during segment production and do not require match-casting of the pier-head segments, and span completion is not on the critical path. Precast or cast in-place high-strength mortar spacers are used to lock the span in place at the closure joints, and 5–10% of the second-phase post-tensioning is applied in order to avoid involuntary joint displacements. The closure joints are cast at the end of the work shift, the continuity tendons are fully stressed the following day, and the jacks are finally released and removed. Construction loads on the previous spans are often prohibited during curing of the closure joints to avoid damage due to movements or vibrations. This construction method requires wide pier caps to support the spans during construction, additional diaphragms in the end segments to anchor the first-phase tendons, a larger number of anchorages, and more complex tendon layouts (Meyer, 2011).
3 The deck is built as a sequence of continuous spans, with wet joints at the front joint of the leading pier-head segment. The curing time of the wet joint is on the critical path, but

prestressing is simpler and less expensive as the span becomes self-supporting in the final structural configuration. Splicing internal tendons by overlapping avoids the finishing work of web couplers. The rear joint of the leading pier-head segment may be glued to the first span segment if match-cast.

4   The deck is built as a sequence of continuous spans with wet joints at the rear quarter/fifth of the span. The gantry takes support on the front deck cantilever, the shorter distance between the supports lightens the gantry and diminishes the rear support reaction, and negative bending at the leading pier diminishes the time-dependent stress redistribution of staged construction within the deck. The gantry erects the new pier-head segment and takes support on it during launching. Gantry operation is simpler and faster when continuity is achieved at the pier-head segments, and closure joints along the span are therefore rarely used.

Span-by-span erection of precast segmental bridges is a fast construction method that minimises traffic disruption on existing highways and railways (ASBI, 2008). This method is very efficient in LRT applications, where the weight of the span is relatively low, but it has also been applied successfully to highway bridges. LRT bridges are erected in congested urban areas, where design constraints often require tight plan radii. Transporting short deck segments is easier than handling long precast beams, segment erection is also easier, and site activities are minimised. Precast segmental bridges may be erected using one of four span-by-span techniques.

1   With shoring falsework that supports the segments. The segments are delivered on the ground and loaded onto the falsework with a ground crane. Sliding header beams are used to adjust the geometry of the segments. When the piers are tall, the span may be assembled (or cast) on the ground or barges and strand jacked into position with two lifting frames or an overhead gantry after application of prestressing.

2   With a twin-girder overhead gantry that suspends an entire span of segments with hangers and spreader beams. The gantry carries an overhead travelling crane that picks up the segments from the ground or the deck and moves them into position.

3   With a single-girder overhead gantry that suspends the segments with hangers and suspension frames. Most of these gantries use underslung winch trolleys to handle the segments. The telescopic gantries comprise a rear main girder and a front self-launching underbridge to allow operation within tight plan and vertical radii.

4   With a twin-girder underslung gantry that supports box girder segments under the side wings. The segments may be loaded onto the gantry with a ground crane, a portal crane carried by the gantry, or a crane or a lifting frame placed on the leading span of the completed bridge. Articulated girders with hydraulic hinges allow operation within tight plan radii.

Span erection on shoring falsework (AASHTO, 2008a, 2008b; CNR, 1985b; CSA, 1975) is a cost-effective option in small bridges with short piers and accessible areas. It is often used for start, end and odd spans of balanced cantilever bridges, and for short spans with tight plan radii. Shoring falsework may also be used to support the gantry during erection of the starter span (Figure 3.7).

The use of commercially available scaffolding and cranes minimises the investment in special equipment; this method, however, is labour intensive (ground preparation, construction of tower bases, tower erection, tower adjustment and bracing, installation of header beams and

Figure 3.7. Gantry erection of a 32.5 m, 460 ton starter span (Photo: BWM)

jacks, and dismantling and relocation after assembly), and is much slower than gantry erection, unless multiple spans of falsework are provided (Homsi, 2012).

In most cases the only real alternative is between an overhead and an underslung gantry (Meyer, 2011). Both types of gantry can be loaded from the ground or from the deck. The width of the segments may amply exceed 20 m, although it should be less than one-third of the span length to facilitate rotation during loading (Meyer, 2011). The gantry supports an entire span of segments during erection and weight, cost and complexity of the gantry increase with span length and deck width.

Span-by-span erection is mostly used for 40–50 m spans in highway bridges and 35–45 m spans in dual-track LRT bridges. On longer highway spans, balanced cantilever erection soon becomes less expensive in spite of higher prestressing costs, due to the cost and complexity of the erection gantry for span-by-span construction. Longer spans are rarely used in LRT bridges due to deflections and train-induced resonance in simple spans. On shorter spans the method competes with precast beams and cast in-place deck slab, and becomes cost-effective only in very long bridges.

Some first-generation cable-stayed gantries for balanced cantilever erection of 100–120 m spans were able to suspend a 50–60 m span of segments for span-by-span erection (Figure 3.8). The deviation tower was placed at the leading pier of the bridge, and the front cable-stayed canti-lever suspended the segments for the new span. Counterweights and tie-downs applied to the rear end of the gantry controlled overturning in combination with a light front auxiliary leg. In spite of their versatility, the cable-stayed convertible gantries have found limited application because of their complexity and labour demand (Rosignoli, 2002). Varying-depth trusses are also rarely used in span-by-span construction, as they are difficult to reuse on different span lengths.

Figure 3.8. First-generation cable-stayed convertible gantry

The gantry supports an entire span of segments during erection. The span weighs 200–2000 ton and the gantry weighs 200–1000 ton, and load imbalance during construction often governs pier design. The overhead gantries take support on the pier-head segments, most underslung gantries take support on pier brackets, and the design of precast segmental bridges for span-by-span erection therefore involves several technological requirements.

3.3.   Twin-girder overhead gantries

A twin-girder overhead gantry comprises two trusses, braced I-girders or box girders supported on crossbeams (Figure 3.9). Trusses are lighter and facilitate access by and mobility of workers, while braced I-girders and box girders are more stable and solid and allow for robotised welding. Field splices designed for fast assembly develop member strength to permit alternative assembly configurations that take advantage of the modular nature of the design.

Overhead winch trolleys or portal cranes bridging between the girders are used to lift and move the segments into position. When the segments are delivered on the deck, the crane picks them up at the rear end of the gantry, moves them out within the clearance between the girders and above the support crossbeams, lowers them beneath the girders, and rotates them by 90°. When the segments are delivered on the ground, the crane reaches down to them, rotates them to the orthogonal position, and lifts them up to the deck level. Hanger bars or ropes are used in combination with spreader beams to suspend the segments from the gantry so that the crane can be released for another cycle.

The overall geometry is similar to that of a twin-truss beam launcher, but the gantry is designed to take the weight of the entire span, and most components are therefore heavier and stiffer. The beam launchers carry two winch trolleys, while only one crane is used for span-by-span erection.

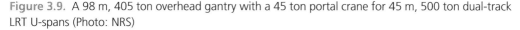

Figure 3.9. A 98 m, 405 ton overhead gantry with a 45 ton portal crane for 45 m, 500 ton dual-track LRT U-spans (Photo: NRS)

Portal cranes are often preferred to flat-frame winch trolleys in new-generation machines, as they are easier to reuse in different applications and are also able to handle the segment hangers to avoid conflicts with the crossbeams during launching.

The overhead gantries are not much affected by ground constraints, straddle bents, C-piers and variations in span length and deck width and geometry. These machines, however, are more complex to design, assemble and operate than are underslung gantries, and are also more expensive and slower in erecting the segments.

Trussed extensions are applied to the main girders to prevent overturning during launching. The length of the gantry is 2.1–2.3 times the typical span, but this is rarely a problem because the overhead machines do not interfere with the bridge piers. Launch noses and tails are heavier than those for underslung gantries, because the tails are used to pick up the segments when delivered on the deck, the noses are designed for placement of the pier-head segments, and both extensions are also used to reposition the support crossbeams with the winch trolley during launching. Therefore, the launch noses and tails of most overhead gantries have the same depth as the main girders in order to extend the operation range of the overhead crane. Launch noses and tails are shorter and lighter than the noses for launched bridges (Rosignoli, 1999).

An underbridge supported at the leading pier of the span to erect and the next pier has been used in some machines in combination with a rear rectangular frame comprising a front crossbeam sliding over the underbridge during launching, two main girders, and a rear C-frame rolling over the completed span. These telescopic gantries are fit to bridges with tight plan and vertical radii. The total length of the main frame and the underbridge is 2.1–2.3 times the typical span, but the articulation between the underbridge and the main frame offers unrivalled flexibility on complex bridge alignments.

As a drawback, the telescopic gantries are tall and sensitive to lateral wind. They must be anchored to the bridge with great care, as the support frames are less stable than the crossbeams of twin-girder overhead gantries and the central articulation further diminishes the torsional stiffness of the machine.

3.3.1 Loading, kinematics, typical features

Hangers and spreader beams are used to align and hold the segments in position during gluing (Figure 3.10). The spreader beams are applied to the segments with stressed bars crossing the top slab to avoid involuntary load displacements and shear loads in the bars. The lifting beam of the winch trolley picks up the spreader beams for lifting. Hanger bars or ropes are used to suspend the spreader beams from the gantry to release the hoist for a new cycle.

Adjustable spreader beams are used to shift the segments of curved spans to the offset location while keeping the hangers aligned with the gantry (Figure 3.11). The hangers are suspended with adjustable anchorages from the bottom chords of the truss. After releasing the span, the spreader beams are removed from the segments and hangers are lifted to avoid conflicts with the support crossbeam of the gantry during launching; the winch trolley is often equipped with monorail hoist blocks to assist these operations.

Figure 3.10. Hangers and spreader beams

Figure 3.11. Segment suspension systems

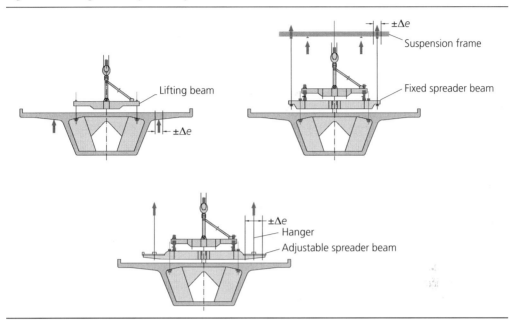

The hangers for fixed spreader beams follow the plan curvature of the bridge and are anchored into adjustable suspension frames spanning between the top chords of the gantry. One suspension frame and two hangers are used for every segment. Dual-track U-segments for LRT bridges can be suspended from the top flange without any spreader beams (Figure 3.12).

The main disadvantage of fixed spreader beams is that the suspension frames interfere with the operation of overhead travelling cranes. Underslung winch trolleys spanning between the bottom chords of the trusses are often used in these machines. Top working platforms are necessary to adjust the geometry of hangers and suspension frames; the working platforms may be lifted over the main girders to create the operation clearance for an overhead winch trolley (Figure 3.13).

When the segments are delivered on the deck, they are moved out with the long side parallel to the gantry to avoid interference with the hangers of the previously placed segments. The segments are rotated during final lowering. When a winch is used to lift the segments, rotation is achieved with a hydraulic slewing ring applied to the lifting beam. When pull cylinders or hoist bars are used for lifting, the hoist is rotated within the crab. Segment rotation above the deck requires a longitudinal clearance between the hangers that is longer than the long side of the segment. Segments wider than one-third of the span are, therefore, delivered on the ground, rotated beneath the deck, and lifted into position.

Typically all segments but one are suspended from the gantry before the gluing of joints in order to avoid deflections during joining. Epoxy is applied to groups of segments; even with slow-set epoxies three or four segments is the practical limit for pressing the group of segments together within the pot life of the epoxy. Finally, the remaining segment is installed.

Figure 3.12. Direct suspension of U-segments (Photo: Systra)

Figure 3.13. Lifted working platform for operation of an overhead winch trolley (Photo: VSL)

The solutions of continuity with the pier-head segments of continuous spans are locked with concrete shims and light tensioning of a few permanent tendons before pouring the in-place stitches and completing post-tensioning. The tendons are grouted after repositioning the gantry to remove these operations from the critical path.

3.3.2   Support, launch and lock systems

Many twin-girder overhead gantries have two main supports and two auxiliary legs. Crossbeams with adjustable support legs are used for the main supports, and articulated frames are used for the auxiliary legs. The auxiliary legs connect the trusses at the ends and provide some degree of torsional restraint.

1   For the erection of simple spans, the front crossbeam frame takes support on the front half of the leading pier cap, and the rear crossbeam takes support on the pier diaphragm of the last span. An entire span of segments can thus be erected without conflicting with the support systems of the gantry. The rear crossbeam provides the longitudinal restraint to the gantry.

2   For the erection of continuous spans with in-place stitches at both pier-head segments, two crossbeams take support on the pier-head segments, and the span is released onto jacks for stitching and completion of prestressing.

3   For the erection of continuous spans with in-place stitches at the rear pier-head segment, both crossbeams take support on the pier-head segments, and the span is released after stitching and partial application of prestressing.

4   For the erection of continuous spans with in-place stitches at the rear quarter/fifth of the span, the front crossbeam takes support on the leading pier-head segment, and the rear crossbeam takes support on the front cantilever of the completed deck. Also in this case, the span is released after stitching and partial application of prestressing.

The prestressing tendons are fabricated and tensioned from a stressing platform applied to the leading end of the span. The platform may be suspended from the gantry or supported on brackets anchored to the front face of the pier. The stressing platform is wide enough to thread the stressing jack over the strand tails extending from the anchor heads, and it is designed to allow insertion of construction materials and equipment into the box cell. Span-by-span erection requires many activities to be performed within the box cell, and these activities require tools, equipment and materials. After completing the span and repositioning the gantry, it will be difficult to store bulky materials such as HDPE ducts, strand and gluing bars for the next span in the box cell. Handling construction materials is much easier when the crews can be assisted by the winch trolley of an overhead gantry. Likewise, waste materials should be disposed of as each span is completed, rather than waiting until the end of the project (ASBI, 2008).

The leading edge of the deck changes constantly during span-by-span erection, and it is not unusual for site conditions to change drastically within one shift. It is therefore indispensable to enforce strict procedures for communicating hazards and protecting site personnel. Personnel working on individual segments during handling and gluing are typically tied off.

The amount of space available on the pier cap for the support systems of the gantry is often limited. Thin piers often require front brackets to support the pendular leg of the gantry during placement of the pier-head segment and repositioning of the front crossbeam (Figure 3.14). These load conditions often govern pier design, and may cause instability in tall

Figure 3.14. Front pier brackets (Photo: VSL)

and flexible piers. The couple due to the eccentric vertical load causes forward deflection of the pier cap, the displacement of the pier cap inclines the pendular leg, the inclined leg now also applies a forward horizontal load to the pier cap, the displacement amplifies increasingly, and the system may become unstable. All potential sources of pier flexibility (foundations, tolerances in verticality, positioning tolerances of the pendular leg on the pier cap, P-delta effect, cracking of cross-sections) should therefore be accounted for in the analysis (ASBI, 2008).

Auxiliary legs support the gantry during repositioning of the crossbeams and provide some extent of torsional end restraint to the trusses. The winch trolley moves the front crossbeam to the new pier with the gantry supported at the front leg and at the rear crossbeam. The rear crossbeam is then moved to the front end of the completed deck with the gantry supported at the front cross-beam and at the rear leg. Some gantries have a third auxiliary crossbeam for additional support during launching (Figure 3.15). No ground cranes are used during repositioning of supports: the gantry is also able to apply the front pier brackets if necessary, even if additional operations lengthen the cycle time.

The front pendular leg is also used to support the gantry during placement of the pier-head segments of continuous spans. The segment is supported on four jacks and anchored to the pier cap, and the front crossbeam is moved onto the segment and anchored to it. Front pier brackets are sometimes necessary to stabilise the assembly on narrow pier caps. The auxiliary legs are not used during span erection.

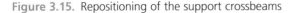

Figure 3.15. Repositioning of the support crossbeams

The pendular legs are applied to the trusses with sliding pinned connections to set the frame vertical when the trusses follow the bridge gradient and to rotate the frame about the vertical axis to the local plan radius. Long-stroke double-acting cylinders are used to take support on stepped pier caps, to adjust the frame length in bridges with vertical curvature, and to recover the deflection of the cantilever trusses when landing at the new pier. Pendular extensions are used to shorten the frame during assembly and decommissioning of the gantry behind the abutment, and in case of reverse launching. Lateral cylinders are used to position the support cylinders over the webs of curved bridges. In the presence of so many systems for geometry adjustment, the auxiliary legs are among the most delicate components of a gantry and are prone to specific forms of instability.

The crossbeams support the gantry with articulated saddles that permit launching, traversing and rotations in the horizontal and longitudinal planes. The crossbeams comprise I-girders connected by lateral braces and cross diaphragms; long crossbeams with four lines of supports and crossfall shims are used when adjacent decks are erected simultaneously to shift the gantry from one alignment to the other, (Figure 3.16). Simultaneous erection of adjacent bridges provides fast erection rates, but requires two complete sets of hangers and spreader beams (Figure 3.17).

The crossbeams are equipped with two lines of double-acting support cylinders that provide geometry adjustment on stepped pier caps and decks with gradient and crossfall. The support lines are spaced longitudinally to resist horizontal loads applied by the launch saddles. Because of the critical role of the support cylinders in assuring torsional stability to the gantry, the locknuts

Figure 3.16. Long crossbeams for simultaneous erection of adjacent decks (Photo: VSL)

Figure 3.17. Simultaneous erection of adjacent bridges (Photo: VSL)

Figure 3.18. Support saddle with vertical geometry adjustment (Photo: VSL)

are kept tightened during most operations. The crossbeams are anchored to the deck and pier cap with stressed bars that resist uplift forces. Long overhangs are necessary in the crossbeams to create the central working clearance between the trusses and to traverse the gantry when launching along curves, and significant uplift forces may therefore arise in the anchor systems.

The launch saddles comprise transverse bogies that roll along the crossbeam, and longitudinal bogies that support the bottom chords of the truss. Cast-iron rolls are assembled on articulated equalising beams to balance the load in the rolls. The top assembly of launch bogies rotates about a vertical pivot, and is articulated to follow flexural rotations of the trusses and the launch gradient. Vertical double-acting cylinders lodged within the launch saddles lift the gantry to the span erection elevation and lower it back onto the launch bogies to release the span in one operation. In some gantries the vertical cylinders have been integrated in the launch saddles to provide geometry adjustment also during launching (Figure 3.18).

Rectangular rails welded to the bottom flanges of the trusses and to the top flanges of the cross-beams facilitate dispersal of the support reactions within the webs guide, which are anchored to the roll flanges to transfer lateral loads, and keep the webs aligned with the bogies. One line of launch saddles lodges longitudinal restraint systems for the gantry, and both crossbeams are equipped with transverse restraint systems.

Most first-generation gantries are launched with capstans. In more modern machines, long-stroke double-acting cylinders lodged within the launch saddles push the trusses forward by acting within perforated rails fixed to the trusses (Rosignoli, 2000). Two launch cylinders are used for

Figure 3.19. Twin launch cylinders acting into a perforated rail

redundancy and to restrain the gantry with one cylinder during repositioning of the other (Figure 3.19).

Capstans or double-acting cylinders are also used to shift the launch saddles laterally along the crossbeams. When anchored into subsequent holes, the shift cylinders are paired to hold the saddle in place with one cylinder during repositioning of the other (Rosignoli, 2000). The saddles may be connected to keep the distance between the trusses constant (Figure 3.20). The shift cylinders also act as transverse restraints during gantry operation.

3.3.3   Performance and productivity
Twin-girder overhead gantries are the first-choice solution where the clearance under the bridge is limited. They are able to erect varying-width decks, transition spans and spans supported on or integral to straddle bents and L-piers. These gantries do not require ground cranes for repositioning of supports but, because of the higher level of automation, they are heavier and more expensive than underslung gantries. They also need hangers and spreader beams to suspend the segments, and site assembly and launching are also more complex.

The main limitation of these machines is their height above the deck, which may cause clearance issues when passing under existing bridges or power lines. The height of crossbeams and launch saddles is dictated by structural requirements and the need for access for inspection and maintenance, the trusses are tall to increase the structural efficiency, and the overhead crane adds height to the gantry.

Figure 3.20. Connection beam between launch saddles for heavy loads (Photo: HNTB)

These gantries perform all segment erection operations from the deck level. When the segments are delivered on the ground, they may be stored under the gantry to shorten the cycle time. A typical 45 m simple span with epoxy joints is erected in 2–3 days in four or five work shifts. Suspending 15 segments from the gantry takes the first shift, aligning and gluing the segments takes the second shift, prestressing takes the third shift, and releasing the span and launching the gantry takes one or two shifts depending on the complexity of the gantry and the plan curvature of the bridge. Span erection is faster with dry joints.

Continuous spans with in-place stitches take 3 days and six work shifts to erect. Compared to a simple span, two additional shifts are necessary to position the new pier-head segment and to cast the wet joints. When wet joints are used at both ends of the span, continuity is achieved after repositioning the gantry, so this approach does not interfere with the cycle time. When one wet joint is used at the rear pier-head segment or within the span, the cycle time is typically 3 days, as the span can be released only after completion.

Continuous 45 m spans can also be erected as balanced cantilevers, with a cycle time of 4 days in eight shifts (Figure 3.21). Positioning the pier-head segment takes the first shift, erecting seven pairs of segments takes 3.5 shifts, casting and curing the midspan closure takes 1.5 shifts, continuity prestressing takes one shift, and launching the gantry takes the eighth shift. Span-by-span erection is twice as fast and prestressing is less expensive, but the gantry is more expensive as it is designed to sustain the weight of the entire span.

Figure 3.21. Balanced cantilever erection of short spans (Photo: VSL)

3.3.4   Structure–equipment interaction

All the segments for the span are suspended from the gantry prior to gluing the joints, and the deflections of crossbeams, trusses, hangers and spreader beams are not design-controlling factors. These gantries are, therefore, flexible and lively, which amplifies structure–equipment interaction at load transfer.

The upward deflection that the application of prestressing generates in the span is a small portion of the deflection accumulated by the gantry during loading of segments. Because of this residual deflection, the gantry supports most of the weight of the span after application of prestressing, and tensioning all the tendons is often impossible (Figure 3.22). Application of prestressing is divided into two phases: a first set of tendons is tensioned to make the span self-supporting, and the remaining tendons are stressed after releasing the span. This shortens the cycle time but the second set of tendons must be tensioned from the rear anchorages within the box cell not to interfere with the erection of the new span.

The span is released by lowering the gantry to avoid load redistribution among the hangers and to accelerate the operation. Support cylinders at the launch saddles lift the trusses from the launch bogies at the end of launching and lower them back onto the bogies to release the span. The support cylinders are designed for the total load (weight of gantry and span) and are equipped with a mechanical locknut to limit the hydraulic operations to lifting and lowering. In simple spans and continuous spans with wet joints at both ends, the support jacks of the span may also be used to lift the span and release the gantry. Two jacks are operated individually and

Figure 3.22. Residual support action

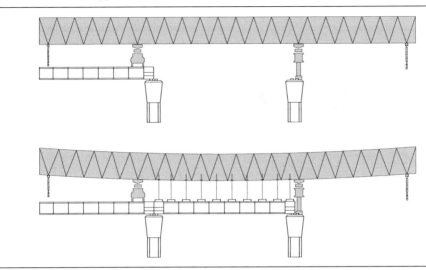

the two jacks at the other pier are connected hydraulically to create a torsional hinge that avoids span twisting.

Most overhead gantries take support on the pier-head segments, and the vertical load applied to the piers is centred or slightly eccentric. These gantries are relatively heavy, and their weight, added to the weight of the span, may affect the design of the bridge foundations. In continuous spans with wet joints at the rear quarter/fifth of the span, the rear crossbeam takes support on the front cantilever of the completed deck. This shortens and stiffens the gantry and improves the time-dependent stress redistribution of staged construction within the deck; releasing the span, however, modifies the stress distribution in and camber of the bridge.

Repositioning the front crossbeam during launching may require the use of a front pier bracket, which induces longitudinal bending in slender columns. Repositioning the rear crossbeam requires the rear auxiliary leg to take support on the deck; the load applied to the deck during this operation is rarely prohibitive, and its location can be adjusted.

Lateral wind load on the gantry should always be considered in the design of tall piers. The taller the piers, the bigger the advertising banners that contractors will want to put on their machines. Gantries and piers should be designed for wind loading on most of the solid area between the chords, which minimises one of the advantages of trusses versus I-girders.

For the erection of curved spans, the gantry is offset outward at the piers to diminish the eccentricity of midspan segments. Additional eccentricity may be necessary for traversing. Load eccentricity generates transverse bending in crossbeams, piers and bridge foundations, and may require stronger tie-downs between crossbeams, pier-head segments and pier caps.

The rear C-frame of telescopic gantries applies a constant load to the deck during launching. Geometry adjustment is necessary only when the deck crossfall varies along the span. The launch bogies of the C-frame roll on rails anchored to the deck over the webs.

3.4.   Single-girder overhead gantries

A single-girder overhead gantry includes a main girder, a launch nose and a rear C-frame. A box girder, two braced I-girders or a rectangular truss may be used for the main girder, in combination with a light launch nose. The bending-to-shear ratio, load application points, cost of hand and robotised welding, access to the working areas, and handling and shipping requirements govern the choice between plate girders and trusses.

Lateral bracing connects trussed webs or I-girders for their entire length. Bracing includes crosses or K-frames, connections designed to minimise displacement-induced fatigue, field splices designed for fast assembly, and sufficient flexural stiffness to resist vibration stresses. Cross-diaphragms connected to flanges and chords at the same locations as lateral bracing, or crossbeams framed into webs and flanges by vertical stiffeners, distribute torsion and provide transverse rigidity. Connections are often designed to develop member strength.

The front nose controls overturning during launching. No rear tail is necessary, as the rear C-frame rolls on the new span during launching, so these machines are shorter than the twin-girder gantries and better adaptable to curved bridges. A rear cantilever is necessary to pick up the segments when delivered on the deck, but it does not interfere with launch kinematics.

A winch trolley travels along the gantry to move the segments into position. When an underslung winch trolley runs along the bottom flanges of the main girder, the hangers for the segments are suspended on either side of the main girder from a working platform supported on the top flanges. Overhead cranes may also be used to pick up the segments with side brackets when these are delivered on the ground.

The gantry suspends an entire span of segments, and most structural elements are heavily stressed. The gantries for the heaviest spans may be equipped with prestressing tendons: sensor-controlled prestressing has been successfully tested on new generation MSSs (Pacheco et al., 2011), and camber control is less critical in precast segmental construction.

Heavy single-girder gantries have been designed for simultaneous erection of adjacent decks (Figure 3.23). These machines are expensive and complex to assemble and operate, and have found limited application. Conventional twin-truss overhead gantries can be easily shifted from one alignment to another simply by lengthening the pier crossbeams, and, even if the cycle time is longer, these gantries are much lighter and are easier to reuse.

Telescopic single-girder overhead gantries have been designed to erect LRT simple spans with tight plan radii (Figure 3.24). An underslung winch trolley suspended from the main girder handles precast segments delivered on the ground or the completed bridge through a rear C-frame. Hanger bars or ropes are suspended from a top working platform. In order to cope with tight plan and vertical radii, the gantry comprises a main girder and a front underbridge. A turntable with hydraulic controls for translation, tilt and plan rotation supports the main girder on the underbridge.

During the first phase of launching, the turntable pulls the main girder over the underbridge. When the turntable reaches the new pier, the underbridge is launched forward to clear the area under the main girder for erection of the new span. The main girder also slides on the turntable to operate the gantry on shorter spans, with the rear C-frame supported on the pier segment of the

Figure 3.23. Single-girder overhead gantry for simultaneous erection of adjacent decks (Photo: De Nicola)

last span (Figure 3.25). The underslung winch trolley cannot be operated beyond the turntable, and these gantries are therefore unfit for erecting continuous spans with wet joints at the rear quarter/fifth of the span. Because of the presence of multiple support systems at the leading pier during launching, these gantries are generally used only for erection of simple spans.

The underbridge has a rear support frame and a mobile auxiliary leg. The auxiliary leg supports the front end of the underbridge during launching of the main girder, and supports the turntable during launching of the underbridge. When the rear support frame of the underbridge reaches the leading pier, it is inserted under the turntable to support the main girder during erection of the new span, and the auxiliary leg is moved to the next pier. The rear support frame of the underbridge has broad geometry adjustment capability, and may take support on a crossbeam for additional lateral shifting.

3.4.1 Loading, kinematics, typical features

The operations of a single-girder overhead gantry are similar to those of a twin-girder machine. If the segments are delivered on the deck, the winch trolley picks them up at the rear end of the gantry, moves them throughout the rear C-frame, rotates them by 90° and lowers them down to the deck level. If the segments are delivered on the ground, the hoist reaches down to them and lifts them. These gantries are typically designed for both operations for enhanced use flexibility. The rear C-frame is wider than the short side of the segments to allow the segments

Figure 3.24. A 41 m, 96 ton underbridge and a 47 m, 132 ton main girder for 37 m, 340 ton dual-track LRT spans (Photo: Deal)

to pass through. Support cylinders and launch bogies of the C-frame are located over the deck webs, and many C-frames therefore require a bottom crossbeam to control transverse bending.

The single-girder gantries are taller than the twin-girder overhead machines because the segments are handled beneath the main girder rather than within the clearance between the two girders. Underslung winch trolleys require clearance for crane bridge, upper and lower sheaves, lifting beam and spreader beams. The bottom crossbeam of the rear C-frame requires additional clearance, and the main girder is tall to enhance flexural efficiency and torsional stiffness. In the telescopic gantries, the depth of the underbridge and turntable is therefore minimised to avoid lifting the main girder further. The height of these machines may pose problems of lateral stability during launching, and may govern pier design in light LRT bridges.

Hangers and spreader beams are used to align and hold the segments in position during assembly. Lateral offset in curved spans is achieved with adjustable spreader beams and aligned hangers, or with fixed spreader beams and offset hangers. In either case, the hangers are located outside the main girder to avoid conflicts with bracing and to create sufficient clearance for the winch trolley to move out the segments within the two planes of hangers. A top working platform simplifies the geometry adjustment of the offset hangers.

3.4.2 Support, launch and lock systems

A single-girder overhead gantry is equipped with a rear C-frame that rolls over the new span during launching, a main support frame at the centre of the machine, and a front auxiliary leg. The main frame can be a tower–crossbeam assembly anchored to the deck during launching, or a rigid frame integral with the gantry.

Figure 3.25. Gantry operation on shorter span (Photo: Deal)

When the deck has simple spans, the main frame takes support on the front half of the leading pier cap, and the rear C-frame takes support on the pier segment of the last span. Continuous spans with wet joints at the pier-head segments are erected with the main frame and the rear C-frame supported on the pier-head segments of the span. For continuous decks with wet joints at the rear quarter/fifth of the span, the main frame takes support on the leading pier-head segment and the rear C-frame takes support on the front cantilever of the deck.

The front auxiliary leg is used to control overturning during launching. In continuous bridges, the winch trolley positions the new pier-head segment, with the gantry supported at the front leg and the main crossbeam. The winch trolley then moves the crossbeam onto the segment, with the gantry supported at the front leg and the rear C-frame. Because of the complexity of these operations, crossbeams are rarely used for the erection of continuous spans. When the piers are short and accessible, the pier-head segments can be erected with ground cranes to shorten the cycle time and move activities out of the critical path of launching.

For the erection of simple spans, the winch trolley moves the tower–crossbeam assembly onto the new pier cap, with the gantry supported at the front leg (often on a front pier bracket) and the rear C-frame. The gantry is then launched forward, with the C-frame rolling over the new span. Most single-girder gantries are fully self-launching and do not require ground cranes.

In the most sophisticate machines, the main support legs are rigidly framed within the main girder and cannot be used for launching. These machines are light, stable, fast to launch and do not require ground cranes, but they need special launchers that support the gantry during repositioning. Launching is achieved by friction, taking advantage of the support reaction that the

Figure 3.26. Friction launcher on adjustable support tower

gantry applies to the launchers (Rosignoli, 2000). Two synchronised friction launchers are used to reposition the gantry; Figure 3.26 shows a launcher assembled on a support tower.

The launcher supports the bottom flanges of the main girder. A support block is used under each flange. The support block has a polished stainless-steel bottom surface that slides along an oscillating arm under the force imparted by a longitudinal double-acting cylinder. PTFE plates are applied to the top flange of the oscillating arm for low-friction sliding of the support block, and lateral guides prevent transverse displacements. Two vertical jacks at the ends of the oscillating arm lift and lower the main girder onto the support block.

The working sequence is similar to operating a friction launcher for launched bridges (Rosignoli, 2000).

1   The launch cylinders push the support blocks forward along the low-friction surfaces of the oscillating arms. The thrust force is transferred to the gantry by steel-to-steel friction between the support blocks and the bottom flanges of the main girder.
2   When the launch cylinders are fully extended, the jacks at the ends of the oscillating arms lift the gantry from the support blocks.
3   The launch cylinders pull the support blocks back to the initial position.
4   The jacks lower the gantry onto the support blocks to start this cycle again.

The oscillating arms cope with the flexural rotations of the main girders and the launch gradient. In some gantries the launch cylinders drive lifting jacks equipped with a bottom polished stainless-steel surface, and fixed support blocks are used at the ends of the oscillating

Figure 3.27. Two friction launchers stored under the launch nose

arm. The working principle of these launchers is the same, but the jack is heavy and difficult to extract for maintenance; the jack also transfers major shear forces during launching (the seals of the hydraulic cylinders typically are not designed for shear) and requires hydraulic hoses instead of rigid connections (Rosignoli, 2000).

Linear position sensors applied to launch cylinders and lifting jacks provide operation feedback to a programable logic controller (PLC) for full automation of the launch cycle. Pressure sensors control the hydraulic systems to stop the launch sequence if pre-set limit pressures are reached. The rear C-frame of the gantry provides most of the torsional restraint during launching.

The two oscillating arms of the launcher are assembled at the ends of a crossbeam supported on a PTFE plate fixed to the top crossbeam of the support tower. A central vertical pivot keeps the crossbeam aligned with the tower and allows rotations to traverse the gantry. Low-friction surfaces between the support tower and a base frame anchored to the deck are used for lateral shifting of the launcher.

All the joints are equipped with sliding clamps to pick up the assembly with the winch trolley or to suspend it from the main girder. Two synchronised launchers are typically used to reposition the gantry. The launchers are suspended from the main girder, moved along the gantry with hydraulic motors, and stored under the launch nose during loading of segments (Figure 3.27). Erecting continuous spans also requires an auxiliary hoist at the tip of the launch nose to lift the pier-head segment from the ground. The support towers of the launchers lodge independent hydraulic systems supplied by the PPU of the gantry through bus-bars or suspended cables.

During application of span prestressing, the rear launcher is moved over the leading pier-head segment. The span is released by retracting the support cylinders at the front legs and at the rear C-frame; during this operation the launcher lands on the deck and is anchored to the pier-head segment by stressed bars. The main support legs of the gantry overhang from the main girder to avoid conflicts with the launchers (Figure 3.28).

In the first phase of launching, the gantry is supported at the rear launcher and at the rear C-frame, which rolls along the new span under the pull imparted by the launcher. When the launch nose reaches the new pier, the vertical cylinders of the C-frame are retracted to lift the front cantilever and create the clearance for insertion of the front auxiliary leg. The front leg takes support on the pier cap, and the front hoist positions the pier-head segment onto four jacks at the corners. After anchoring the segment to the pier cap, the front launcher is moved onto the segment.

When the deck has simple spans, the front launcher is moved over the new pier and anchored to the pier cap, and no front hoist is necessary. The front auxiliary leg controls negative bending and overturning during these operations. Both launchers are devoid of vertical cylinders because the main support systems of the gantry provide adjustment of vertical geometry.

In the second phase of launching, both launchers drive the gantry. The load that the rear C-frame applies to the deck is controlled hydraulically because excessive reduction in the support reaction at the rear launcher would cause slippage (Rosignoli, 2002). When the front support legs reach the front launcher at the new pier-head segment, the main support cylinders of the gantry are extracted to lift the gantry and disengage the launchers, which are moved to the storage area.

When integral with the gantry, the main support frame includes an L-leg overhanging from the main girder on either side, and a central diaphragm that connects the two legs within the main girder. The bottom portions of the legs are not braced to each other to allow unrestricted movement of the main girder over the launchers. When the bridge has construction joints along the span, the clearance between the bottom portion of the legs is also necessary for the winch trolley to pass through and feed the segments for the front cantilever. The legs typically have a box section configuration to provide longitudinal and transverse stiffness. High-tonnage double-acting cylinders with mechanical locknuts are lodged within the bottom portion of the legs to provide geometry adjustment and to lower the gantry to release the span in one operation.

The rear support of these gantries is a complex C-frame. Transverse double-acting cylinders shift the frame relative to the main girder when launching along curves for the launch bogies to follow rails anchored over curved deck webs. The lateral cylinders are also used to rotate the gantry about the rear launcher to align the launch nose with the new pier at landing. Longitudinal cylinders and low-friction sliding surfaces are used to rotate the C-frame in the horizontal plane to the local radius (Figure 3.29). Telescopic connections and vertical cylinders are used to control the load that the C-frame applies to the deck in the three-support configurations of the second phase of launching and to cope with varying deck crossfall.

The rear C-frame of the underbridge of a telescopic gantry is subjected to similar geometry adjustment requirements, but it also supports the main girder and the turntable, and is therefore heavily

Figure 3.28. Support leg overhanging from the main girder

stressed. When the underbridge takes support on a narrow pier cap, its rear C-frame provides most of the lateral stability of the gantry, and the anchor bars are typically oversized.

The front auxiliary leg is applied to the launch nose with sliding connections to set the leg vertical and to rotate it about the vertical axis. Long-stroke double-acting cylinders are used to take support on stepped pier caps and to adjust the length of the leg in bridges that have a vertical curvature.

3.4.3 Performance and productivity

Light custom-design single-girder overhead gantries are the first-choice solution in long bridges comprising a number of spans sufficient to depreciate the investment and with limited clearance under the bridge. These machines are also used for curved bridges, which can benefit from the superior adaptability of a telescopic gantry. Spans of constant length simplify the operations, but these gantries are immediately usable on shorter spans. The different number of segments in the span does not greatly affect the cycle time.

Figure 3.29. Telescopic C-frame with low-friction rotation surfaces during assembly

These machines allow erection of bifurcations and varying-width decks, as well as spans supported on or integral to straddle bents and L-piers; telescopic gantries are also able to cope with tight plan radii. For their adaptability within ground constraints and complex geometries, these gantries have found frequent application in urban LRT bridges. Most single-girder overhead gantries are able to reposition the support systems without any need for ground cranes.

The main limitation of these machines is their height above the deck, which may cause clearance issues when passing under existing bridges or power lines. Single-girder gantries are taller than twin-girder overhead machines, and telescopic gantries are even taller.

The productivity of these machines is not much different from that of a twin-girder overhead gantry. Launching is faster and requires less labour because of the higher level of automation; shipping and site assembly are also less expensive. Most overhead gantries are more expensive than underslung gantries, due to their higher level of sophistication and adaptability.

3.4.4   Structure–equipment interaction
The structure–equipment interaction during application of prestressing and span release is similar to that of a twin-girder overhead gantry; the effects, however, are more marked, as single-girder gantries are more flexible. Also in these machines all the segments for the span are suspended before the application of epoxy, span prestressing is often divided into two phases, and the span is released by lowering the gantry or by jacking the span.

The segments are suspended from hangers and spreader beams. When the span is curved, both types of suspension (adjustable spreader beams with aligned hangers and fixed spreader beams with offset hangers) generate torsion in the gantry and transverse bending in the piers.

In continuous decks with wet joints at the rear quarter/fifth of the span, the rear C-frame takes support on the front deck cantilever. Releasing the span diminishes the load applied to the cantilever, which modifies the stress distribution and camber in the bridge. Gantries for the erection of simple spans are always supported at the piers, and structure–equipment interaction is therefore less marked.

The rear C-frame rolls on the new span during launching. The load applied to the deck diminishes as the launch progresses. Moving the front launcher forward may diminish the rear support reaction in the first phase of launching; these gantries, however, are relatively light, and the load applied to the deck is rarely governing. Structure–equipment interaction is more demanding when second-hand gantries previously used for highway bridges are used to erect light LRT spans.

Lateral wind on the gantry should always be considered in the design of tall piers. Single-girder overhead gantries with integral support frames cannot be shifted outward during launching on curved spans, and transverse bending in piers and foundations is more demanding. The gantries supported on tower–crossbeam assemblies are easier to traverse. The support legs of telescopic gantries may take support on crossbeams at the leading pier for some traversing capability, although this further complicates the operation of these machines.

Telescopic gantries are tall and sensitive to lateral wind, and must be anchored with great care. This is often difficult with LRT box girders, as the pier caps are narrow and provide a short lever arm to the anchor bars. The piers of LRT bridges, on the other hand, are rarely very tall.

3.5. Underslung gantries

Many precast segmental box girders have been erected with twin-girder underslung gantries. These machines operate beneath the deck with one girder on either side of the pier. The gantry supports the precast segments under the wings, with adjustable carts that roll along the girders. The main girders have front and rear extensions to control overturning during launching. Noses and tails are lighter than those of the overhead gantries, as they are not used to handle precast segments (Figure 3.30).

Underslung gantries are simple to design, assemble and operate. They allow for fast segment erection, and facilitate access to the segments via working platforms applied along the main girders. These gantries are typically supported on pier brackets that are repositioned with ground cranes or barge-mounted cranes on water. Intermediate props from foundations may be used to increase the load capacity of the gantry when the piers are short.

The gantry may carry a portal crane spanning between the main girders. The crane picks up the segment from the ground, rotates it to longitudinal, lifts it up, rotates it to orthogonal, moves it along the gantry and lowers it onto the segment carts. Cantilever runways for the carts permit unrestricted operations of the portal crane; the segments, however, are typically erected in a forward fashion (Figure 3.31). The last segments are lifted from a side overhang of the portal

Figure 3.30. A 76 m, 325 ton underslung gantry with a 40 ton crane for 35 m, 490 ton dual-track LRT spans (Photo: NRS)

crane or within the main girders beyond the leading pier; this avoids anchoring the crane but makes the launch nose more expensive.

A lifting frame over the leading span of the completed bridge may also be used to load the segments onto the rear end of the gantry. Lifting frames with side overhangs hoist segments delivered on the ground alongside the bridge, while bidirectional lifting frames pick up the segments from haulers on the deck, move them forward throughout the frame, rotate them to orthogonal, and load them onto the gantry. In both cases the segments are rolled into position with adjustable carts that provide lateral and vertical adjustment for setting of cambers, crossfall and plan curvature.

Wheeled or crawler cranes may also be used on the deck to load the gantry. Commercially available cranes are more expensive than lifting frames, and are much heavier, because swinging requires counterweights, but their cost is easier to depreciate as they are multi-purpose machines. Cranes on the deck lift the segments alongside the bridge, and this process has two weak points.

1   Crane operation (weight of crane and counterweights, hoist load, net hoist couple and dynamic allowance) increases the couple applied to the leading pier cap in the longitudinal plane due to the missing weight of the next span. Crane operation is also demanding in continuous decks with wet joints at the quarter/fifth of the span because the crane is placed on the front deck cantilever.

Figure 3.31. Cantilever runways for the segment carts (Photo: Strukturas)

2   The torque applied to the span in the transverse plane may be critical. For normal
    operations, basic and car-body counterweights are fine-tuned with the hoist load, load
    radius and boom configuration to apply an equal and opposite torque to the span prior to
    and after picking up the segment. For the assessment of span stability in case of accidental
    release of the segment, however, many design standards (AASHTO, 2003, 2012; BSI, 2006)
    require application of a vertical dynamic allowance $\varphi_V = -2$ to the hoist load. The crawler
    facing the load may lift from the deck when the negative hoist couple is combined with the
    counterweight couple, the weight of the crane is suddenly transferred to the outer crawler,
    and the sudden load redistribution can destabilise crane, span or both if the span support
    jacks are closely spaced or the span is short or curved.

When the segments are delivered on the ground or the water, the area under the gantry is accessible and the piers are not tall, the segments may be lifted with a crane operating on the ground or
mounted on a barge. Access must be maintained throughout bridge erection, and this may be
expensive when working on the water due to the cost of barges and tugboats. Weather and sea
conditions and the distance from the docking facility also affect delivery and lifting of segments
(ASBI, 2008). The crane is placed beyond the leading pier and the segments are loaded onto the
gantry and rolled backward into position. A crane travelling alongside the gantry can load the
segments directly into position, but the width of the segments after rotation requires a long
boom, and therefore a powerful crane (Homsi, 2012).

Figure 3.32. A 91 m, 333 ton articulated gantry with a 40 ton crane for 36.7 m, 392 ton dual-track LRT spans (Photo: Strukturas)

When launch noses and tails are applied to the main girders to control overturning, the length of the gantry is more than twice the typical span. These machines are unfit for bridges with tight horizontal curves because the nose and tail of the inner girder conflict with the piers and the completed deck, and traversing the outer girder becomes too complex. Articulated girders with hydraulic hinges have been used to overcome this limitation, although these gantries are more expensive and complex to operate (Figure 3.32). Combining a varying-span portal crane with an articulated underslung gantry permits erection of varying-width spans and bifurcations (Figure 3.33).

The gantry projects beneath the deck, which may cause clearance issues when passing over existing highways or railways, and difficulties at the end spans as the abutment walls are wider than the piers. This problem is solved by applying the launch tails after erecting the first span, with the gantry propped from foundations (Figure 3.34). For the same reason, the launch noses are dismantled prior to moving the gantry to the last span. The abutment walls must also be taller than the gantry so as not to prevent operation on the first and last span.

Most underslung gantries are incompatible with decks with varying width or devoid of side wings. When straddle bents, L-piers, hammerheads of aerial stations, and other custom substructures of LRT bridges make the clearance beneath the box-girder wings insufficient for gantry operation, the spans are erected at a raised elevation, released onto jacks and lowered after repositioning the gantry.

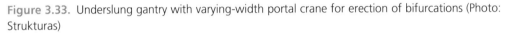

Figure 3.33. Underslung gantry with varying-width portal crane for erection of bifurcations (Photo: Strukturas)

Self-launching underbridges supporting a sliding crossbeam that connects the front end of the main girders have been used on a few curved bridges. Rear C-frames rolling over the new span during launching have been used in these telescopic machines to shorten the gantry and to avoid pier brackets. These gantries require a V-shaped design of the pier caps to create a central clearance for underbridge operation.

Underslung gantries comprising a central truss supported on the pier caps, and two edge trusses supported on pier brackets have been designed for simultaneous erection of twin box girders (Figure 3.35). These machines are very specialised and have found limited application. The vast majority of new-generation underslung gantries include two independent girders equipped with launch noses and tails, supported on pier brackets, and launched individually. Hydraulic articulations are more and more frequently used to connect noses and tails to the main girders to provide some geometry adjustment when working on curved bridges.

3.5.1 Loading, kinematics, typical features

The segments are mostly loaded onto the gantry with ground cranes or lifting frames. When the segments are delivered on the deck, the lifter is placed at the rear end of the gantry. When the segments are delivered on the ground or on water, the crane is placed at the front end of the gantry. In both cases the segments are loaded onto the gantry close to the lifter and rolled into position.

Figure 3.34. Prop from foundation in the first span

Adjustable carts are used to align and hold the segments in position during assembly. Lateral shifting for geometry adjustment and offset in curved spans is achieved with steel–PTFE skids and adjustment screws. Vertical adjustment for camber and crossfall is achieved with hydraulic jacks housed within the carts. The carts include two bottom rollers that run along the top flange of the main girder or special cantilever runways; winches are typically used to roll the segments into position. Two carts support the segment on one side and one cart is used on the other side so as not to twist the segment during handling and gluing.

Working platforms along the main girders provide access to the segments for the entire length of the span. A front stressing platform spanning between the girders is used for fabrication and tensioning of the prestressing tendons.

Typically all segments but one are loaded onto the gantry before gluing the joints so that no deflections can occur during joining. Epoxy is applied to groups of segments, which are then pressed together with prestressing bars. Even with slow-set epoxies, three or four segments may be the practical limit in order to avoid setting of the epoxy before the group is pressed together (Hewson, 2003). Finally, the remaining segment is installed.

Figure 3.35. Three-girder underslung gantry for twin box girders (Photo: Freyssinet)

Underslung gantries can be used to erect simple and continuous spans. The wet joints of continuous spans are locked with concrete shims and light tensioning of the prestressing tendons prior to casting the closure pours. Forming the external surfaces of the closure pours interferes with the gantry and lengthens the cycle time when narrow piers prevent span release onto temporary jacks (Meyer, 2011).

Compared to an overhead gantry, with an underslung gantry the pier-head segments can be erected at any time because the gantry is not supported on them.

3.5.2 Support, launch and lock systems
Most underslung gantries are supported on pier brackets. Articulated saddles support the gantry at the brackets for launching and some extent of lateral shifting. The pier brackets have long side overhangs to traverse the gantry in curved bridges. The two girders of the gantry are independent

Figure 3.36. Propped pier bracket of LRT bridges (Photo: Strukturas)

and are launched and traversed individually. The stressing platform is removed from the gantry prior to launching and reapplied at the end of repositioning.

The pier brackets include two horizontal I-girders connected by lateral bracing and cross-diaphragms. Shear keys take support into recesses in the pier wall to transfer the vertical load applied by the gantry. Transverse bending is resisted by means of inclined props and horizontal stressed bars that clamp the two brackets together.

When the piers are too slender to form wide support recesses, crossbeams are suspended from the top of the pier cap to support the launch saddles, or, more frequently, the brackets are propped from foundations (Figure 3.36). Suspended crossbeams are difficult to remove after span erection, and require tall bearing pedestals or span jacking to create sufficient clearance between the pier cap and the deck soffit. Ground cranes are mostly used to apply the gantry support systems to the piers and to remove them when disengaged.

The typical launch saddle is an articulated assembly of two bottom bogies that shift along the pier bracket, and two top bogies that support the truss. Equalising beams are used to balance the load in the cast-iron rolls and to follow the flexural rotations of the trusses and the launch gradient. In some new-generation gantries the bottom bogies have been replaced with PTFE skids. Double-acting cylinders drive lateral shifting and provide the lateral restrain to the gantry during operation.

Rectangular rails welded to the flanges of trusses and pier brackets facilitate load dispersal and guide the roll flanges for transfer of lateral loads. PTFE skids load most of the flange width, and lateral guides transfer the load to the flange edges. Double-acting vertical cylinders lift the gantry from the launch saddles at the end of launching and lower it back to release the span after application of prestressing. Locknuts are tightened to provide mechanical support to the gantry during span erection.

Not all underslung gantries are self-launching. Some cheap first-generation trusses are repositioned individually with a ground crane that hangs and pulls the launch nose forward while the rear tail rolls on a pier bracket (ASBI, 2008). Dragging the trusses with ground cranes is a delicate and risky operation due to side loading of the crane boom (Homsi, 2012). These launch methods have been abandoned over time, and capstans and deck-mounted winches have also been progressively abandoned. In new-generation machines, redundant double-acting cylinders lodged within the launch saddles push or pull the truss forward by acting within perforated rails fixed to the truss.

Some pier brackets have been designed to be suspended from the truss. Hydraulic motors drive the brackets along the truss, and the transverse shift systems of the brackets are used to insert the shear keys into the pier recesses. Self-launching brackets carry independent hydraulic systems fed by the PPU of the gantry; launch noses and tails are also more expensive, as they are designed for the additional negative moment. Self-launching brackets can be a cost-effective solution when the segments are delivered on the deck or loaded with lifting frames; when a ground crane is used to load the segments onto the gantry, the crane may also be used to reposition the pier brackets off-shift or during the application of span prestressing.

3.5.3   Performance and productivity

Underslung gantries are the first-choice solution for box girders where the clearance above the bridge is limited, the area beneath the bridge is free from clearance restrictions, and the bridge is rectilinear or slightly curved. Articulated underslung gantries are compatible with tight plan radii. However, these gantries are expensive, slow to launch and difficult to operate.

Underslung gantries require side wings in the precast segments and are not really compatible with varying-width decks and spans supported on or integral to straddle bents and L-piers, which often require erecting the span with the gantry propped from foundations. The gantry is then lowered onto the ground, moved forward and lifted with ground cranes.

The use of self-launching brackets avoids the need for ground cranes. When ground or floating cranes are used to reposition the pier brackets, the same cranes may also be used to load the segments onto the gantry. Many LRT bridges have been erected by repositioning the pier brackets overnight with ground cranes, and by loading the gantry during the day with a lifting frame on the deck or a portal crane on the gantry.

Underslung gantries are lighter than overhead gantries, and are easier to design, assemble and operate. They are also less automated and therefore less expensive. Hangers and spreader beams are not necessary, and a 40 m simple span with epoxy joints may be erected in 2-day cycles. With dry joints, 1-day cycles have been achieved with two working shifts. Launching is particularly fast when the pier brackets are repositioned with ground cranes, as these operations are removed from the critical path.

The location of the precasting facility and its distance from the bridge site often become the bottleneck in terms of productivity (Homsi, 2012). If continuous delivery of segments with a reasonable number of haulers cannot be maintained, the segments are delivered off-shift and stored near the erection site, which doubles the handling cost. The distance that the haulers have to back up and the availability of crossover areas also affect the cycle time when the segments are delivered on the deck.

3.5.4 Structure–equipment interaction

Structure–equipment interaction during the application of prestressing and span release is similar to that with the other types of gantry. All the segments for the span are loaded onto the gantry prior to gluing the joints, prestressing may be divided into two phases, and the span is released in one operation, either by lowering the gantry or by jacking the span.

Reinforcement and transverse post-tensioning in the side wings of the segments are checked for the positive moment generated by supporting the segments under the wings. The temporary support stresses may be demanding in the short and heavy end segments of the span. The support points shift transversely in curved bridges because the gantry is aligned with the chord between the piers. Unless the main girders are articulated to follow the curve, tight plan radii are often incompatible with eccentric transverse post-tensioning in the top slab.

The pier brackets are wider in curved bridges to avoid conflicts between the launch tail of the inner truss and the completed deck. A wider transverse distance between the trusses also increases the positive moment in the top slab of the segments.

Application of span prestressing requires particular care. When the span becomes self-supporting, the side wings of the end segments transfer part of the span weight to the gantry. It is therefore necessary to support the end segments with hydraulic jacks prior to the application of prestressing in order to avoid cracking in the wings. The support jacks on the pier cap slide on PTFE surfaces such that they do not counteract span shortening induced by prestressing.

Most underslung gantries are supported on pier brackets and the vertical load applied to the piers during span erection, and launching is centred longitudinally and slightly eccentric transversely in curved bridges. Staggered launching of the trusses increases transverse bending in piers and foundations, especially in curved bridges with wide piers.

In addition to the longitudinal bending due to friction and the launch gradient, vertical load imbalance prior to erecting the next span applies a couple to the pier cap in the longitudinal plane. Although this load condition is common to all gantries and MSS for span-by-span construction of simple spans, the lighter weight of the gantry may require additional reinforcement in the columns. Loading segments with lifting frames on the deck increases longitudinal bending in the leading pier, and the loads of counterweighted swinging cranes on the deck may be particularly demanding.

Some underslung gantries have a rear C-frame that rolls over the new span during launching. Launching may require three-support arrangements where the flexibility of the deck and gantry governs the load distribution. In such cases, the load applied by the C-frame may be controlled with the vertical cylinders of the launch bogies, although the support reactions of such light gantries rarely affect the bridge design.

REFERENCES

AASHTO (American Association of State Highway and Transportation Officials) (2003) *Guide Specifications for Design and Construction of Segmental Concrete Bridges*. AASHTO, Washington, DC, USA.

AASHTO (2008a) *Construction Handbook for Bridge Temporary Works*. AASHTO, Washington, DC, USA.

AASHTO (2008b) *Guide Design Specifications for Bridge Temporary Works*. AASHTO, Washington, DC, USA.

AASHTO (2012) *LRFD Bridge Design Specifications*. AASHTO Washington, DC, USA.

ASBI (American Segmental Bridge Institute) (2008) *Construction Practices Handbook for Concrete Segmental and Cable-supported Bridges*. ASBI, Buda, TX, USA.

BSI (2006) BS EN 1991-3: 2006. Eurocode 1: Actions on structures – Part 3: Actions induced by cranes and machinery. BSI, London.

CNR (Consiglio Nazionale delle Ricerche) (1985a) CNR 10021. Strutture di acciaio per apparecchi di sollevamento. istruzioni per il calcolo, l'esecuzione, il collaudo e la manutenzione. Consiglio Nazionale delle Ricerche, Rome, Italy.

CNR (1985b) CNR 10027. Strutture di acciaio per opere provvisionali. istruzioni per il calcolo, l'esecuzione, il collaudo e la manutenzione. CNR, Rome, Italy.

CSA (Canadian Standards Association) (1975) *Falsework for Construction Purposes*. Design. CSA, Toronto, Canada.

FEM (Fédération Européenne de la Manutention/European Federation of Materials Handling) (1987) *Règles pour le Calcul des Appareils de Levage*. FEM 1.001. FEM, Brussels, Belgium.

Harridge S (2011) Launching gantries for building precast segmental balanced cantilever bridges. *Structural Engineering International* **21(4)**: 406–412.

Hewson NR (2003) *Prestressed Concrete Bridges: Design and Construction*. Thomas Telford, London.

Homsi EH (2012) Management of specialized erection equipment: selection and organization. *Structural Engineering International* **22(3)**: 349–358.

Kamaitis Z (2008) Field investigation of joints in precast post-tensioned segmental concrete bridges. *The Baltic Journal of Road and Bridge Engineering* **3(4)**: 198–205.

Meyer M (2011) Under-slung and overhead gantries for span by span erection of precast segmental bridge decks. *Structural Engineering International* **21(4)**: 399–405.

Pacheco P, Coelho H, Borges P and Guerra A (2011) Technical challenges of large movable scaffolding systems. *Structural Engineering International* **21(4)**: 450–455.

Podolny W Jr (1985) The cause of cracking in post-tensioned concrete box girder bridges and retrofit procedures. *PCI Journal* **30(2)**: 82–139.

Rombach GA (2004) Dry joint behavior of hollow box girder segmental bridges. FIB Symposium 'Segmental Construction in Concrete', New Delhi.

Rombach GA and Abendeh R (2004) Temperature induced deformations in match-cast segments. In *Metropolitan Habitats and Infrastructure*. IABSE Report 88. International Association for Bridge and Structural Engineering, Zurich, Switzerland.

Rosignoli M (1999) Nose optimization in launched bridges. *Proceedings of the Institution of Civil Engineers, Structures and Buildings* **134**: 373–375.

Rosignoli M (2000) Thrust and guide devices for launched bridges. *ASCE Journal of Bridge Engineering* **5(1)**: 75–83.

Rosignoli M (2001) Deck segmentation and yard organization for launched bridges. *ACI Structural Journal* **23(2)**: 2–11.

Rosignoli M (2002) *Bridge Launching*. Thomas Telford, London.

Vion P and Joing J (2011) Fabrication and erection of U-through section bridges. *Structural Engineering International* **21(4)**: 426–432.

Bridge Construction Equipment
ISBN 978-0-7277-5808-8

ICE Publishing: All rights reserved
http://dx.doi.org/10.1680/bce.58088.079

Institution of Civil Engineers

publishing

Chapter 4
Launching gantries for macro-segmental construction

4.1. Technology of macro-segmental construction

Handling and transportation requirements govern the length and weight of precast segments. Lengths up to 3.5 m are often transportable on public roads without excessive restrictions. If the precasting facility is close to the erection site and no transportation constraints exist along the route, the segments are made as long as is practicable, but they rarely exceed 4.5 m.

Long segments accelerate construction, there being a smaller number of segments to cast and fewer joints to close, but they do pose transportation challenges (ASBI, 2008). Macro-segmental construction is based on casting long deck segments beneath the bridge or over the deck in order to avoid transportation. The segments are extracted from the casting cell and moved into position with a self-launching gantry. Macro-segmental decks are erected span-by-span or as balanced cantilevers; in both cases the deck is erected directionally from one abutment toward the opposite one.

Wet joints between the segments avoid the geometry control requirements of match-casting. Compared to in-place casting, the use of multiple segments reduces the cost of the casting cells, makes construction more repetitive, and improves risk management, with parallel tasks for carpenters, ironworkers and post-tensioning crews. The erection gantry is also less expensive than a MSS, as the segments are handled individually.

With span-by-span erection, each span comprises four macro-segments. A longitudinal joint at the deck centreline divides the box girder into two halves, and transverse joints at the span quarters divide each half of the deck into a sequence of pier-head segments and midspan segments. The segment length rarely exceeds 45 m, and the maximum span achievable is therefore 90–95 m. The deck will have varying depth on such long spans, and the use of single-cell box sections simplifies lateral extraction of the inner form from the segment. Special twin-girder over-head gantries are used for span-by-span erection, as the hoist load is excessive, both for ground cranes and for the gantries for balanced cantilever erection of standard segments.

The segments may be cast behind the abutment and transported along the deck; in most cases, however, the segments are cast directly over the leading span of the bridge, beneath the rear section of the erection gantry. Casting cells rolling along the deck are more expensive than fixed casting cells behind the abutment, but avoid the need for heavy lifters and special transportation means, as the gantry picks up the segment directly from the casting cell. The forming systems include a casting cell for the midspan segments and a casting cell for the pier-head segments. Both casting cells are as wide as one-half of the deck and as long as one-half of the span (Figure 4.1).

Figure 4.1. General view of the two erection lines

Span-by-span macro-segmental construction is used for the simultaneous erection of two adjacent decks. The gantry cannot rotate such long and heavy segments, and the segments are therefore cast in their final alignment. After segment extraction, the casting cell is moved back to the abutment, rotated by 180°, adjusted to the geometry of the conjugate segment, fed with the prefabricated cage of the new segment, and moved back to the gantry (Figure 4.2). The load of the casting cell is applied over the deck webs, and each deck can therefore accommodate only one casting cell.

The same operations are performed on the adjacent deck for the second casting cell. The construction cycle for two adjacent spans includes casting and positioning four pier-head segments and four midspan segments. All the pier-head segments are cast over one deck, all the midspan segments are cast over the adjacent deck, and the gantry moves the segments into position by shifting from one bridge to the other.

Two powerful winch trolleys span between the top chords of the trusses. The pier-head segments include negative-bending tendons and are picked up at the centre, while the midspan segments include positive-bending tendons and are picked up at the ends. Integrative tendons are fabricated after joint closure; bending is low at the span quarters and the integrative tendons are therefore a small part of the total prestressing. In-place stitching at the span quarters minimises geometry adjustment of the casting cells after 180° rotation, as both types of segment are symmetrical. Transverse prestressing is unusual in adjacent decks because the top slab is rarely wide.

Figure 4.2. Cage handling for the pier-head segment

Span-by-span macro-segmental construction requires a temporary pier at the front quarter of every span; this construction method is therefore primarily used for bridges with short piers located on good soils. The temporary pier supports the front end of the midspan segments and stabilises the pier-head segments against overturning about the leading pier. The rear end of the midspan segments is suspended from the front cantilever of the continuous deck by steel brackets equipped with vertical cylinders for geometry adjustment (Figure 4.3). The brackets are anchored to the bottom slab by vertical bars crossing the webs.

The pier-head segments are placed first. The gantry extracts the segment from the casting cell, moves it out over the rear crossbeam, shifts it laterally to the final alignment, lowers it down to the final elevation, and inserts it longitudinally under the front support frame. Two winch trolleys are used to handle the pier-head segments: when the front hoist reaches the support frame (Figure 4.4) the segment is temporarily supported on the pier cap and the front hoist is moved beyond the support frame to pick up the segment at a more advanced location for final placement. Each pier-head segment is therefore equipped with three lines of anchor bars. After positioning the second pier-head segment, the two midspan segments are lowered into the clearance between the pier-head segments and the front end of the completed deck.

The same operations are repeated on the second deck during casting of the closure pours on the first deck. Placing all left segments first and all right segments second requires rotation of the casting cells on alternate segments, while completing one span prior to erecting the adjacent one allows earlier prestressing and shortens the cycle time.

Balanced cantilever macro-segmental construction involves casting the long deck segments on the ground beneath the bridge (Matyassy and Palossy, 2006) or tugging them under the bridge with barges. Lifting frames or heavy gantries are used to lift the segments and hold them in position during connection to the deck. In prestressed concrete decks this involves coupling the top-slab

Figure 4.3. Rear bracket with hydraulic geometry adjustment

ducts, casting 1.0–1.5 m reinforced closure pours, and fabricating and tensioning the top-slab tendons. In steel cable-stayed decks this involves closing field splices or welding the adjoining segments, and fabrication of the new pair of stay cables (Gimsing and Georgakis, 2012).

Long deck segments may be cast on the ground beneath the bridge (Figure 4.5). Bottom form tables on scaffolding shoring support the segments during curing, and the rest of the casting cell is used to cast other segments. When a twin-girder overhead gantry is used to lift the segments, short pier tables are cast in place at the end of pier erection to accelerate and simplify launching of such heavy machines. Closure pours between the segments avoid the geometry requirements of match-casting and allow bridge design for partial prestressing.

A gantry for balanced cantilever erection of macro-segments operates like a suspension-girder MSS, where the casting cells have been replaced with lifting platforms for strand jacking. The gantry stabilises the hammer during erection, and the two types of machine are actually interchangeable and easily convertible (Matyassy and Palossy, 2006). Compared to span-by-span erection of macro-segments, prestressing is more expensive and the area under the bridge is heavily affected, but the absence of temporary piers allows deck erection on tall piers and over water.

Lifting frames carried by the cantilevers are used frequently for the handling of macro-segments, even if lifting the hoists on the pier table and removing them from the deck after midspan closure

Figure 4.4. Insertion of the left pier-head segment within the front support frame

takes time (ASBI, 2008). When the bridge comprises multiple 100–120 m spans, using a self-launching gantry for directional erection of the hammers from abutment to abutment accelerates construction and avoids disruption of the erection cycle for placement and removal of the lifters.

The distinction between segments and macro-segments is somewhat arbitrary in balanced cantilever construction. Typically, the macro-segments are not match-cast and are connected with in-place stitches. Narrower pier tables are cast in place at the end of pier erection, and support the gantry during operations to avoid crossbeams.

4.2. Twin-girder overhead gantries for span-by-span erection of adjacent macro-segmental decks

A self-launching gantry for simultaneous erection of two continuous macro-segmental decks comprises two overhead trusses supported on a front portal frame and a rear crossbeam. Two overhead winch trolleys span between the trusses to handle the macro-segments within the clearance between the trusses and over the rear crossbeam. The segment is moved to the adjacent deck by shifting the gantry along the front portal frame and rear crossbeam, and by shifting the hoist crabs within the winch trolleys. The rear crossbeam is supported on both decks while the front portal frame spans between the pier caps.

The design of these gantries is subject to specific technological requirements. The segments are very heavy, the pier-head segments typically being heavier than the midspan segments due to their varying depth, the thicker bottom slab, and the additional weight of the pier diaphragm.

Figure 4.5. Strand-jacking of a 8.0 m, 650 ton segment (Photo: ThyssenKrupp)

The weight of a 45 m pier-head segment may exceed 650 ton, and winch trolleys for such loads are among the most powerful hoists used in special bridge construction equipment – that is, they are also very heavy. The length and weight of the segments require a rear auxiliary W-frame to support the gantry during segment extraction from the casting cell, (Figure 4.6). The W-frame is kept lifted during insertion of the casting cells under the gantry and during segment casting. After shifting the gantry over the segment, the W-frame is lowered to take support on the deck for lifting.

The gantry is supported at the leading piers, on the front cantilevers of the two decks, and at the rear W-frame. The rear cantilever of the gantry is 55–60% of the span length, and the total length of the trusses is about 1.8 times the typical span. Compared to a gantry for balanced cantilever erection of standard segments, the front cantilever is shorter, as the winch trolleys operate

Figure 4.6. A 46 m, 640 ton pier-head macro-segment placed with a 162 m, 1280 ton gantry

close to the front portal frame for positioning of the pier-head segments, and the total length of the trusses is therefore similar.

The trusses are designed to handle the pier-head segments and the launch stresses imposed by such heavy machines on long spans. The winch trolleys are heavy and work close to each other, and the top chords and diagonals are therefore heavily stressed. The lighter midspan segments are suspended at the ends, truss loading is less localised, and load displacement along the gantry is also shorter.

Such heavy machines lead to demanding launch loads on such long spans. Portal frames and rear crossbeams designed to shift a loaded gantry from one bridge to another are also heavy and cumbersome. The main supports are repositioned using the winch trolleys, and these gantries therefore have four supports: a front pendular leg which is used during repositioning of the portal frame; the portal frame; the rear crossbeam; and the rear W-frame.

The trusses are braced at the ends and cannot be launched individually. Control against overturning on such long spans may require moving the winch trolleys to the rear end of the gantry during launching in order to suspend counterweights, which further increases the launch stresses in the front support region and generates specific conditions of out-of-plane buckling.

4.2.1   Loading, kinematics, typical features
Two articulated lifting beams permanently connected to the winch trolleys are attached to the segment with through bolts. The articulations cope with gradient and crossfall in the top slab.

Figure 4.7. Double-winch crab

Each winch trolley carries two hoist winches or strand jacks to lift the segment at four points. The winches of one trolley can share the hoisting rope to create a torsional hinge that avoids twisting the segment during handling.

Two winches or strand jacks are assembled on a wide hoist crab that rolls laterally along the crane bridge (Figure 4.7). Shifting the hoist point within the winch trolley provides additional geometry adjustment capability and shortens the side overhangs of the portal frame and rear crossbeam. Capstans, motorised wheels or double-acting cylinders drive the crab along the crane bridge. Auxiliary winches control payload oscillations with diagonal ropes and anchor the trolley to the rear crossbeam in out-of-service wind conditions.

Modular temporary piers at the third-quarter of the span support the midspan segments and stabilise the pier-head segments against overturning about the leading pier. The vertical load applied to the temporary piers is markedly eccentric. Sliding saddles have been used to support the segments in an attempt to allow thermally induced deck displacements after joint stitching. In most cases, however, the temporary piers are too flexible to overcome the breakaway friction of sliding saddles, and fixed saddles provide a more predictable response and a more reliable control of out-of-plane buckling.

4.2.2 Support, launch and lock systems
During segment handling, the gantry takes support at the front portal frame and the rear crossbeam. The rear W-frame is lowered to pick up the segment and to move it beyond the

Figure 4.8. Placement of the second pier-head segment

rear crossbeam, and is then lifted to shift the gantry to the segment erection alignment. The front auxiliary leg is used only during repositioning of the portal frame.

The front portal frame spans between the two pier caps, while the rear crossbeam has four transverse support lines over the webs of the two box girders. Both supports have long side overhangs for placement of the outer segments. Additional load eccentricity is achieved by shifting the hoist crabs along the crane bridges, which increases the load applied to the outer truss and overhangs.

The crossbeam of the portal frame is a stiff box girder supported on two tall legs. One of the two legs may be shifted laterally along the crossbeam to be anchored at the pier centreline when two bridges diverge. A diagonal prop provides lateral stability during insertion of the pier-head segments of the first deck (Figure 4.8). After inserting the first segment of the second deck, the crossbeam is anchored to the segments and the prop is removed to insert the last segment. Shuttling such heavy machines along long overhangs generates high uplift forces, and many bolts and stressed bars are necessary to anchor the support legs to the crossbeam and pier caps.

During launching of the gantry, the crossbeam of the portal frame is detached from the support legs and stored on the deck. The legs are extracted from the deck and repositioned on the new pair of pier caps, and the crossbeam is placed over them. The winch trolleys reposition the front portal frame and rear crossbeam during launching, without any need for ground cranes.

The rear crossbeam comprises two braced I-girders or a box girder. The crossbeam is supported on four transverse lines of adjustable legs, and is therefore lighter than the crossbeam of the front

Figure 4.9. Rear crossbeam during gantry assembly behind the abutment

portal frame. The rear crossbeam provides the longitudinal restraint to the gantry during operations, and the two longitudinal lines of hydraulic legs are therefore spaced for better control against overturning. High-tonnage double-acting cylinders with a mechanical locknut at the eight support points cope with gradient and crossfall in the deck slab (Figure 4.9).

The winch trolleys reposition the portal frame and rear crossbeam during launching. The crossbeams are not rotated during repositioning and the gantry is therefore launched between the two bridges to minimise transverse load eccentricity. A front auxiliary leg controls overturning during these operations by bridging between the two pier caps.

The pier-head segments of 90–100 m spans are tall, and the front auxiliary leg is therefore very tall. A shorter leg configuration is used during gantry assembly behind the abutment and to take support on the deck for maintenance of the crossbeams. Long-stroke double-acting cylinders rotate a wide-bottomed pendular extension that bridges between the pier caps during launching. Double-acting support cylinders adjust the frame geometry to different pier cap elevations and recover the gantry deflection during launching. In curved bridges, lateral cylinders shift the support cylinders along windowed crossbeams to align them with the deck webs. Longitudinal cylinders at the connection of the auxiliary leg to the trusses are used to rotate the frame to the local plan radius. A box top crossbeam may be used to brace the trusses and to transfer the vertical load from the leg. Work platforms are provided at all articulations.

A simpler pendular W-frame supports the rear end of the gantry during lifting of segments and repositioning of the rear crossbeam (Rosignoli, 2007). The W-frame is lifted to insert the casting cell beneath the gantry and during casting of the segments. The windowed bottom crossbeam of the W-frame is equipped with lateral cylinders that shift double-acting support cylinders over the webs of curved decks (Rosignoli, 2010).

Figure 4.10. A 32-roll articulated saddle for a 635 ton service load

Large assemblies of cast-iron rolls with conventional geometry are used for the articulated support saddles. Such loads often require long eight-roll bogies comprising seven equalising beams under each chord of the truss (Figure 4.10). Four-roll bogies are often sufficient for the hoist crabs of the winch trolleys. Capstans are preferred to double-acting cylinders for driving lateral shifting because of the higher speed in such frequent operations; capstans and shift cylinders also provide the lateral restraint during operations.

The support saddles house double-acting launch cylinders that are anchored to twin perforated rails fixed to the trusses (Figure 4.11). Vertical plates with two through pins are used to transfer the launch force from the cylinder to the twin rails. Rectangular rails welded to the flanges of trusses and crossbeams facilitate dispersal of the support reactions in the webs and guide to the roll flanges to transfer lateral loads. Launching is complicated by the weight of these machines, the need to respect precise support locations on the deck, and the poor longitudinal stability of the front portal frame.

4.2.3 Performance and productivity

Span-by-span macro-segmental construction is used for continuous single-cell box girders with 80–100 m spans. In addition to balanced cantilever segmental precasting with epoxy joints and incremental launching of steel girders followed by segmental in-place casting of the concrete

Figure 4.11. Launch cylinder acting into twin perforated rails

slab (Rosignoli, 2002), several alternative methods are available for in-place casting of prestressed concrete decks in this span range.

1   Balanced cantilever casting of 4–5 m segments with form travellers.
2   Balanced cantilever casting of 9–12 m macro-segments with a suspension-girder MSS.
3   Balanced cantilever lifting of 9–12 m macro-segments cast beneath the bridge or delivered on barges.
4   Span-by-span casting of half spans with a propped underslung MSS.
5   Span-by-span casting of entire spans with a single-truss overhead MSS.

Balanced cantilever casting of 4–5 m segments with form travellers is the most versatile method. The span length may amply exceed 200 m, the deck may have a tight plan radius or a wide top slab, and the hammers may be erected randomly as soon as the piers become available. Weekly cycle times for a pair of travellers result in slow erection rates, but the special equipment is relatively inexpensive and multiple pairs of travellers can be used at the same time. Poor access to the working areas and expensive handling of materials affect productivity, and temporary pier cap stabilisation gear is necessary at every hammer when the deck is supported on bearings.

Balanced cantilever casting of 9–12 m macro-segments with a suspension-girder MSS is a less versatile method (Matyassy and Palossy, 2006). The span length governs the design of the MSS, and 100–110 m spans are rarely exceeded. The deck must be almost rectilinear and the hammers must be erected directionally from one abutment to the other. The erection rate is almost three times faster than with a pair of form travellers but cannot be increased further; easy access to the working areas and delivery of materials on the deck are additional advantages. The MSS is also used to stabilise the hammers during construction.

Balanced cantilever lifting of 8–10 m macro-segments cast beneath the bridge (Matyassy and Palossy, 2006) or delivered on barges (ASBI, 2008) is an even less versatile method. Load imbalance discourages the use of lifting frames, and handling the segments with a gantry implies the same restrictions as a suspension-girder MSS. Productivity is similar, and these gantries also offer easy access to the working areas and can be used to stabilise the hammers. The main disadvantage of this construction method is disruption of the area beneath the bridge over the entire bridge length.

An underslung MSS for 40–60 m spans can be supported with a temporary pier at the third-quarter of every span to cast a sequence of half-span segments with joints at the span quarters (Rosignoli, 2011). The rear end of the MSS is suspended from the completed deck, and the temporary pier supports the front end of the MSS and the deck. Compared to span-by-span macro-segmental construction, the casting cell is designed for an entire half-span segment, forming the varying-depth box girder is more expensive, and a special carrier is necessary to deliver the prefabricated cage. Also, this construction method requires a temporary pier at every span.

The main advantage of a single-truss overhead MSS for full-span casting is that there is no need for temporary piers. This is a significant advantage, as long spans are typically chosen because of inaccessible terrain or a need for tall piers. At the current state of practice, however, these MSS have reached 70 m spans, with weekly casting cycles. Spans of 80–90 m may probably be reached in narrow bridges or by casting the box core with the MSS and then the side wings with deck-supported forming carriages (Pacheco et al., 2011).

The construction methods are compared in Table 4.1. Full-span casting with a long MSS and span-by-span macro-segmental construction are the fastest methods for in-place casting. Macro-segmental construction is used for simultaneous erection of two adjacent decks, and two temporary piers are necessary in every span. When a half-span underslung MSS is used to cast two adjacent decks, the first deck is completed prior to reverse launching of the MSS for the second deck, and the temporary piers may be shifted from one alignment to the other to save costs.

Macro-segmental construction requires less labour and is faster than in-place casting with a half-span MSS, as the segments are cast with a weekly cycle and erecting two adjacent spans (4 + 4 segments) takes 4 weeks. The cycle time of a half-span MSS is typically 2 weeks per half-span segment due to the complex form adjustment.

The main limitation of span-by-span macro-segmental construction is the cost and level of special-isation of the erection equipment. An underslung MSS does not require huge investments, the machine can be reused for shorter spans, and temporary piers are necessary in either case. Gantries and casting cells for macro-segmental construction are highly specialised pieces of equipment, can rarely be reused in different ways, and require long twin bridges to depreciate the investment and the costs of shipping and site assembly. If the dimensions of the project justify the investment, however, construction is faster and the quality is typically better.

Table 4.1. Alternative erection methods for medium to long spans

Feature	In-place casting						Match-cast segmental precasting with epoxy joints		Composite construction
	Span-by-span macro-segmental construction	Balanced cantilever casting of 4–5 m segments with form travellers	Balanced cantilever casting of 9–12 m macro-segments with a suspension-girder MSS	Balanced cantilever lifting of 8–10 m macro-segments	Span-by-span casting of half spans with a propped underslung MSS	Full-span casting with a single-truss overhead MSS	Balanced cantilever erection with a full two-span gantry	Balanced cantilever erection with two lifting frames	Incremental launching of the steel girder and segmental casting of the concrete slab with a forming carriage
Span length	80–100 m	80–200 m	90–130 m	90–130 m	80–120 m	70–80 m	50–120 m	80–180 m	60–90 m
Tight plan radius	No	Yes	No	No with lifting frames and shorter segments	Possible but complex	No	No	Yes	Possible but complex
Pier height	Short	Any	Any	Any	Short	Any	Any	Any	Any
Disruption of the area beneath the bridge	Partial	No	No	Complete	Partial	No	No	Partial	No
Out-of-sequence erection	No	Yes	No	No	No	No	No	Yes	No
Possibility of acceleration	Poor	High	Poor	Poor	Poor	Poor	Medium	Medium	High
Access to work areas	Easy	Complex	Easy	Easy	Easy	Easy	Easy	Complex	Medium
Mechanisation of casting processes	High	Low	Medium	Low under the bridge	Medium	Medium	High	High	Medium
Logistics follows erection equipment	Yes	Yes	Yes	No	Yes	Yes	No	No	Yes
Handling of materials	Easy	Complex	Easy	Medium	Easy	Easy	Easy	Easy	Medium
Need for match casting	No	No	No	No	No	No	Yes	Yes	No
Labour demand for equipment assembly and decommissioning	High	Low	High	High	Medium	High	Medium	Medium	Low
Labour demand for deck production	Low	High	Medium	Medium	Medium	Low	Low	Low	Low
Financial exposure for erection equipment	High	Low	High	High	Medium	Medium	High	Medium	Low
Re-usability of equipment	Poor	High	Medium	Medium	High	Medium	High	Medium	High
Pier cap gear for hammer stabilisation	No	Yes	No	Yes	No	No	No	Yes	No
Temporary piers	Two per span	No	No	No	One per span	No	No	No	No
Monthly erection rate for one set of equipment	160–200 m	25–30 m	80–110 m	90–150 m	80–180 m	280–320 m	250–400 m	120–200 m	260–400 m

4.2.4   Structure–equipment interaction

Gantry and casting cells interact with the bridge in several ways. Even though the front portal frame and rear crossbeam spread the load to the two bridges, when such a heavy gantry is shifted over the casting cell to pick up the segment, the weight of the gantry, casting cell and macro-segment is applied to only one bridge. The rear end of the midspan segments is also suspended from the front deck cantilever, which further increases the negative moment at the leading pier.

Casting the closure pours and application of continuity prestressing make the deck continuous. The different flexibility and thermal inertia of permanent and temporary piers and the launching loads of gantry and casting cells may overload young in-place stitches. On the positive side, the load deflections of the gantry do not affect the deck, and application of prestressing does not affect the gantry.

The casting cells roll on rails anchored to the deck and the number of bogies is minimised to avoid distortion in the presence of varying gradient and crossfall in the deck. Multiple lines of double-acting cylinders lift the casting cells from the rails prior to filling, to provide uniform support to the frame and to spread the load applied to the deck; in both configurations the load is applied over the deck webs. The casting cells are stiff, deflections are minimal and load redistribution at the application of prestress is easily controlled. Cambers are applied initially and adjusted when necessary. During filling, concrete pumps and mixers apply additional load to the deck.

The macro-segments are designed taking into consideration the handling requirements and temporary support conditions. The cross-sections are asymmetrical prior to casting the longitudinal closure pour, and rotated principal axes complicate the analysis of staged construction as time-dependent deflections also occur in the horizontal plane. The outer form for the front transverse closure pour is inserted between the two lines of supports of the temporary pier; suspended forms are used for the rear closure and the longitudinal joint.

The rear W-frame takes support on the deck prior to picking up the segment. This applies a localised load to the deck that increases at segment lifting, while the distributed load of the casting cell diminishes. Localised loads are also applied here and there during launching and for repositioning the front portal frame and rear crossbeam. These gantries are particularly heavy, and launching is complex and requires many operations. The front auxiliary leg and rear W-frame are used only at specific locations during launching, and the gantry is mostly supported at the portal frame and crossbeam.

The temporary piers are affected by specific forms of structure–equipment interaction. The rear columns of braced modular towers are heavily loaded, the front columns are almost unloaded, and out-of-plane buckling also depends on the longitudinal deflections induced by thermal deck displacements (Rosignoli, 2011). The axial load due to prevented thermal pier deformations after casting the closure pours also affects the buckling factor. The load on the temporary piers is released by retracting the top support systems prior to launching the gantry or moving the casting cells.

4.3.   Twin-girder overhead gantries for balanced cantilever macro-segmental construction

Balanced cantilever macro-segmental decks are erected with lifting frames (Figure 4.12) or twin-girder overhead gantries. Lifting frames apply significant loads to the hammer and require

93

Figure 4.12. Lifting frames for strand jacking of 7.6 m, 25.0 m wide, 725 ton segments for 160 m spans (Photo: SDI)

oversized stabilisation gear at the pier cap and powerful cranes to lift the hoist onto the pier table and to remove it from the deck after midspan closure. Lifting frames are typically used to erect 3–3.5 m segments in curved prestressed concrete bridges with continuous piers and long spans.

Lifting frames are also used to erect macro-segments of prestressed concrete or steel cable-stayed bridges where the stay cables and pylons resist most of load imbalance. Composite cable-stayed decks are typically erected with lighter self-launching derrick platforms. The lifting frames are discussed in Chapter 7.

A twin-girder overhead gantry for balanced cantilever erection of macro-segments takes support on the leading pier and the front deck cantilever. Two lifting platforms spanning the main girders lift the segments from the ground level. The segments may be precast and delivered with barges and tugboats, or may be cast on the ground beneath the bridge. After erecting the hammer and achieving midspan continuity, the gantry is launched to the new span.

Such payloads are demanding on long spans in pure cantilever configurations. The gantry comprises box girders, braced I-girders or rectangular trusses for enhanced lateral stability of the compression chords. Hybrid assemblies of different types of cross-section are often used to lighten the gantry. The lifting platforms are also particularly heavy, and often use strand jacks for reasons of weight and overall economy.

Gantry erection is fit to rectilinear or slightly curved bridges with spans up to 120–130 m and length sufficient to depreciate the investment. The length of the gantry is about 1.3 times the maximum span and 1.5 times the typical span. Despite their cost, these machines offer numerous advantages. The gantry permits access to the working areas from the completed deck for workers and materials. If the deck is supported on one line of bearings, the gantry stabilises the hammers during erection and avoids temporary stabilisation gear at the pier caps. Hammer stabilisation is also useful when the deck is continuous with tall piers to minimise longitudinal bending in the piers due to load imbalance. Less top-slab prestressing is used in the deck because no erection equipment is applied to the cantilevers.

Short pier tables are cast in place at the end of pier erection to support the gantry during operations and repositioning (Figure 4.13). Wet joints with through reinforcement avoid the tight geometry tolerances of match-casting and allow design for partial prestressing. The

Figure 4.13. Lifting of the starter 11.3 m, 680 ton segment with a 158 m, 1600 ton gantry for 125 m balanced cantilever spans (Photo: ThyssenKrupp)

gantry hangs two segments at the ends of the hammer during casting of 1.0–1.5 m closures, curing, and the application of top-slab prestressing. The segments are then released to move the lifting platforms out to lift another pair of segments. This construction method requires the use of wide reinforced closures to ensure duct coupling with reasonable wobbles in case of tendon misalignment in the adjacent segments.

These gantries can be easily reconfigured into suspension-girder MSS for in-place casting of macro-segments, and vice versa (Matyassy and Palossy, 2006). The lifting platforms are replaced with suspended casting cells that roll along the main girders, and the rest of the machine does not require major modifications.

4.3.1   Loading, kinematics, typical features

These gantries operate like short 1.5-span gantries for balanced cantilever erection when standard segments are delivered on the ground. The macro-segments are delivered on barges or cast beneath the bridge. When the piers are tall, the segments are strand jacked into position. Two spreader beams, four stand jacks and hydraulic supply lines controlled by PLCs are used to equalise the load among the strand cables and for geometry adjustment prior to casting the closure pours.

The segments are lifted and held in place during stitching. The lifting platforms may be supported on PTFE skids to avoid the cost of multiple articulated bogies designed for such loads. Auxiliary winch trolleys span between runways overhanging from the main girders to assist the operations of the casting cells for the closure pours. The auxiliary winch trolleys are also used for delivery of materials from the deck, handling of stressing platforms, and repositioning of the launch saddles during launching.

Short casting cells roll along the gantry to form the outer surface of the closure pours, and the corresponding tunnel forms are moved from joint to joint to form the inner surfaces. Concrete is delivered at the front end of the completed deck, pumped to the casting cell through pipes supported on the main girders, and fed with a counterweighted distribution arm. Each lifting platform works in tandem with a casting cell and carries a concrete distribution arm.

Short pier tables are cast in place at the end of pier erection. A tower crane assists pier operations, and assembling a small casting cell on the pier cap is not very labour intensive. The segment is cast on the permanent bearings and stabilised with four jacks at the corners; the tie-downs are minimal, as the gantry stabilises the hammer during construction. The pier table is slightly wider than the pier cap to simplify setting of the casting cells for the first closure pours.

Casting the pier tables at the end of pier erection shortens the cycle time of the gantry and mitigates the risks with parallel tasks out of the critical path. Avoiding a tall front support frame for the gantry balances the additional labour cost for the pier tables, and launching the gantry is also faster.

4.3.2   Support, launch and lock systems

The gantry is supported on the leading pier table and the front deck cantilever. The support saddles are anchored to the deck through concrete pedestals cast over the webs to cope with deck crossfall. Minor traversing may also be necessary in rectilinear bridges to cope with tolerances of verticality in the piers. Flat crossbeams comprising two PTFE skids over the support saddles and a central connection beam are shifted by double-acting lateral cylinders for minor traversing of the gantry. The main launch saddles are assembled over the ends of the flat crossbeams.

The launch saddles transfer the vertical load directly to the bottom support saddles, and the flat crossbeams are used only for geometry adjustment. The assembly is equipped with bidirectional restraints to be repositioned with an auxiliary winch trolley.

Such heavy machines often have three mobile supports: two main supports are used during operations, and a front auxiliary support controls overturning and negative bending during launching. Compared with the use of a fixed front auxiliary leg, a mobile support permits repositioning of the main supports with different configurations of the main girders, and launching is also faster. Long-stroke double-acting support cylinders are used to adjust the support geometry when the deck has a vertical curvature.

These gantries are never lifted from the launch saddles, which are therefore designed for self-weight and full payload (Rosignoli, 2011). Launch saddles designed for such loads are often based on PTFE skids. Friction launchers (Rosignoli, 2000) or redundant double-acting cylinders acting within perforated rails provide the launch force and restrain the gantry during operations.

4.3.3   Performance and productivity

The productivity of these gantries is governed by the lifting rate of the hoists and the cycle time of the closure pours between segments. Strand jacks are slower than hydraulic winches with reeved hoisting ropes but can be used with any pier height. The use of wet joints has a substantial impact on the cycle time, as the closure pours must be cured prior to stressing the top-slab tendons. Although the closure pours are subjected to uniform stresses and the curing time may therefore be short, the typical cycle time for a pair of macro-segments is 3–4 days.

The use of wet joints adversely affects productivity (see Table 4.1). The erection rate is about one-third that achieved with balanced cantilever erection of standard precast segments with epoxy joints. When the segments are cast under the bridge, the advantages are no precasting facility, no need for storage and transportation of segments, less stringent geometry control, and bridge design for partial prestressing.

When the segments are precast in a docking facility for delivery on barges, match-casting with both short- and long-line techniques dramatically accelerates the cycle time due to the use of epoxy joints. In this case, the choice between erection by gantry or lifting frames is governed by the length of the spans, the radius of the plan curvature, and the pros and cons of directional erection of the hammers.

4.3.4   Structure–equipment interaction

Balanced cantilever erection of macro-segments requires powerful and heavy gantries, and the loads applied to the bridge are a major concern. These gantries are flexible, but this is not a major problem as the segments are handled individually and supported by the hammer after application of top-slab prestressing. The casting cells for the closure pours are anchored to the adjoining segments, and can therefore be very wide without problems of excessive form flexibility.

The front support of these machines loads the leading pier with minor or no eccentricity. The rear support on the front deck cantilever is repositioned after midspan closure, and erection stresses are therefore locked into the bridge. The gantry is used to stabilise the hammer during erection, which relieves most of load imbalance of staged construction in the piers.

The weight of the gantry may affect the design of piers and foundations, and may cause differential settlement of foundations, especially when the piers are short and the weight of deck and gantry is therefore a major component of the foundation load. As the deck is erected directionally and made continuous during the process, progressive settlement of foundations may generate locked-in stresses (Rosignoli, 2011).

The macro-segments are lifted at four points for geometry adjustment prior to casting the closure pours. Synchronous lift systems controlled by PLCs use feedback from linear position sensors and load transducers to control lifting, lowering and positioning of the macro-segments, regardless of the weight distribution. Synchronous lifting minimises the risk of warping, tilting and load shift between the lift points, and minimal erection stresses are therefore locked into the deck.

REFERENCES

ASBI (American Segmental Bridge Institute) (2008) *Construction Practices Handbook for Concrete Segmental and Cable-supported Bridges*. ASBI, Buda, TX, USA.

Gimsing NJ and Georgakis CT (2012) *Cable Supported Bridges*, 3rd edn. Wiley, New York, NY, USA.

Matyassy L and Palossy M (2006) Koroshegy Viaduct. *Structural Engineering International* **16(1)**: 36–38.

Pacheco P, Coelho H, Borges P and Guerra A (2011) Technical challenges of large movable scaffolding systems. *Structural Engineering International* **21(4)**: 450–455.

Rosignoli M (2000) Thrust and guide devices for launched bridges. *ASCE Journal of Bridge Engineering* **5(1)**: 75–83.

Rosignoli M (2002) *Bridge Launching*. Thomas Telford, London.

Rosignoli M (2007) Robustness and stability of launching gantries and movable shuttering systems – lessons learned. *Structural Engineering International* **17(2)**: 133–140.

Rosignoli M (2010) Self-launching erection machines for precast concrete bridges. *PCI Journal* **2010(Winter)**: 36–57.

Rosignoli M (2011) Bridge erection machines. In *Encyclopedia of Life Support Systems*, Chapter 6.37.40, 1–56. United Nations Educational, Scientific and Cultural Organization (UNESCO), Paris, France.

Bridge Construction Equipment
ISBN 978-0-7277-5808-8

ICE Publishing: All rights reserved
http://dx.doi.org/10.1680/bce.58088.099

publishing

Chapter 5
Movable scaffolding systems (MSSs) for in-place span-by-span casting

5.1. Technology of span-by-span casting

Span-by-span in-place casting of prestressed concrete spans ranging from 30 m to 70 m is a popular alternative to precast segmental construction in many countries. Solid or voided slabs with pier haunches are used for 30–40 m continuous spans, double-T-ribbed slabs are rarely used for spans longer than 50 m, and box girders are used, for different reasons, over the entire range of spans (Rosignoli, 2011a).

Simple spans are preferred in LRT and high-speed railway (HSR) bridges because of the favourable stress distribution in the continuous welded rail. Continuous spans are preferred in highway bridges because of structural efficiency and durability, and the smaller number of bearings and expansion joints.

In very short bridges the casting cell is supported on ground-based falsework. After application of prestressing, the outer form is suspended from the deck, the falsework is moved to the next span, and the form modules are lowered to the ground and placed on the new falsework. Casting on falsework is labour intensive, but also requires minor investments in construction equipment; falsework is, therefore, the typical choice in emerging countries where there is cheap labour and small projects that do not require rapid construction (Harridge, 2011).

In bridges of length sufficient to depreciate the investment and comprising multiple spans with constant length and large plan radius, the outer form is assembled on a self-launching frame to create a MSS. The frame supports the form during span casting and launches itself to the next span for a new casting cycle.

MSSs are among the heaviest, and in some cases the most expensive, construction machines for prestressed concrete superstructures. They were developed after the introduction of post-tensioning technology to shorten the cycle time and to reduce the labour demand of prestressed concrete bridges with enhanced mechanisation of in-place casting (Däbritz, 2011). The MSSs are normally used for spans up to 50–60 m, but new-generation MSSs have consistently reached 70 m spans and can potentially reach 80–90 m spans (Pacheco *et al.*, 2011).

Although an MSS is expensive, mechanical transfer of the casting cell to the next span takes hours instead of weeks, and this often results in weekly cycle times instead of monthly ones. Safety of workers and cost-effective production add to no ground constraints and minimised use of ground cranes. Factory production enhances quality and facilitates quality control. Minor site disruption and easy maintenance of highway and railway traffic are additional advantages (Figure 5.1).

Figure 5.1. A 110 m, 1490 ton overhead MSS with a 60 ton portal crane for 57.4 m, 1380 ton single-track HSR spans

The disadvantages include the investment required for the MSS, the costs of shipping, site assembly and final decommissioning, the high level of technology needed for bridge design and construction, and the need for casting the deck directionally from abutment to abutment. Without some form of cage prefabrication, the time schedule is rigid due to multiple activities at the only work location (Povoas, 2012). The demand on batching plant and concrete delivery lines is also very imbalanced when long spans are cast in one phase.

In-place casting with an MSS competes with incremental launching on the same range of spans, and the cycle time is also similar (Rosignoli, 2001). Both construction methods are linear and repetitive and cannot be accelerated. Key differences are the ability of an MSS to cope with complex deck geometries, the higher cost of labour and equipment, and the need for logistics to follow the MSS throughout the bridge (Rosignoli, 2002). The absence of a maximum deck length is another potential advantage, even if by the time that the 1200–1500 m limit length of incremental launching is reached, the MSS may often be replaced by precast segmental techniques on economic grounds (ASBI, 2008; Homsi, 2012).

An MSS comprises three major components: the casting cell, the support frame, and the support and launch systems (Däbritz, 2011). The MSSs are distinguished from the position of the support frame relative to the casting cell. Underslung MSSs support the casting cell from underneath and provide full clearance for feeding of materials; they are the lightest, most adaptable and most common type of MSS. Overhead MSSs suspend the casting cell, minimise the need for clearance beneath the bridge, are fully self-launching, facilitate sheltering of the casting cell, and are simpler to decommission at the end of bridge construction. Both types of MSS can cast simple and continuous spans. Reversible MSSs can cast the deck in both directions to avoid reverse launching in the construction of parallel bridges (Povoas, 2012).

An MSS is typically used to cast an entire span, and its capability is expressed in terms of casting span (distance between the supports during span casting) and load capacity of the casting cell. The self-weight of the casting cell often exceeds 20% of the weight of the prestressed concrete span and is a considerable portion of the total load carried by the self-launching frame.

Steel or wood forms are used for the casting cell. The choice depends on the number of spans to cast, plan curvature, framing system of the casting cell, required finishing of concrete surfaces, and cross-section geometry (Däbritz, 2011). Steel forms are used in applications that do not require changes to geometry, while wooden forms are more adaptable and are the typical choice for varying-depth bridges.

The framing system of the casting cell is connected to the self-launching frame with shims, adjustment screws or jacks for setting of camber and crossfall. The framing system is divided into two halves and the outer form is opened prior to launching to avoid conflict with the leading pier (Figure 5.2). Splice design influences the deflections of the casting cell during filling and the duration of launching. The interaction between forms and framing system stiffens the casting cell, and is sometimes accounted for in the design in order to reduce the amount of bracing required and to lighten the MSS (Däbritz, 2011).

Trusses, braced I-girders or box girders are used for the self-launching frame. Plate girders provide a large wind drag area but permit robotised welding, and the loads of the casting cell and launch saddles can be applied anywhere. Triangular and rectangular trusses facilitate access by workers but require hand welding, and the loads should be applied at the panel nodes. Light launch noses and tails are used in most applications to control overturning during launching. Hybrid noses and tails are also used where I-girders are stiffened by lateral trusses. One or two main girders are used in the overhead machines, while underslung MSSs are always based on two girders.

Long overhead MSSs have been tested on 70 m spans and may open new perspectives for 80–90 m spans over water or difficult terrains where incremental launching and half-span casting with a propped MSS require temporary piers, and balanced cantilever casting is not yet cost-effective (Pacheco et al., 2011). Short and light MSSs have been used for span-by-span casting of the deck of arch bridges, but also in this case the incremental launching method is often preferred because of the inexpensive equipment, simple logistics and less load imbalance on the arch.

If the piers are short and the area under the bridge is free from obstructions, the span-by-span method can be used on spans longer than 100 m. The span is divided into two halves, with construction joints at the span quarters, and a temporary pier supports the leading end of the midspan segment prior to casting the next pier segment. Underslung MSSs are used for these applications because the temporary pier also supports the MSS (Figure 5.3). Props from foundations may further increase the load capacity of the casting cell.

Figure 5.2. Opened outer form during launching (Photo: AP-BCS)

Simple spans are cast full-length prior to the application of prestressing. For continuous decks, span-by-span construction involves casting a deck length between vertical joints positioned along the span. The abutment span is cast with a front cantilever, and the subsequent spans extend out over the piers. The length of the front cantilever is 20–25% of the typical span, and the abutment span is therefore 75–80% of the typical span to avoid modifications of the casting cell.

When an overhead MSS is used to cast continuous spans, the rear end of the machine is supported on the front cantilever of the completed deck. When an underslung MSS is used, the rear end is suspended from the cantilever. In both cases a shorter casting span lightens the MSS, and the negative bending generated by the front cantilever of the casting cell relieves stresses and deflections in the MSS and the rear support reaction applied to the deck.

Supporting the MSS on the front deck cantilever also avoids casting cell deflections at the construction joint and improves the time-dependent stress redistribution of staged construction. This is a critical issue in span-by-span construction, as prestressing is applied at short curing. Negative bending due to the rear support reaction of the MSS, however, may require additional post-tensioning in the deck. The farther the construction joint from the leading pier, the lighter the MSS and its rear support reaction. The optimal joint position varies between 20% and 25% of

Figure 5.3. Launching of a half-span MSS to the pier segment configuration

the span length depending on the weight of the deck and MSS and the use of constant- or varying-depth sections for the deck.

Solid, voided and ribbed slabs are cast in one phase, while the box girders may be cast in one or two phases. When a box girder is cast in one phase, the inner tunnel form remains within the span during repositioning of the MSS. After inserting the new reinforcement cage into the casting cell, the tunnel form is pulled forward and reopened within the cage. One-phase casting is the fastest way to cast a box girder, and a weekly cycle time is soon achieved with full or partial cage prefabrication. One-phase casting is frequent in HSR bridges, where the owners require no horizontal joints and the depth of the box girder facilitates access to the inner cell.

Operating a full-span tunnel form is labour intensive due to the number of articulations and hydraulic systems necessary to collapse the forms, to avoid interference with tendon blisters and deviators and folded form sectors, and to extract the assembly out through the front pier diaphragm. Stripping and forming the inner cell, on the other hand, is very fast.

Variations in cross-section geometry, tendon blisters, deviators and pier diaphragms govern the modularity of the form segments (Figure 5.4). Form design may be simplified with diaphragms of consistent geometry and thickness throughout the deck (piers and abutments) and by widening the opening of the diaphragms as much as possible (CNC, 2007). Tunnel forms carried by

Figure 5.4. Inner tunnel form (Photo: AP-BCS)

full-span trusses are extracted more easily through V-openings: the truss rolls on saddles supported on the bottom slab of the rear span, and is received by a second set of saddles supported on the bottom form table of the casting cell. Independent form modules assembled on adjustable towers are handled with hydraulic manipulators and require openings at the base of the diaphragm.

Additional geometry constraints derive from the need to support overhead MSSs on the pier caps without conflict with the bridge bearings. The pier diaphragms may also be cast after extracting the inner form, although this requires a large number of bar couplers and increases the labour demand (CNC, 2007).

The webs and the bottom slab may be thickened at the pier to stiffen diaphragms with large openings. Thicker webs are also necessary at the construction joints when the tendons are coupled (Figure 5.5). Splicing the tendons by overlapping avoids the need for thicker webs, but the blisters are difficult to access prior to extracting the inner form.

When conventional concrete is used, the top surface of the bottom slab is not formed and the slab is cast using concrete with a lower slump to avoid reflow during web filling. Concrete pouring starts at the bottom web-slab nodes and the pouring zones are staggered longitudinally with a sequence designed to avoid cold joints. The use of self-compacting concrete requires a top

Figure 5.5. Tendon couplers at the construction joints (Photo: BERD)

shutter for the bottom slab and air-drainage systems to minimise air bubbles; shutters and anti-floating tie-downs are designed for full hydrostatic load.

Box girders for highway bridges are typically cast in two phases. The bottom slab, webs and pier diaphragm are cast first, and the top slab is cast after removal of the inner forms (Figure 5.6). In some applications the top slab strip between the webs has been cast in the MSS and the side wings have been cast with forming carriages after launching the MSS. This casting method reduces the design load of the MSS but complicates the analysis of the time-dependent stress redistribution within the cross-section.

In most cases, the top slab is cast full-width in the MSS. The central strip is cast on a form table supported on brackets bolted to the inner side of the webs or on modular towers supported on the bottom slab. After stripping the inner forms for the first-phase segment, the form table is extracted from the previous span for completion of the cross-section. Form tables supported on web brackets require full-width openings at the top of the pier diaphragm, while hydraulic manipulators for support towers require a flush bottom slab at the pier diaphragm.

The main concern with two-phase casting with horizontal joints at the top slab level is cracking of the first-phase U-span due to MSS deflections. The weight of the form table and top slab generates

Figure 5.6. Inner web forms for two-phase casting (Photo: BERD)

additional deflections in the MSS and, as the U-span is stiffer than the MSS, it tends to resist most of the additional load. The total deflection of the MSS may be significant on long spans, and the typical limit of 1/400 of the casting span may be insufficient to avoid cracking of the young concrete. Tensioning some tendons in the U-span is often necessary to avoid cracking and, although staged prestressing increases the labour costs and slows the cycle time, it reduces the design load of the MSS.

Another potential disadvantage of two-phase casting is the shrinkage differential between the two segments. Some design standards require that the time between casting the U-span and the top slab must not exceed 3 days, even though creep is high in these first curing stages and shrinkage tension fades quickly.

When the horizontal joints are at the bottom slab level, the inner form is similar to a tunnel form for one-phase casting. The inner form is more expensive, but staged prestressing is not necessary because the bottom slab is more flexible than a U-span and can follow the deflections of the MSS without cracking. Hydraulic manipulators for modular segments of the inner form are very fast along the two-span strip of bottom slab. Mid-depth horizontal joints are sometimes used in the webs of wide decks with a ribbed top slab.

Figure 5.7. Full-span cage prefabrication with precast anchor blocks

Two-phase casting and staged application of post-tensioning reduce the design load of the MSS, simplify pier diaphragms and extraction of inner forms, and reduce the quantity of concrete processed daily. One-phase casting of a 70 m span requires handling 600–700 m^3 of concrete in a few hours and exerts a substantial demand on batching plant and concrete delivery lines. Horizontal joints at the top slab level also avoid horizontal cracks arising from settlement of fresh concrete in the webs.

The reinforcement cage is often prefabricated behind the abutment and delivered on the deck. The cage includes front bulkhead and post-tensioning hardware. Precast anchor blocks may be used to shorten the cycle time and remove activities from the critical path (Figure 5.7). The strand is inserted into the ducts during curing, and tendon splicing with couplers therefore requires temporary working recesses in the webs. Overlapped tendons are less labour intensive but the tendons are longer. In both cases the tendons are tensioned from a front stressing platform.

Partial cage prefabrication has also been tested successfully. The webs are the most complex portion of the cage as they contain the tendon ducts (Figure 5.8). In some overhead MSSs, cage segments for the webs have been picked up from trucks on the deck and moved into the casting cell. In other applications, full-length web cages have been assembled over the new span during curing, suspended from the MSS during launching, and lowered into the casting cell after closure (Figure 5.9).

Concrete is poured using either conveyor belts or pumps feeding distribution arms. In the longitudinal direction, the casting cell for continuous spans is filled in one of two alternative sequences.

Figure 5.8. Segmental web cage prefabrication

1 Concrete is poured starting at the leading pier, proceeding symmetrically until the front overhang is filled; then the rear span section is filled backwards toward the construction joint. This filling sequence is preferred when two lines of supports are used for the MSS at the leading pier. Forward filling from the construction joint is never used due to the risk of joint cracking with increasing deflections of the MSS. This sequence would also overload the MSS and front deck cantilever due to there being no negative bending from the front overhang of the casting cell, and would lead to large midspan sags and upward deflections in the front overhang (Däbritz, 2011).

2 Concrete is poured starting at the front bulkhead and proceeding backward toward the construction joint. This sequence requires less labour and facilitates finishing of the top slab, but the flexural rotations at the leading pier are larger. Backward filling is preferred when one line of supports is used at the leading pier and when the MSS is propped from foundations. Simple spans are filled from bulkhead to bulkhead in either direction.

The difference in temperature between adjoining spans should be controlled in the first curing stages. In particularly severe climates or when cements with high hydration heat are used, it may be necessary to maintain the temperature in the previous span or to heat its front end, and also to cool the setting concrete of the new span (Rosignoli, 2002).

Precast anchor blocks, local thermal cycles at tendon anchorages, and steam curing of the entire span have been used to shorten the cycle time (Rosignoli, 2001). Increasing the design strength of

Figure 5.9. Suspended web cages for ribbed slabs (Photo: Strukturas)

concrete is also cost-effective, and reduces the influence of unexpected events and time-dependent phenomena.

5.2. Advancing shoring based on ground falsework

Span-by-span advancing shoring based on ground falsework may be cost-effective in short bridges with complex geometry, short spans, short piers, good soil and, most of all, cheap labour. Disruption of the area under the bridge and high crane demand are additional limitations; in some countries; however, labour is so cheap that ground falsework is hardly beatable when construction duration is not a major issue (AASHTO, 2008a, 2008b; BSI, 1996, 2008a, 2008b; CSA, 1975; FIB, 2009; Harridge, 2011).

Ground falsework (BSI, 1996, 2008a) may lead to optimal quantities of concrete and prestressing because construction does not cause permanent stresses in the bridge. In spite of savings in materials, however, ground falsework is so labour intensive that in industrialised countries it is used only for specific tasks such as uneven spans within bridges built with other methods, abutment closures of balanced cantilever bridges, and curved spans that prevent the use of a gantry.

Several proprietary shoring systems are available (ACI, 2004). Towers, adjustment screws, turn-buckles, diagonals, braces, diaphragms and joints have different shapes and features, and

Figure 5.10. Modular trusses on pier towers

typically are not interchangeable from system to system. The same shoring system can be used to support precast segments or a casting cell for in-place casting (BSI, 2008b). The towers have adjustment screws at the bottom and at the top; the bottom screws are used to set the towers vertical and to cope with irregularities in the support surface, and the top screws are used to set precast segments or the casting cell to gradient and crossfall (FIB, 2009).

Modular trusses on pier towers offer several advantages over shoring tower systems (Figure 5.10). The pier towers are supported on the bridge foundations, the outer form is lowered with a few sand boxes for stripping, and the trusses are rolled laterally out of the span to be picked up with two ground cranes. The trusses are assembled into blocks of two or three braced units, and the entire span is assembled and dismantled with three or four lifts. Turnbuckles in the bottom chords of the trusses are used for setting the camber.

The increase in labour cost with the height to the ground is less marked than with shoring towers, and the cycle time is shorter. The main limitation of modular trusses is their flexibility: spans longer than 25–30 m require the use of a central temporary pier, with additional costs of foundations and labour.

5.3. Twin-girder overhead MSSs

A twin-girder overhead MSS comprises two main girders supported on two sets of pier cross-beams. For in-place casting of continuous spans, the MSS takes support at the leading pier and on the front deck cantilever. In configurations for simple spans, the rear support is placed on the pier diaphragm of the previous span.

Hanger bars suspend the casting cell from the main girders (Figure 5.11). After application of prestressing, the MSS is lowered to release the span in one operation. A portal crane bridging

Figure 5.11. Casting cell suspended with hangers

between the main girders lowers the outer form modules to the ground, and the hangers are lifted so as not to interfere with the front support crossbeams during launching of the MSS to the next span. Flat frame winch trolleys are rarely used in these machines because side overhangs are necessary in the crane bridge to hoist the outer form modules beyond the main girders.

Twin-girder overhead MSSs are easily adaptable to shorter spans and small variations in deck geometry. They are less affected by ground constraints than are underslung MSSs, and they are easier to operate in curved bridges. These MSSs, however, are heavier and more complex to design, assemble and operate than underslung MSSs in typical conditions. They are also slower in casting the span, they cannot be propped from foundations, and the need for lowering the outer form to the ground prior to launching adds activities to the critical path, slows the cycle time and affects the area under the bridge.

Launch noses and tails are applied to the main girders to control overturning during launching and the total length of these machines is 2.1–2.3 times the typical span. This is rarely a problem, as the overhead MSSs do not interfere with the piers or the completed deck and have minimal impact on the clearance beneath the bridge. The height of the main girders, however, may cause clearance issues when passing under existing bridges or power lines.

Full-span cage prefabrication requires multiple portal cranes for cage insertion, and is complicated by the tower–crossbeam assembly that supports the MSS at the leading pier. However, the cycle time of these machines is so long that full-span cage prefabrication is rarely worthwhile. Prefabricated cage segments are easily delivered on the ground, lifted beyond the front bulkhead, and moved backwards into the casting cell.

5.3.1 Loading, kinematics, typical features

A twin-girder overhead MSS suspends the outer form with hanger bars (Rosignoli, 2011a). Square trusses with light front and rear extensions to control overturning during launching are the typical choice for the main girders; braced I-girders and box girders are rarely used due to the difficult access to the hangers. The hangers are suspended from rails supported onto spreader beams spanning between the truss chords for geometry adjustment.

Square trusses often include three planar trusses: two adjacent trusses support the spreader beams of the hangers, and an outer auxiliary truss controls out-of-plane buckling and carries the runway for the portal crane (Figure 5.12). One of the main trusses is interrupted at the ends of the casting cell, and the other two trusses are extended into a front nose and a rear launch tail that allow full-length operation of the portal crane (Rosignoli, 2010). Full-length working platforms along the trusses facilitate access to the hangers and maintenance and inspection.

Some square trusses include long welded modules that can be used in both overhead and underslung configurations (Figure 5.13). Shipping is more expensive, but handling, site assembly and decommissioning are very fast. The field splices in the chords are typically pinned for fast assembly.

The modules of the outer form comprise a form table supported on a bottom frame and adjustable shutters for walls and side wings. Each module of the bottom frame includes two crossbeams suspended from the hangers and two edge girders for structural continuity with the adjacent

Figure 5.12. Adjustable anchor rails supported onto spreader beams

Figure 5.13. 4.0 m tall, 2.5 m wide, 21.0 m long welded module of square truss

frame modules. The bottom form table and auxiliary cross braces provide in-plane stiffness (Däbritz, 2011). The bottom crossbeams support lateral working platforms at both edges, provide contrast for geometry adjustment of the rib frames, and anchor the tie-downs that prevent floating of the inner form during filling.

Conical spindles are the typical choice for connecting form modules. Hammering spindles into position is faster than inserting and tightening bolts and also drives the form panels into position. Conventional bolted connections are used between the form panels that do not require disconnection for stripping.

After application of prestressing, lowering the MSS releases the span in one operation. The crane bridge of the portal crane has side overhangs to pick up the crossbeams of the bottom frame at the ends, and the form modules are lowered to the ground and stored on trailers. After launching the trusses to the next span, the form modules are driven under the trusses and lifted up to the deck level for progressive reassembly of the outer form.

The bottom frame of the casting cell is anchored to the leading pier cap and the front deck cantilever to transfer lateral loads and to stabilise the front overhang of the casting cell. Structural continuity between form modules also ensures uniform camber and avoids steps in the formed surfaces at the joints.

Curved spans are formed by shifting the rib frames transversely along the bottom crossbeams. The inner bottom node of each rib frame is pinned to a clamp that slides along the crossbeam, and adjustment screws and diagonal turnbuckles are used for form setting. Geometry adjustment at the bottom frame level keeps the hanger bars aligned with the main trusses. Camber and crossfall are set by jacking the hangers at the top anchorages.

The prestressing tendons are fabricated and tensioned from a stressing platform at the front bulk-head. The stressing platform is also lowered to the ground prior to launching. Tendons are grouted after launching to remove these operations from the critical path.

5.3.2 Support, launch and lock systems

Most twin-girder overhead MSSs have three main supports. Two tower–crossbeam assemblies support the MSS at the piers of the span to be cast, and a third assembly receives the trusses at the next pier during launching. The three tower–crossbeam assemblies have the same load capacity and are anchored to the pier caps, or to the bridge foundations when the piers are short. Supporting the MSS from the bridge foundations simplifies forming due to there being no interference with deck diaphragms and bearings.

The tower–crossbeam assemblies are erected using ground cranes and comprise modular towers, crossbeams and high-tonnage double-acting cylinders that provide geometry adjustment and may be interconnected to generate hydraulic hinges. Casting continuous spans also requires a rear auxiliary frame. The portal crane places the frame on the front deck cantilever, and the frame lifts the trusses from the rear pier support for casting the span and lowers them back at span release for launching.

In light applications, longitudinal support frames connect the ends of two braced crossbeams (Figure 5.14). The support frames carry double-acting vertical cylinders and articulated launch bogies; the cylinders lift the MSS from the launch bogies for span casting and lower it back after application of prestressing to release the span in one operation. The locknuts of the cylinders are tightened to provide mechanical load transfer during span casting. The support frames can be shifted along the crossbeams to traverse the MSS in curved bridges. The support cylinders apply localised loads to the main girders, and vertical stiffeners welded to the web panels control local buckling (Rosignoli, 2002). The MSS is relatively light during repositioning, and the launch bogies are short and sometimes devoid of equalising beams. Guide rolls act against the edges of the bottom flange to transfer lateral loads (Rosignoli, 2000).

In heavier applications, two pairs of braced crossbeams support the trusses under the panel nodes (Figure 5.15). This solution reduces the load applied to the support saddles and permits the use of commercially available shapes for the crossbeams. Double-acting cylinders at the top of the tower are used to lift and lower the crossbeams, and the support saddles are designed for the total load of the MSS and span (Rosignoli, 2007). The cylinders may be interconnected during launching to generate a hydraulic hinge that controls longitudinal bending in the tower and equalises the load applied to the crossbeams. After lifting the MSS to the span casting elevation, the hydraulic hinge is locked by tightening the locknuts of the cylinders.

The cost of articulated support saddles grows with the design load, and PTFE skids often replace the launch bogies in heavy applications. Articulated bogies are better suited to light loads and fast launching; and PTFE skids are better suited to heavy loads and slow launching, but the friction

Figure 5.14. Tower–crossbeam assembly for a light MSS

coefficient is typically more than double. The use of polished stainless steel is unusual and the trusses are supported directly on greased PTFE pads. The flexibility of the crossbeams reduces the transfer of longitudinal bending to the tower, and hydraulic hinges are rarely necessary, and if so only for launching (Rosignoli, 2011a).

PTFE skids load most of the flange width to reduce the contact pressure. Lateral guides act against the edges of the flange to transfer lateral loads (Rosignoli, 2002). The bottom flanges of the trusses and the top flanges of the crossbeams have flush contact surfaces (Rosignoli, 1998) of constant width.

The girders of light MSSs are sometimes launched with winches and reeved ropes anchored to the deck or the next pier; launching with ground cranes is not recommended (Figure 5.16). In heavier units, redundant double-acting cylinders operating into perforated rails push and pull the girder forward. Twin launch cylinders are safer than winches when launching along inclined planes, although launching is slower. Winches cannot be used for downhill launching (Rosignoli, 2000).

Figure 5.15. Tower–crossbeam assembly for a heavy MSS with tie-downs at the support cylinders

The girders are launched and traversed individually. In the span-casting configuration, the girders are located at the tips of the crossbeams to suspend the bottom frame of the casting cell outside the outer forms. When the MSS does not carry a portal frame, the girders may be shifted inward prior to launching in order to control transverse bending in the tower–crossbeam assemblies. Shifting adds operations to the launch cycle, and in most cases the tower–crossbeam assemblies are designed for the full load imbalance of staggered launching. The use of counterweights to control overturning further increases load imbalance (Figure 5.17). Long crossbeams are prone to uplift and overturning, which may limit the launch differential during staggered launching.

Long-stroke double-acting cylinders are typically used to shift the main girders transversely along the crossbeams. Paired cylinders add redundancy (Rosignoli, 2007) and hold the saddle in place during staggered repositioning of the anchor pins. When the MSS carries a crane, the launch saddles may be connected to keep the gauge of the crane runway and to use only one set of cylinders for traversing. The shift cylinders also act as transverse restraints during launching and operations.

The pier towers are anchored to pier caps and bridge foundations with stressed bars that resist uplift forces. The crossbeams have long side overhangs and significant uplift forces may arise in the anchor systems (Rosignoli, 2011a). When the crossbeams are supported on hydraulic

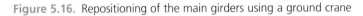

Figure 5.16. Repositioning of the main girders using a ground crane

cylinders, adjustable tie-downs are often necessary also at the cylinders (see Figure 5.15). Longitudinal bending due to friction and the launch gradient causes additional uplift.

The bridge bearings for simple spans are located at the corners of the pier cap, and this often requires the use of narrow pier towers to avoid interference. Narrow pier towers further increase uplift in the tie-downs. The top diaphragm of hollow piers often houses a manhole, and through tie-downs can be anchored at the bottom surface of the diaphragm. Some twin-girder overhead MSSs are so heavy that the pier diaphragms are designed for the loads applied by the MSSs. The rear auxiliary frame is also anchored with stressed bars and equipped with geometry-adjustment systems to cope with deck crossfall.

Continuous spans have only one line of bearings at the pier cap but the four legs of the tower are all embedded in the span. The legs are protected from concrete with left-in-place pipes and are stiffened in both directions as the bottom panel of the tower cannot be braced (see Figure 5.14).

Figure 5.17. Staggered launching with counterweights

5.3.3   Performance and productivity

Some twin-girder overhead MSSs are not many steps ahead of ground falsework in terms of the scale of mechanisation of production. They are heavy and slow to operate and require ground cranes for repositioning of supports. The need to lower the outer form to the ground before launching is another major limitation, and several activities affect the area under the bridge.

These machines are designed to cast short bridges with large-plan radius where the vertical clearance under the deck is limited and disturbance to part of the area can be tolerated for short periods of time. These MSSs have also been used successfully on large-scale HSR projects for in-place casting of special crossings with spans longer than those achievable in a precasting facility for the full-span method (Rosignoli, 2011b).

In such cases, the long cycle time of a twin-girder overhead MSS is not a major issue, as construction of the special crossings can start as soon as the piers are available. Continuous three-span bridges, with a central span 1.5–1.7 times longer than the simple spans of the approaches and 20% shorter side spans, provide similar dynamic performance (train-induced resonance is less marked in continuous spans) and can use the same cross-section. In-place casting of a three-span continuous bridge may take 3–4 months, including the assembly and decommissioning of the MSS (Rosignoli, 2012).

Site assembly and decommissioning of such heavy and cumbersome machines are also complex. When these MSSs are used to cast special crossings within HSR viaducts built by full-span precasting, the approach spans are not yet available and the MSS is assembled from the ground. A starter module is placed on two temporary piers and new modules are attached to and supported on temporary piers. The two girders are kept close to each other during assembly to minimise transverse bending in the towers. After application of launch noses and tails, the girders are launched over the pier towers and shifted out to the tips of the crossbeams, and the portal crane is placed on the girders to lift the casting cell. Decommissioning follows an inverse sequence of operations. Lighter girders may be lifted onto the pier towers in one lift using two cranes.

When the MSS is used in different zones of the project, partial dismantling adds labour and crane costs. Reverse launching for casting adjacent bridges also takes time because of slow operations and the low level of automation of these machines.

5.3.4 Structure–equipment interaction

Twin-girder overhead MSSs are typically used to cast heavy box girders on a few 40–60 m spans. The main concerns with two-phase casting are related to the MSS deflections during casting of the top slab. The load path of these machines includes multiple flexible systems (bottom crossbeams, hangers, main girders, tower–crossbeam assemblies), and the total deflection of the casting cell may therefore be pronounced. One-phase casting with retarding admixtures mitigates the risk of cracking, but is demanding on batching plant and concrete delivery lines due to the large quantity of concrete to be processed in a few hours.

Outer and inner forms are stripped prior to the application of prestressing to support the span only with the bottom form table. This operation reduces the prestress losses into the forming systems at tendon stressing. The span is released by lowering the MSS to avoid load redistribution among the hangers, to accelerate the operation and to save labour.

Tensioning all the tendons prior to releasing the span is often impossible due to the residual support action exerted by the MSS. Application of prestressing is often divided into two phases: a first set of tendons is stressed to make the span self-supporting, and the remaining tendons are tensioned after span release. This also allows earlier repositioning of the MSS. Two-phase casting of heavy box girders often requires stressing some tendons in the U-span prior to casting the top slab, which further complicates the application of prestressing.

Continuous spans are cast with the rear auxiliary frame of the MSS supported on the front deck cantilever to shorten the casting span. The new span is released after achieving structural continuity with the previous span, and repositioning the MSS removes the load applied by the rear auxiliary frame. Stresses, deflections and time-dependent effects of load redistribution may be significant due to short curing, full prestressing, and the absence of superimposed dead loads.

Most twin-girder overhead MSSs are supported at the piers during launching. This simplifies the analysis of structure–MSS interaction, as the vertical loads are applied with minor eccentricity. Some of these machines are very heavy and their weight may affect the design of piers and foundations. The longitudinal launch loads may also affect pier design, especially when the use of PTFE skids is combined with uphill launching (Rosignoli, 2002). Tall piers may require realignment upon launch completion.

When casting curved bridges, the main girders are offset outward at both piers to reduce the lateral eccentricity of the casting cell at midspan. Additional eccentricity may be necessary for traversing during launching. This generates additional transverse bending in piers and foundations.

Lateral wind loading on the MSS may be demanding on tall piers. Compared to an underslung MSS, the wind drag area is larger and higher on the ground. Vertical and lateral loads on the leading pier are more pronounced in continuous spans due to the longer tributary length of the MSS.

5.4. Single-girder overhead MSSs

A single-girder overhead MSS for simple spans takes support at the leading pier of the span to be cast and on the front pier diaphragm of the completed deck. For casting of continuous spans, the rear support of the MSS is moved forward on the front deck cantilever to shorten the casting span of the MSS (Povoas, 2012) and to better control the time-dependent stress redistribution of staged construction within the bridge (Figure 5.18).

Modular machines based on a single truss are not highly mechanised but are easy to reconfigure. Two braced I-girders are used in the most specialised MSSs in combination with a launch nose that controls overturning during launching (Figure 5.19). A rear C-frame rolls on the new span during launching, and these machines are therefore relatively short and better suited to use on curved bridges. The main girder is stiffer than a typical truss, and the support systems are also stiffer.

Hanger beams suspend the outer form from brackets overhanging from the main girder. After application of prestressing, the MSS is lowered to release the span. The outer form is divided

Figure 5.18. Single-truss overhead MSS (Photo: Strukturas)

Figure 5.19. Single-girder overhead MSS on the starter span

into two halves by releasing central splices in the bottom crossbeams, and the two form shells are rotated hydraulically to clear the leading pier during launching (Figure 5.20). After repositioning, the MSS is closed and realigned in a reversed sequence of operations.

The design of the central splice in the bottom crossbeams influences the deflections of the casting cell, the cycle time and the labour demand, because of the number of splices to open and reclose in every casting cycle. The central splice is rarely designed as a slip-critical connection due to the large number of bolts to be removed and reassembled. A transverse pin at the bottom flange and two longitudinal shear bolts at the top of a frontal splice ensure transfer of positive moment and shear, fast release, and insensitivity to vibrations (Figure 5.21). Large conical pins minimise geometry tolerances and force the splice into position, but may require hydraulic extractors for handling.

The single-girder overhead MSSs suspend the outer form during launching and are designed to reposition the pier supports. When a box girder is cast in two phases, the inner shutters of webs and pier diaphragms are also hung from the girder during launching. In addition to a cycle time substantially shorter than with a twin-girder overhead MSS, these machines do not require the use of ground cranes. These advantages often drive the choice of this type of MSS in long bridges over water or with tall piers.

Figure 5.20. Open casting cell during launching (Photo: Strukturas)

Figure 5.21. Transverse pins and top shear bolts in the central splice of the bottom crossbeams

Figure 5.22. Walkways and working platforms (Photo: Strukturas)

A drawback of the full self-launch capability is that these machines require numerous walkways and working platforms: full length on the main girder to reach the next pier through the launch nose, at the tip of the form brackets for adjustment of the camber and crossfall of the casting cell, at the end of the side wing shutters for worker protection during concrete finishing, on the bottom crossbeams of the casting cell for adjustment of the rib frames, and beneath the casting cell for opening and closing the outer form (CNC, 2007). A stressing platform is also necessary at the front bulkhead (Figure 5.22).

Winch trolleys run along the main girder to assist the operations of the casting cell and insertion of the prefabricated cage. Underslung winch trolleys span between the bottom flanges so as not to interfere with the form brackets, and the main girder may therefore have varying depth. Cage insertion is fast for simple spans, but more complex for continuous spans due to conflicts with the MSS's support legs at the leading pier (Figure 5.23).

These MSS have different levels of automation depending on the length of the bridge. The most sophisticated machines are custom designed for bridges of length sufficient to depreciate the investment. They are easily adaptable to shorter spans and are able to cope with variations in deck geometry, plan and vertical radii, and transverse crossfall. They are less affected by ground constraints, and typically are lighter and structurally more efficient than the other types of MSSs (Figure 5.24).

Figure 5.23. Support legs at the leading pier of continuous span

Figure 5.24. High structural efficiency of a single-truss overhead MSS (Photo: AP-BCS)

These machines are also simple to assemble, operate and decommission. Fast launching and form setting save labour and permit weekly casting cycles. These MSSs are therefore the first-choice solution for long viaducts on difficult terrains with on-deck delivery of materials (Rosignoli, 2011a). The casting cell is easily sheltered for cold-weather applications, where the only limitation is the availability of fresh concrete.

5.4.1   Loading, kinematics, typical features

The main girder is the backbone of these machines, and bracing has a fundamental role in ensuring satisfactory performance. The girder must be stiff in the vertical plane to limit the deflections of the casting cell, in the horizontal plane to resist wind loading with minor vibrations and lateral deflections, and about the longitudinal axis to transfer the torque generated by lateral wind and asymmetrical filling of the casting cell to the support systems.

Lateral bracing includes crosses or K-frames, connections designed for fast assembly and to minimise displacement-induced fatigue, and sufficient flexural stiffness to resist vibration stresses. Diaphragms connected to chords and flanges at the same locations as lateral bracing and form brackets distribute torsion and provide transverse rigidity. K-frames with flexural connections to vertical web stiffeners are used for the diaphragms of MSSs that carry underslung winch trolleys (Figure 5.25). Permanent connections and field splices are often designed to develop member strength.

Figure 5.25. K-frame diaphragms with flexural connections to vertical web stiffeners create the vertical clearance for operations of underslung winch-trolleys and support bus-bars for power supply

Figure 5.26. Form brackets with hydraulic adjustment of telescopic connections with the form hangers

Tied form brackets cantilever out from the main girder to suspend the hangers for the casting cell. The brackets are aligned with the diaphragms of the main girder to create full-width planar transverse trusses. Telescopic connections allow vertical adjustment of the form hangers for setting of camber and crossfall. Some MSSs have hydraulic systems for lateral shifting of the connections between brackets and hangers (Figure 5.26). Full-length working platforms are provided along the brackets for access to the geometry adjustment systems.

Form brackets and diagonal ties may be pinned to the main girder for fast assembly, inspection and dismantling, and to provide insensitivity to vibrations. The brackets support a covering for the casting cell. A working platform over the main girder closes the covering and facilitates access to the new pier through the launch nose (Figure 5.27). Lateral enclosures are rarely used around the casting cell due to the large wind drag area.

The MSS may carry a hydraulic concrete distribution arm to feed the casting cell from the top through multiple windows in the covering. Concrete is delivered on the deck and a concrete pump at the rear end of the MSS feeds the distribution arm through pipes supported on the top working platform to provide for easy inspection and cleaning.

Longitudinal edge girders connect the tips of the form brackets. Auxiliary lateral bracing stiffens the panels between the main girder, edge girder and form brackets to increase the first lateral frequency of the MSS for better control of wind vibrations (Pacheco *et al.*, 2011). Auxiliary lateral bracing is very effective, as the horizontal truss thus generated is wider than the deck (Figure 5.28). Vertical crosses brace the form hangers in the longitudinal plane. Along with the

Figure 5.27. Access walkway along the launch nose

Figure 5.28. Working platform for geometry adjustment over edge girder and top auxiliary bracing

Figure 5.29. High-flow-rate hydraulic systems open and close the casting cell in 1–2 hours (Photo: AP-BCS)

in-plane stiffness of the bottom frame, vertical crosses and the full-width horizontal truss provide torsional stiffness to the MSS.

The connections between form brackets and hangers transfer transverse shear and bending to resist wind loading during operations and to stabilise the open form shells during launching (Figure 5.29). The bottom crossbeams are hinged to the hangers to open and close the outer form and to set the crossfall of the casting cell. Opening and closing the outer form are on the critical path of the casting cycle, and some MSSs are equipped with synchronised double-acting cylinders and high-flow-rate hydraulic systems to perform these operations in 1–2 hours and to leave the winch trolleys available for other operations (Povoas, 2012).

The central splices in the bottom crossbeams are opened and reclosed at every repositioning, and these operations are on the critical path of the casting cycle. One transverse pin connecting over-lapped plates welded to the bottom flanges and two frontal shear bolts at the top flange are the typical solution for transfer of positive moment and mild shear. The pins are heavily loaded during concrete vibration and tend to seize with time, which may suggest the use of hydraulic extractors. The pins are tied to the crossbeams and suspended during launching.

The bottom crossbeams may be designed for the full load of the casting cell or may be suspended from form brackets and diaphragms with hanger bars crossing the casting cell. This solution reduces the weight. the cost of the crossbeams and the deflections of the casting cell, but increases the labour costs for forming and sealing holes in the top slab.

The bottom frame of the casting cell is anchored to the front deck cantilever and the leading pier cap to prevent displacements during filling. The pier cap anchorages are particularly stiff in order to stabilise the front overhang of the casting cell for continuous spans. The pier cap anchorages are removed prior to the application of prestressing to release the sliding bearings and to not counteract span shortening.

The outer form comprises a bottom form table fixed to the crossbeams, and side shutters fixed to adjustable rib frames connected to the crossbeams with sliding clamps, adjustment screws and diagonal turnbuckles. Bottom bracing connects the two half-crossbeams of every form module and does not extend to the adjacent modules, to allow individual opening and closing. Steel shutters are spindled at the joints between form modules, while wooden forms are spliced using proprietary hardware. The bottom form table is not spliced at the joints. The inner form is similar to the systems used in the other types of MSSs.

Camber and crossfall are set with telescopic connections between form brackets and form hangers, or by shimming the bottom form table. The side forms are adjusted by shifting and tilting the rib frames relative to the crossbeams. Form setting at the bottom frame level allows keeping the form hangers aligned with the main girder in curved bridges. Electric or compressed-air concrete vibrators are applied to the inner and outer forms.

The tendons are fabricated and tensioned from a stressing platform at the front bulkhead. The inner and outer side forms are stripped prior to application of prestressing to avoid prestress losses in the forming systems. Light winch trolleys or monorail hoist blocks assist bulkhead stripping and handling of the stressing jacks. The stressing platform is opened into two halves prior to launching, and provides access to the bottom walkway for pin extraction. Tendons are grouted after launching the MSS to the next span.

Full-span cage prefabrication is not common due to the cost of the cage carrier and multiple winch trolleys for cage handling. However, cage prefabrication shortens the cycle time, creates a special-ised working area for parallel tasks, and improves risk mitigation. Prefabricated cages for simple spans often include the prestressing tendons. The cages for continuous spans include the front bulkhead and prestressing hardware and ducts, but the strand is inserted into the ducts during curing of the span to simplify tendon splicing, to lighten the cage during handling, and for better risk mitigation and labour rotation.

The cage is assembled behind the abutment on a full-length motorised carrier. A tower crane assists storage and handling of rebars, and the prefabrication yard can be sheltered. The carrier is equipped with a full-length lifting frame (Figure 5.30). Side jigs are used to align the bars to the deck geometry. The carrier and lifting frame comprise hinged modules to cope with plan and vertical curves during transportation along the deck. The lifting frame is equipped with non-motorised hydraulic legs that roll along the carrier. After completion of the cage, the cage is tied to the lifting frame and the carrier is driven to the rear end of the MSS.

Three or four underslung winch trolleys are stored on a rear cantilever of the main girder over-hanging the end span of the deck (Figure 5.31). The front end of the carrier is moved under

Figure 5.30. Cage carrier with full-length lifting frame

Figure 5.31. Storage area for the winch trolleys used for cage insertion

the front winch trolley, and the legs of the lifting frame are extended to lift the cage from the carrier platform. The front winch trolley picks up the first module of the lifting frame and moves forward, thus pulling lifting frame and suspended cage along the carrier. The second winch trolley picks up the second module of the lifting frame, and so on until the cage is fully suspended. The lifting frame and cage are moved out and lowered into the casting cell. The legs of the lifting frame are retracted to release the cage onto the outer form, and the frame is detached from the cage and moved back to the carrier in the reverse sequence of operations.

The front legs of the MSS take support on the leading pier. When the deck is integral with the pier and has expansion joints along the span, reinforced-concrete columns are used to support the MSS at the top slab level. When continuous spans are supported on one line of bearings, the MSS takes support on the pier cap and the four tower legs are protected from the concrete and extracted during repositioning of the MSS. Braced reinforced-concrete columns supported on the bridge bearings and permanently embedded into the span have also been used to support MSSs.

The front support legs of the MSSs for continuous spans prevent full cage prefabrication for the front overhang. Web cages and prestressing ducts are prefabricated full length, and the slabs are completed within the casting cell. Full-span cage prefabrication is easier in the simple spans, as the front legs of the MSS are beyond the front bulkhead and do not interfere with cage handling. In both cases the rear C-frame of the MSS is designed to allow full-width cage insertion. When the C-frame is too wide to resist the full casting load, pendular legs are lowered after cage insertion in order to support the MSS on the front deck diaphragm (see Figure 5.31).

5.4.2 Support, launch and lock systems

Some single-truss overhead MSSs for 45–55 m spans are shortened configurations of modular MSSs for 70 m spans, and their support, launch and lock systems are discussed in Section 5.5.2. Alignment wedges are often used at the tip of the launch nose to force deflection recovery during launching.

Custom single-girder overhead MSSs designed for full-span cage prefabrication are equipped with a rear C-frame that rolls over the new span during launching, a main support frame at the leading pier, and a front auxiliary leg. The main support frame may be a tower–crossbeam assembly with articulated launch saddles, or a narrow C-frame integral to the main girder.

When the bridge has simple spans, the main frame takes support on the front half of the leading pier cap, the rear C-frame takes support on the front pier diaphragm of the previous span, and the new span is cast without conflicts with the support systems of the MSS.

Continuous spans may be framed to the piers or supported on bearings. Seat expansion joints are used at the quarters/fifths of the span in the first case, and match-cast construction joints with through reinforcement are used at the same locations in the second case (Rosignoli, 2002). In either case the main support frame of the MSS takes support on the leading pier cap, and the rear C-frame is supported on the deck cantilever to shorten the casting span of the MSS and to improve the time-dependent stress redistribution of staged construction within the deck.

The tower–crossbeam assembly is anchored to the pier cap and the support legs cross the new span at the pier diaphragm (Figure 5.32). Repositioning the frame requires extracting the support legs

Figure 5.32. Front tower–crossbeam assembly (Photo: VSL)

from the completed span. The tower is spliced at the top slab level and the winch trolley removes the top portion of the assembly, stores it on the deck, and extracts the legs for repositioning on the next pier cap.

When the support systems are integral to the main girder, the MSS is repositioned with friction launchers supported on adjustable towers. These MSSs are much lighter than the twin-girder MSSs, and are very stable. They are fast to launch and reconfigure, and weekly casting cycles with one shift per day can be achieved. They do not require ground cranes, the casting cell is sheltered, and production is unaffected by meteorological conditions.

The friction launchers are heavier than the systems used for single-girder overhead gantries for erection of precast segments because the main girder is heavier in order to control deflections, and the MSS also carries the casting cell and technological plant during launching. In several applications the launchers have been suspended from the main girder and stored under the launch nose during span casting.

After application of prestressing, the rear launcher is moved over the leading pier and aligned with the anchor bars. The span is released by retracting the support cylinders of front legs and rear C-frame, and the launcher lands on the deck. The anchor bars are stressed, the outer forms are opened, and the cylinders of the front legs are fully retracted in preparation for launching.

In the first phase of launching the MSS is supported at the rear launcher and the rear C-frame, which rolls over the new span under the pull imparted by the launcher. When the launch nose reaches the new pier, the front launcher is positioned on the new pier cap for launch completion. The front auxiliary leg controls negative bending and overturning during this operation, and lowers the front launcher on the pier cap by retracting its vertical support cylinders.

Control of overturning is a major issue as these machines are short – typically 1.7–1.8 times the design span. These MSSs are mostly used for continuous spans and when the construction joint is at the quarter of the span, only three-quarters of the casting cell are behind the leading pier at the start of launching. After 25% of launching, the centre of gravity of the casting cell reaches the leading pier and the weight of the casting cell becomes destabilising. The PPU and fuel tanks are applied to the rear end of the MSS to enhance stability, and for the same reason the winch trolleys are often stored at the rear end of the MSS prior to launching.

In the second phase of launching the two launchers are synchronised electronically (Rosignoli, 2011a). Position sensors and pressure transducers provide feedback to a PLC-controlled launch computer that drives the launchers by means of actuators. The load that the rear C-frame applies to the deck is controlled hydraulically. When the front support legs reach the new pier, the support cylinders are extracted to lift the MSS to the span-casting elevation and the launchers from the deck. The launchers are moved to the storage area under the launch nose, and the outer form is reclosed.

When integral to the main girder, the front support of the MSS is a narrow C-frame comprising two side legs overhanging from the main girder and a central diaphragm connecting the two legs. The bottom portion of the legs is unbraced to allow unrestricted movement of the main girder over the friction launchers and of the winch trolleys along the MSS.

The legs of the C-frame have box section to provide longitudinal and transverse stiffness. High-tonnage double-acting cylinders housed within the legs provide vertical geometry adjustment, lift the MSS to the span-casting elevation, and release the span after application of prestressing in combination with the support cylinders of the rear C-frame. The locknuts of the support cylinders are released only for the vertical movements, being kept tightened the rest of the time.

The legs of the C-frame may be designed as telescopic assemblies locked by large-diameter pins to shorten the design stroke of the main cylinders. Auxiliary long-stroke double-acting cylinders move the telescopic leg to the next lock position for the pin. The pins are inserted and removed with hydraulic extractors.

A complex C-frame supports the rear end of these machines. Long-stroke double-acting cylinders shift the frame transversely when launching along curves for the launch bogies to follow curved rails anchored to the deck over the webs. The transverse cylinders are also used to rotate the MSS about the rear launcher in the launch plane to align the tip of the launch nose with the new pier for landing. Longitudinal cylinders rotate the C-frame about the vertical axis to the local radius. Vertical cylinders provide torsional stability during launching, control the support reaction applied to the deck in the second phase of launching, and adjust the frame geometry to deck crossfall.

The rear C-frame is designed for insertion of a full-width prefabricated cage into the casting cell. Because of its geometry, the frame is subjected to high flexural and shear stresses; the load conditions are very complex and the frame is often overdesigned. Auxiliary pendular legs hinged to the crossbeam of the C-frame are lifted during cage insertion and lowered prior to span casting to take support on the front deck diaphragm for direct load transfer; in this case the C-frame is designed for the launch stresses and the weight of the prefabricated cage.

The front auxiliary leg is adjustable laterally and may be equipped with a support crossbeam to amplify traversing capability. The front leg has broad vertical adjustment to recover the deflection of the launch nose and to cope with vertical curves in the bridge.

5.4.3 Performance and productivity

Some single-girder overhead MSSs are among the most complex machines used for bridge construction. They are compact, light, fast to operate and typically custom designed for casting several tens of spans of equal length. They have full self-launching capability to avoid the need for ground cranes and to leave the area under the bridge unaffected. Launching is fast and safe, and synchronised friction launchers minimise the loads applied to the piers. The outer form is suspended from the MSS also during reverse launching to reposition the MSS for casting a second deck.

Camber and crossfall are easy to set, and long-stroke cylinders in the support legs provide general control of elevation. The length of the casting cell may be adapted to shorter spans by leaving outer form modules open or by hugging the front deck cantilever with the casting cell.

The casting cell is sheltered and steam curing may be easily applied in cold weather. The typical cycle time for one-phase casting of rectilinear box girders and ribbed slabs with cage prefabrication is one span per week with one 10–11 hour shift per day, despite the complexity of the inner tunnel form; the cycle time may lengthen to 8–9 days in curved spans. Two-phase casting of box girders may require 10 days when the first-phase U-span is prestressed prior to casting the deck slab.

The critical path tasks of the casting cycle for a ribbed slab or a solid or voided slab include repositioning of the MSS, cage insertion, duct splicing, concrete pouring, cleaning of anchorages, tendon fabrication and application of prestressing. Short-span box girders are often cast in one phase, which adds extraction and repositioning of the inner tunnel form to the activities on the critical path.

One-phase casting puts a high demand on batching plant and concrete delivery lines, especially when concrete is supplied on the deck with a great number of mixers. Concrete is kept fluid with retarder admixtures to avoid cracking due to the progressive deflection of the casting cell, and this requires design of the shutters for full hydrostatic pressure (FIB, 2009; SAA, 1995). One-phase casting typically provides the best possible finish of outer concrete surfaces.

When a box girder is cast in two phases, the cage for the top slab is assembled within the casting cell. The form table for the central strip of the top slab is easier to extract and reposition than is an inner tunnel form. When the first-phase U-span is not prestressed, the casting cycle is weekly and requires a similar number of workers (Povoas, 2012).

The pier diaphragm of the next span may be cast in place prior to repositioning the MSS to extract a complex activity from the critical path. When a single-truss overhead MSS for 45–55 m spans is a shortened configuration of a modular MSS for 70 m spans, lengthening the launch nose implies minor costs. These MSSs use three tower–crossbeam assemblies. The third assembly may be positioned on the next pier cap during span casting in order to avoid repositioning of supports during launching. If the launch nose is as long as the typical span, the form for the pier diaphragm is attached to the third support tower. After inserting the cage into the form, the winch trolley moves the third tower–crossbeam assembly to the new pier. Concrete for the pier diaphragm is fed through the launch nose during span casting (Povoas, 2012).

The span diaphragms of the ribbed slabs with double-T-section may be precast with emerging reinforcement to simplify the forms of the ribs and accelerate opening and closing of the casting cell.

5.4.4   Structure-equipment interaction

The typical span range of a single-girder overhead MSS is 45–55 m. Box girders for highway bridges are often cast in one phase to take advantage of the high level of mechanisation associated with most of these machines. Solid, voided and ribbed slabs are also cast in one phase. Simple and continuous spans are both possible; the MSSs for continuous spans are lighter because the casting span is shorter and the front overhang of the casting cell further relieves positive bending in the main girder.

The main concerns with two-phase casting are related to the deflections of the casting cell. The single-girder overhead MSSs are more flexible than the twin-girder machines and the structure–MSS interaction is therefore more pronounced. When a 2-week cycle with cage assembly in the casting cell is used, the first-phase U-span may be mildly prestressed for better control of cracking during casting of the top slab. In most cases, however, such sophisticated MSSs are used with cage prefabrication and weekly cycles, and the curing time of the U-span is insufficient for two-phase prestressing.

The current design trend for highway bridges is toward continuous spans (AASHTO, 2012). A reduced number of expansion joints improves durability and reduces annoyance to drivers; continuous spans, however, complicate staged prestressing. Even when the tendons are stressed in two phases for earlier span release, prestressing is completed as soon as possible. Large time-dependent deflections can arise when prestressing is applied so early in the process, and the structure–MSS interaction must therefore be analysed with great care. This is particularly true with solid and voided slabs, due to the poor moment of inertia of the cross-section.

Continuous spans are cast with the rear support of the MSS placed on the front deck cantilever. The support reaction generates negative bending at the leading pier, which reduces the time-dependent stress redistribution of staged construction within the deck. At span release, however, the load applied to the cantilever fades, the stiffness of the new span affects load redistribution, and the new bending moment diagram modifies stresses and deck camber.

Single-girder overhead MSSs are used in long bridges, and optimising the structure–MSS inter-action can save prestressing throughout the bridge. Starting from a few-span model of the continuous bridge inclusive of prestressing tendons, the rear support point of the MSS is moved back and forth along the front cantilever, and the corresponding support reaction is calculated

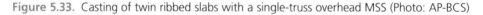

Figure 5.33. Casting of twin ribbed slabs with a single-truss overhead MSS (Photo: AP-BCS)

based on the casting span of the MSS. Time-dependent analysis for different locations of the construction joint is used to minimise the influence of staged construction on the prestress demand.

These MSSs are relatively light and their weight rarely governs the design of the bridge. Applications with large transverse eccentricity, however, require special design of piers and foundations (Figure 5.33).

The time-dependent deflections are rarely a major issue in well-designed simple spans for HSR bridges, as the bending moments from self-weight, superimposed dead load and prestressing are easy to balance. Integrative tendons can also be tensioned after several weeks of curing, in easy and protected operations within spacious box cells.

5.5. Modular single-truss overhead MSSs for long spans

Span-by-span casting with an MSS offers several advantages over balanced cantilever casting with form travellers. Materials are delivered on the deck, access to the working area is simpler and production is more continuous. Prestressing is more effective because of less influence of staged construction, and less expensive because of longer tendons and less anchorages. The number of construction joints is much smaller, and cage prefabrication saves time and removes activities from the critical path.

If proceeding linearly from abutment to abutment is not a problem, span-by-span casting can also be mechanised more easily (Homsi, 2012). Lifting form travellers onto the pier tables requires powerful ground cranes, and assembling the travellers over the pier tables lengthens the cycle time. Different construction equipment is also necessary for pier tables and cantilever segments.

Compared with span-by-span casting, however, the weakest point of balanced cantilever casting is the duration of construction. An MSS can cast a 70–80 m span every week. An 80 m balanced cantilever span would require a 10 m pier table and seven 5 m cantilever segments on either side, plus midspan closure. Casting the pier table may take 2 months, but typically this task is performed with a specific casting cell and therefore is not on the critical path. Including set-up of travellers and midspan closure, balanced cantilever casting of an 80 m span may take 9–10 weeks.

Balanced cantilever casting is typically associated with inaccessible terrains and tall piers. In such circumstances propping a 40 m underslung MSS from ground foundations to cast two half-span segments is often impossible or too expensive. Span-by-span macro-segmental construction also requires a temporary pier in every span, even if the load on the pier is smaller. These spans are too long for full-span incremental launching (Rosignoli, 2002), and the use of temporary piers to halve the launch span involves the same limitations; 70–80 m spans also have varying depth, which is a counterindication to incremental launching.

In such cases a single-truss overhead MSS may be stretched to the limits to extend the application range of span-by-span casting. Bridges over water or with tall piers are situations where span-by-span casting of 70–80 m spans may be faster and less expensive than balanced cantilever construction. Long spans reduce the number of foundations and piers required, and a fully self-launching MSS avoids the need for ground cranes.

Single-truss overhead MSSs may potentially reach 80–90 m spans for highway bridges and 70 m spans for railway bridges (Pacheco et al., 2011). Varying-depth continuous prestressed concrete box girders supported on bearings or integral to the piers are the typical solutions for these span lengths. In spite of simple forming, ribbed slabs with double-T sections are rarely used on spans longer than 50 m because of the poor torsional constant of the cross-section (Povoas, 2012).

The construction joints are at the rear quarter of the span to shorten the casting span of the MSS, to reduce the rear support reaction, and to provide some flexibility in positioning the rear support in order to minimise out-of-node load eccentricity in the truss. However, locating the construction joint at the span quarter may not be the optimal solution in terms of structure–MSS interaction, and the number of spans to cast governs the break-even point between the higher costs of prestressing and a less expensive MSS.

The bridge should be almost rectilinear, because plan curvature requires complex traversing operations on long spans. Torsion in the truss due to the lateral offset of a curved casting cell is another problem on long spans, and the cycle time is longer due to form adjustment. The top slab should be narrow to limit the design load of the MSS; wide box girders may be built by casting the central box core span-by-span and by widening the side wings segmentally with forming trolleys.

A single-truss overhead MSS is the most versatile starting point for long-span applications. These machines are light and structurally very efficient, are easy to launch, and clearance issues are rarely encountered when operating above the deck. The use of trusses instead of braced I-girders lightens the MSS and improves the response to lateral wind on tall piers. A tall and wide central truss supported directly on the deck is also stiffer than two smaller lateral trusses supported on long and flexible crossbeams.

The truss suspends one-quarter of the outer form beyond the leading pier and, at 25% of launching, the weight of the casting cell becomes destabilising. Applying the PPU and counterweights to the rear end of the MSS increases the load applied to the deck and does not relieve negative bending in the truss; the typical solution is, therefore, to lengthen the launch nose for earlier landing on the new pier.

From a structural standpoint, the optimal length of the launch nose is 85–90% of the typical span. Lengthening the nose to full span involves minor additional costs, but provides access to the new pier one week earlier for placement of bearings, casting the pier diaphragm and other preparatory activities (Povoas, 2012). An auxiliary leg may be used at the new pier to control wind-induced vibrations in the long front cantilever.

Negative bending from the long launch nose and the front overhang of the casting cell suggests the use of varying-depth trusses, despite the difficultly of reusing them on shorter spans. In some applications the front support zone of the MSS has been equipped with a tower to deviate inclined tension chords. Similar schemes have been used on several gantries for balanced cantilever erection of precast segments in the past.

Most long-span MSSs are based on braced constant-depth trusses. The modular nature of assembly facilitates reuse of the truss in new projects, but the great number of field splices increases cost and the duration of site assembly. The deflections of the casting cell are often controlled by increasing the height of the truss. This provides ample room for a mid-depth working platform that facilitates access to the new pier through the launch nose and to the side form brackets (Figure 5.34).

The axial force in the chords of constant-depth trusses increases toward the middle of the casting span, while parabolic trusses may be designed for uniform axial force in both chords (Figure 5.35). Uniform tension at the bottom chord suggests the application of longitudinal prestressing tendons that actively control the deflections of the casting cell during filling and application of span prestressing (Pacheco et al., 2011).

An MSS for long spans must be as light as possible in order not to overload tall bridge piers. If the piers are short and the area is free from obstructions, supporting a short MSS on temporary piers to cast half-span segments may be less expensive, and the spans may amply exceed 100 m.

Cross-diaphragms stiffen the truss and resist the loads applied by side form brackets. Form hangers made with modular welded trusses increase the lateral stiffness of the casting cell and save weight. The bottom crossbeams are also replaced with welded trusses or three-dimensional space frames to further stiffen and lighten the casting cell. The number of field splices is still reasonable, but massive recourse to hand welding increases the cost per unit weight of these highly efficient modular assemblies of light members (Figure 5.36).

After application of prestressing, the MSS is lowered to release the span, and the outer form is opened for launching. When the box girder is cast in two phases, the inner web shutters are hung from the form brackets during launching. The inner tunnel form for one-phase casting is extracted from the previous span after cage fabrication for the bottom slab and webs.

Figure 5.34. Mid-depth working platform along the main truss (Photo: AP-BCS)

These MSSs have a low level of mechanisation because of the small number of spans to cast, and the cage is often assembled within the casting cell, with some degree of prefabrication (Figure 5.37). In some applications, the cage for the webs and pier diaphragm has been fabricated on the new span during curing, suspended from the MSS during launching, and lowered into the

Figure 5.35. 140 m, 780 ton overhead MSS for 70 m, 1400 ton spans (Photo: BERD)

Figure 5.36. Modular MSS for 70 m spans (Photo: BERD)

casting cell after closure. Monorail hoist blocks are often sufficient for these operations. The most advanced single-truss overhead MSSs for 70 m spans offer several interesting features (Povoas, 2012).

1 Assistance in preparatory activities for the next pier cap (placement of bearings, forming of soffits, casting of pier diaphragms) through the launch nose, thus removing tasks from the critical path and the need for ground cranes.
2 Fabrication of web cages over the completed span during curing and cage suspension from the MSS during launching.
3 Easy adaptation to different span lengths (a modular truss, however, is typically heavier than a custom design).
4 Reversibility for casting parallel viaducts in the opposite directions (without reverse launching).
5 Minimal storage and easy shipping when members are designed to fit within 40 ft containers.

Modular design facilitates reconfiguration for shorter spans, although a higher level of mechanisation is often preferred for span-by-span casting of 40–50 m spans. These machines are unaffected by ground constraints, but the height of the truss may be a limitation when passing

Figure 5.37. Prefabricated panels of bottom slab reinforcement (Photo: BERD)

under existing bridges or power lines, and the length of the truss limits their use to rectilinear bridges.

5.5.1   Loading, kinematics, typical features

The main truss of these MSSs is highly stressed during operations and launching. Vertical stiffness is necessary to control the deflections of the casting cell on long spans. Lateral stiffness is necessary to resist wind loading on tall piers, and to avoid major vibrations, aeroelastic phenomena and resonance. Torsional stiffness is necessary to stabilise the casting cell under lateral wind and during filling. Stability from out-of-plane buckling is another critical issue, especially when the truss is equipped with prestressing tendons (Pacheco *et al.*, 2007).

Bracing has a fundamental role in ensuring satisfactory performance. Bracing connections are designed for fast assembly and to minimise displacement-induced fatigue. Sufficient flexural stiffness is necessary to resist vibration stresses. Cross- or K-diaphragms connected to the chords at the same locations as lateral bracing and form brackets distribute torsion and provide transverse rigidity (Figure 5.38).

Figure 5.38. Lateral braces and K-diaphragms in a prestressed truss (Photo: BERD)

Tied brackets are applied on either side of the truss to suspend the form hangers. The brackets may be connected to the verticals of the truss instead of to the bottom chords to create additional vertical clearance for handling of prefabricated web cages. The central diaphragm within the truss, form brackets, form hangers and bottom crossbeam of the casting cell create closed planar frames that resist transverse bending during operations and launching.

Form hangers and the bottom frame may be designed for the full load or may be integrated with hanger bars crossing the casting cell. The bottom frame may comprise I-crossbeams, transverse trusses or three-dimensional space frames; typically, the lighter the bottom frame, the larger the number of field splices there are to assemble and inspect. The bottom frame is divided into two halves with pinned connections, and the form adjustment devices are similar to those used in single-girder MSSs for shorter spans.

Full-span cage prefabrication requires long bridges to depreciate the investment in the cage carrier, rebar jig and multiple winch trolleys for cage handling. Long bridges rarely have 70–80 m spans, and the cage is therefore partially prefabricated over the new span and completed within the casting cell. This solution does not require extra investment, and has minor impact on the cycle time.

The most proven approaches for in-place casting of 70 m spans with a non-propped MSS are one-phase casting and two-phase casting with mild prestressing of the first-phase U-span. Both approaches have drawbacks: one-phase casting of such long spans is very demanding for batching plant and concrete delivery lines, and requires expensive inner tunnel forms, while the curing time of the first-phase U-span prior to application of prestressing lengthens the cycle time.

In some MSSs for two-phase casting, prestressing has been applied to the truss of the MSS instead of to the first-phase U-span (Pacheco et al., 2011). Prestressing may be considered for the design of a new MSS and for reconditioning existing machines. Prestressing is not aimed at overcoming cross-sectional deficiencies – the addition of cover plates would be more effective because unbonded tendons do not increase the moment of inertia. The purpose of prestressing is to control the deflections of the MSS with active systems of forces that counteract the loads applied by the casting cell. The high payload-to-weight (P/W) ratio of most of these machines maximises the efficiency of active prestressing, the tendons are immediately reusable for all the spans of the bridge, and they are not affected by friction and time-dependent losses.

The truss of a prestressed MSS is designed for the launch stresses and the loads applied by casting cell and prestressing tendons. The strands are anchored to the truss at the ends of the casting span and are supported on the lateral braces of the bottom chords; rubber plates at the supports damp transmission of vibrations (Figure 5.38). This strand layout is particularly favourable in parabolic trusses because axial tension is constant along the bottom chord.

Because of the high payload-to-weight ratio, the pull in the strands is increased during filling of the casting cell, kept constant during curing, and released progressively during application of span prestressing. The pull may be adjusted in two or three steps, or may vary almost continuously if the stressing jacks are controlled by PLC computers that receive position feedback from displacement sensors applied to the casting cell (Pacheco et al., 2007).

Prestressing provides highly effective control of the deflections of the casting cell: Figure 5.39 shows the midspan deflection and the stroke of the stressing jack (Pacheco et al., 2011) during second-phase casting of a 70 m span using the parabolic MSS shown in Figure 5.35. A 26 mm peak deflection corresponds to $L/2692$, and is 6.7 times smaller than the typical limit of $L/400$.

However, sensor-controlled prestressing has also drawbacks. The automation and active control systems include displacement sensors, data links and communication systems, a PLC, electric boards, hydraulic systems, actuators and special stressing jacks with motorised locknuts. The response of the actuators must be damped via control software to avoid tendon vibrations and to control numerical instability due to the response of the PLC to concrete pouring and vibration. Locked tendons are also exposed to thermal effects during span curing.

System failure would leave the deflections uncontrolled, and unusual risk scenarios (bugs in the PLC software, failure of components, electric blackout, human error, etc.) should be well understood and mitigated. Some weight saving may be expected in the casting span of the truss, but the weight of launch nose, casting cell, working platforms, support and launch systems, and technological systems is unaffected. Prestressing adds costs to the MSS and complexity to operations.

Auxiliary prestressing has been amply tested in the self-launching gantries for balanced cantilever erection of precast segments, and it has progressively been abandoned because of the complexity

143

Figure 5.39. Midspan deflection and jack stroke for PLC-controlled 7.0 MN prestressing

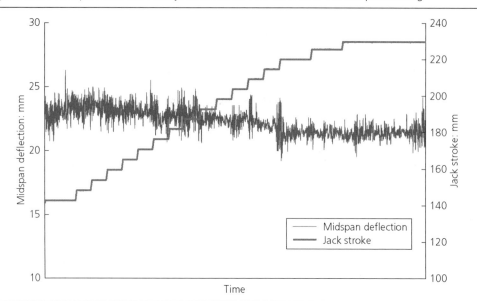

and labour demand of these machines. Real-time control of the deflections of a 70 m casting cell, short reaction times to unforeseen events, and combining sophisticated automation and control systems with the skill and attention typically exhibited by construction workers, do involve some challenges.

On the other hand, however, the virtual absence of deflections in the casting cell may open new perspectives for two- and multiple-phase casting of 80–90 m spans. One-phase casting of 70 m spans and two-phase casting of 70–75 m spans, combined with mild staged prestressing of the U-span, are the current limits of span-by-span technology, and also have some drawbacks, while prestressed MSSs are almost ready for the next step forward.

5.5.2 Support, launch and lock systems

The support and launch systems of long-span MSSs are rarely sophisticated because of the small number of spans to be cast. These machines typically have three supports: two tower–crossbeam assemblies for the casting span, and an auxiliary front support placed on the next pier prior to or during launching.

These MSSs reposition the supports with a winch trolley or a light service crane at the tip of the launch nose. No need for ground cranes is a basic requirement for machines to be used on difficult terrains and tall piers. The main supports comprise a braced crossbeam that sustains the truss, and an adjustable support tower. The crossbeams are relatively short, due to no need to traverse in rectilinear bridges.

The crossbeams support the truss with long articulated bogies that spread the load over long sections of the bottom chords to control local bending. Assemblies of cast-iron rolls on equalising beams cope with flexural rotations of the truss and the launch gradient. Rectangular rails welded to the chords facilitate load dispersal and guide the flanges of the rolls to transfer lateral loads.

Minor geometry adjustment may be provided in the transverse direction with guided PTFE skids sliding over the crossbeams.

The rear support restrains the truss longitudinally to the front deck cantilever during span casting. The front support allows displacement to avoid locked-in stresses due to thermal expansion of the truss and shortening of the prestressed MSSs under the axial load applied by the tendons.

Some MSS launched with capstans or redundant hydraulic winches applied to the rear end of the truss to pull reeved haulage ropes anchored to the deck. Redundant long-stroke double-acting cylinders acting on perforated rails, or hollow cylinders acting on large-diameter prestressing bars, are slower but safer and more robust. Moving couplers along the bars to transfer the load to the hollow cylinders is labour intensive, while insertion of the contrast pins into the perforated rail may be automated easily.

The support towers are adjustable in order to set the crossbeams horizontal. High-tonnage double-acting cylinders under the bottom nodes of the truss lift the MSS to the span-casting elevation and lower it back onto the launch bogies for span release. Locknuts are tightened after lifting the MSS for mechanical load transfer. The towers are anchored to deck and pier caps with stressed bars that resist uplift forces. Lateral wind on such tall trusses can generate significant uplift forces, and bidirectional restraints are often necessary at the support cylinders to control uplift and overturning.

5.5.3   Performance and productivity

The MSSs for 70 m spans are not very automated, as the length of the spans reduces the number of casting cycles. Reinforcement, prestressing and concrete are typically delivered on the deck. With web cage prefabrication on the new span during curing, the casting cycle is typically weekly for both one- and two-phase casting.

In-place precasting of the pier tables simplifies the casting cycle and front support frame of the MSS, and removes activities from the critical path, but increases the labour costs and requires a special casting cell. The cycle time for wide decks can be maintained by casting the box girder core with the MSS, and by constructing the side wings with precast panels or in-place casting with forming carriages (Figure 5.40). Span-by-span casting with a 70 m MSS is compared to the other construction methods for similar span lengths in Table 4.1.

Assembly and decommissioning of a modular MSS are complex operations that take time. The truss is almost three spans long and is composed of a great number of modules, which adds design flexibility and facilitates reuse but complicates assembly. Long trusses must be moved frequently during assembly to balance the load in the support frames. Form brackets, form hangers and the bottom frame of the casting cell are also composed of a great number of braced modules, and many components need to be applied in the air.

Load testing at the end of site assembly is complicated by the entity of the test load. Further complication results from fine-tuning of PLC-controlled prestressing systems. Reverse launching also takes time because of the low level of automation of these machines.

5.5.4   Structure–equipment interaction

The single-truss overhead MSSs for 70 m spans are designed for one- or two-phase casting of rectilinear box girders. The flexibility of the casting cell during filling is not a major issue with

Figure 5.40. Pier tables cast at the end of pier erection and a widened central box core (Photo: BERD)

one-phase casting, provided that retarding admixtures are used to keep the concrete fluid during most of filling. With two-phase casting, the effects of MSS deflections are mitigated by mild prestressing of the first-phase U-span prior to casting the deck slab, or by using a prestressed MSS.

The MSSs for one-phase casting exert a substantial residual support action on the span after application of post-tensioning, and only a part of the tendons can be stressed prior to span release. Prestressed MSSs permit full application of post-tensioning prior to span release, because the tendons of the MSS are progressively released to reduce the support action. Because of the deformability of the truss, small variations in the prestressing force may generate large deflections, and the interaction between span post-tensioning and truss prestressing must be accurately calibrated to avoid concrete cracking or damaging the truss.

When the truss is not prestressed, span post-tensioning for two-phase casting is typically divided into three phases. A first set of tendons anchored within the U-span is stressed prior to casting the deck slab, to stiffen the U-span and to improve crack control. The remaining tendons are stressed after span completion: the second-phase tendons are stressed prior to span release to make the span self-supporting, and the third-phase tendons are stressed from the rear anchorages within the box cell after repositioning of the MSS. Three-phase post-tensioning improves the structure–MSS interaction, shortens the cycle time and removes activities from the critical path, but increase labour costs.

Prior to the application of post-tensioning, the bottom frame of the casting cell is detached from the leading pier cap to release the sliding bearings, and the outer and inner forms are stripped to support the span only at the bottom form table. This operation minimises the friction losses of prestress into the forming systems. The span is finally released by lowering the MSS.

The rear support of the MSS is placed on the front deck cantilever to shorten the casting span and to reduce the time-dependent stress redistribution of staged construction within the deck. Releasing the span and repositioning the MSS modifies the stress distribution in the deck and the camber in both the new span and the preceding one.

Figure 5.41. Long-span MSS on tall piers (Photo: AP-BCS)

The vertical loads applied by an MSS for 45–55 m spans rarely influence the deck design. MSSs for longer spans are more demanding, due to their weight, the quantity of concrete in the casting cell and the long front deck cantilever. A shorter deck cantilever increases the weight, cost and rear support reaction of the MSS, and the break-even point between the higher cost of the MSS and the higher prestressing costs throughout the bridge depends on the number of spans and the local cost of post-tensioning. Easier control of deflections is another advantage of a longer front cantilever.

Long-span MSSs are often used on tall piers, and the large wind drag area may result in significant wind loads, (Figure 5.41). The increasing tributary mass during filling of the casting cell may trigger dynamic responses to mild forcing wind in the leading pier. The mass increase is not accompanied by an immediate increase in stiffness, and the natural frequencies of the pier reduce substantially. This makes tall piers potentially prone to vortex-induced oscillations and transverse galloping (Pacheco *et al.*, 2011).

Stability to wind-induced vibrations is an unexplored field for these machines. Single-truss MSSs are stiffer than twin-truss MSSs and have higher lateral frequency. In the absence of scientific research, a lower-bound frequency of 0.20–0.25 Hz has been proposed for the first lateral mode (Pacheco *et al.*, 2011), based on satisfactory performance of existing machines (Figure 5.42). This should allow spans of about 90 m to be reached with a single-truss MSS.

In addition to vertical and transverse loads, most types of full-span bridge construction equipment apply longitudinal loads to the bridge piers. These loads may be pronounced in a long-span MSS due to the weight of the span and the MSS itself (Povoas, 2012). Non-seismic longitudinal loads are due to multiple causes.

Figure 5.42. First lateral frequency of overhead and underslung MSSs

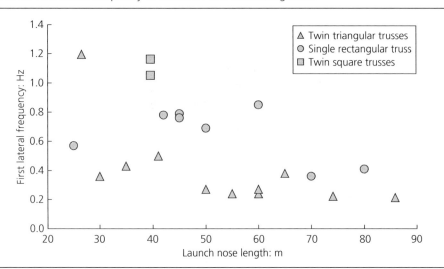

1   *Longitudinal projection of the vertical load due to inclined launch saddles.* The trusses follow the bridge gradient, and inclined launch saddles apply longitudinal loads to the pier caps. When the MSS is lifted from the launch bogies for span casting, the longitudinal load projection is related only to the weight of the MSS. When the MSS is supported on PTFE skids and the vertical cylinders are inserted between support towers and crossbeams, the weight of the span also has a longitudinal load projection. As the weight of the span is about twice the weight of the MSS, the total longitudinal load triples.

2   *Frictional resistance of supports.* Launch bogies have a friction coefficient of 2–5%, while that for PTFE skids may reach 6–10%, which further increases the longitudinal demand associated with this type of launch saddle. The different combinations of bridge gradient, frictional resistance and launch direction are discussed in Section 10.8.

3   *Reaction to launch forces applied to the MSS.* The launch forces are sometimes applied to piers devoid of vertical load – for example, when a winch pulls a haulage rope anchored to the landing pier. The launch force is the sum of longitudinal self-weight projection and the friction resistance, and its effects may be particularly severe in the absence of vertical load.

4   *Braking loads during launching.* The braking loads depend on the launch devices and may be more severe than the launch loads, especially when friction brakes are used. Fast braking may also cause vibrations and dynamic amplification. The use of hydraulic launch cylinders (friction launchers, perforated rails, prestressing bars) minimises braking loads and avoids irregularities in the launch speed.

5   *Wind loading.* Launching takes place in controlled weather conditions, and, although longitudinal wind may increase the launch and braking loads, its influence is rarely marked. Transverse wind in both service and out-of-service conditions may be much more demanding, and may trigger lateral vibrations. Wind-induced resonance may be particularly risky due to the absence of effective damping systems in most of these machines.

6   *Accidental loads.* Most accidental situations affect equipment design more than pier design; however, imbrication of launch bogies and PTFE skids and dynamic load redistribution at failure of a launch winch should be considered in the pier design. Steps in the launch rails

may increase the launch force and overload the pier that restrains the launcher. When two launch winches are used for redundancy, one winch takes the entire load when the other winch collapses, and fast load redistribution amplifies the load applied to the pier.

Piers equipped with sliding bearings are designed for longitudinal loads of 3–5% of the vertical service load (permanent and live load). As the weight of an MSS may exceed 50% of the span weight and the friction coefficient of the launch bogies is higher, the longitudinal load during construction may govern pier design from a 2–3% uphill launch gradient. PTFE skids are typically avoided for launching on tall and flexible piers due to the higher friction coefficient (Däbritz, 2011) and the additional longitudinal load component due to the weight of the span itself (Rosignoli, 2002).

5.6.   Underslung MSSs

A great number of solid, voided and ribbed slabs and box girders have been cast with underslung MSS on spans ranging from 30 m to 50–60 m; 75 m spans have recently been achieved without props from foundations. The casting cell of these scaffolding systems is assembled on two girders supported at the piers of the span to cast. The girders are located alongside the piers, under the side wings of the deck. The 97 m, 690 ton underslung MSS shown in Figure 5.43 was used to cast 44 m, 1190 ton dual-track HSR spans.

In the configuration for continuous spans, a crossbeam suspends the rear end of the MSS from the front deck cantilever. Hydraulic cylinders at the rear crossbeam and the leading pier lift the MSS to the span-casting elevation, and lower it back after application of prestressing to release the span. The rear crossbeam is sometimes replaced with a C-frame that also suspends the MSS during launching by rolling onto the new span.

Box girders and square trusses are used for the main girders, and light noses and tails control over-turning during launching. Triangular trusses are often used for the launch extensions. Hybrid

Figure 5.43. Germany Unity Transport Project 8, VDE 8.1 New Line, Wümbach Viaduct (Photo: DB AG)

Figure 5.44. I-girders with stiffening truss for the launch noses (Photo: Strukturas)

sections comprising a vertical I-girder and a stiffening truss are a cost-effective alternative: the I-girder is designed for the launch loads and robotised welding, and the bracing truss is designed for lateral loads and to control instability (Figure 5.44). Hybrid noses and tails are supported only at the I-girders.

The main girders are typically located below the deck soffit. Alongside MSSs have been designed with the main girders lifted closer to the deck wings, in the typical configuration of an underslung gantry for the erection of precast segments (Däbritz, 2011). Alongside MSSs are more complex to operate and require expensive pier brackets, and have found limited application in spite of the lower clearance demand (Figure 5.45).

Adjustable supports connect the bottom crossbeams of the casting cell to the top flanges of the main girders in short-span applications. Adjustment screws, shims or hydraulic jacks are used to set the camber and crossfall of the casting cell. Sliding clamps at the supports permit lateral shifting of the form shells relative to the main girders, and control transverse overturning during launching. Concrete counterweights are often suspended outside of the main girders to minimise the transverse torque during launching.

The bottom crossbeams are spliced at midspan. After span release, the splices are opened to divide the outer form into two halves. Each half is shifted outwards over the main girder to create a central launch clearance wider than the leading pier cap. The main girders may also be shifted outwards along the pier brackets to further widen the central clearance for launching on curves (Figure 5.46).

The two halves of the MSS are launched to the next span, the casting cell is closed, hangers are applied between the rear suspension crossbeam and the main girders, and the MSS is lifted to the span-casting elevation for another casting cycle. The centre of gravity of the wind drag

Figure 5.45. Alongside MSS for ribbed slabs (Photo: Strukturas)

Figure 5.46. Launching on a curve (Photo: ThyssenKrupp)

area is above the launch supports, the two halves of the MSS are less stable than the closed machine, and the weight of the form shell is eccentric transversely and often requires balancing counterweights. The underslung MSSs are more risky to launch than the overhead machines, and launching overnight or in the presence of wind should be avoided.

In some alongside MSSs the bottom crossbeams of the casting cell have been supported on the bottom chords of the main trusses. Square trusses offer numerous advantages over windowed box girders in these applications. As the crossbeams are close to the pier brackets, the bottom form table cannot be rotated to vertical for launching. These MSSs are opened by shifting the entire truss outwards, which increases the cost and complexity of the pier brackets.

Both types of underslung MSSs are less adaptable to different deck geometries than are overhead machines. As the main girders are located beneath the deck, ground constraints, straddle bents and C-piers complicate the operations much more than with an overhead MSS. The MSS also projects beneath the deck for its entire length, which may cause clearance problems when passing over existing roads or railroads and difficulties in the first and last span of the bridge because of the interference with the abutment walls.

Underslung MSSs are not suitable for bridges with a tight plan radius, as the landing pier is out of line and the casting cell is offset from the chord at midspan. Articulated MSSs have been used for curved bridges, where hydraulic hinges connect launch noses and tails to the main girders for plan rotations. In optimal conditions, an underslung MSS is less complex to design, assemble and operate than an overhead machine. Underslung MSSs facilitate access to the casting cell, and the operations may be assisted using tower cranes anchored to the piers, rolling on rails on the deck, or supported by the MSS itself.

Overturning during launching is controlled by trussed extensions, and the total length of these MSSs is 2.1–2.2 times the typical span. In the first-generation MSSs, launch noses and tails had approximately the same length. In some new-generation MSSs for continuous spans, the launch nose is 85% of the main girder and the rear tail is 20%. The launch nose takes support on the next pier in the span-casting configuration for earlier starting of preparatory activities, while the rear tail just reaches the rear pier of the span to be cast. Hydraulic articulations between the front nose and main girder make these MSSs compatible with plan radii of 1000–1200 m on 70 m spans. When combined with a rear C-frame rolling on the new span during launching, this geometry improves stability by moving the centre of gravity of the machine backward.

When the piers are tall, the MSS takes support on brackets anchored to the sides of the pier by shear keys and stressed bars. Hydraulic cylinders housed within the launch saddles lift the MSS from the launch bogies to the span-casting elevation. After the application of prestressing, the cylinders are retracted to release the span and to lower the MSS onto the bogies for launching. The pier brackets may be designed to be repositioned by the MSS during launching; this solution lengthens the cycle time but avoids the use of ground cranes.

W-frames on longitudinal through girders are used for the heaviest applications. In this case, the support cylinders are placed on the through girders (see Figure 5.53) and lift and lower the W-frame. The MSS is supported on high-capacity PTFE skids designed for the entire load of MSS and span. The W-frames are applied on both sides of the pier to halve the load and for control of overturning of the through girders. Application and removal of these support systems requires ground cranes. The longitudinal loads due to the combined effect of high friction and inclined supports during span casting are a counterindication to uphill launching on tall piers (Rosignoli, 2002).

Props from foundations or reinforced-concrete temporary piers may be used to support an under-slung MSS when the piers are not tall. One temporary pier at the third-quarter of every span is used to cast 100–120 m spans with an MSS for 50–60 m spans. Directional casting from abutment to abutment is faster than balanced cantilever construction and saves prestressing (see Table 4.1); however, to be cost-effective this construction method requires good foundation soil and short piers.

The construction joints are at the span quarters. The temporary pier supports the front end of the casting cell during casting of the midspan segment, while the rear end of the casting cell is suspended from the front deck cantilever. For casting of the pier segment, the temporary pier supports the midspan segment and the rear end of the casting cell, and the MSS cantilevers out from the leading pier. Varying-depth decks with vertical webs are easy to form by shimming the bottom form table (Figure 5.47). This construction method is also compatible with overhead MSSs, but applying and removing the form shims is much easier in an underslung MSS due to the unobstructed access to the casting cell (Rosignoli, 2011a).

Figure 5.47. Form shimming for casting half-span segments of varying depth

Full-span cage prefabrication is less expensive than with an overhead MSS, as no winch trolleys are required for cage handling. A long straddle carrier is used to suspend the cage, and the shutters of the side wings are stiffened to create support routes for the front wheels of the carrier to move over the casting cell.

5.6.1 Loading, kinematics, typical features

Box girders and square trusses are the typical choice for the main girders of these MSSs; braced I-girders are uncommon due to the primary role of torsion in resisting the lateral load imbalance during launching. Three-truss assemblies comprising two twin trusses that support the bottom crossbeams of the casting cell and a lighter outer truss that controls out-of-plane buckling and resists lateral loads are also frequently used (Rosignoli, 2010). Hybrid assemblies comprising box girders under the casting cell and triangular trussed extensions have also been used on many occasions (Figure 5.48).

Triangular trusses are rarely used for the main girders due to the complex control of form overturning during launching. Underslung triangular trusses have a lot of merits and are the first-choice solution for the gantries for span-by-span erection of precast segments to achieve a runway for the segment carts on either side. Underslung gantries and MSSs resist similar loads, but different technological requirements lead the choice between alternative solutions having similar structural efficiency.

Figure 5.48. Triangular launch noses and tails (Photo: NRS)

Lateral bracing connects the chords of the square trusses over their entire length. Lateral bracing includes diagonals or crosses, connections designed for fast site assembly and decommissioning and to minimise displacement-induced fatigue, and sufficient flexural stiffness to resist vibration stresses. Diaphragms connected to the chords at the same locations as the lateral bracing distribute torsion and provide transverse rigidity. Connections and field splices are often designed to develop member strength.

Box girders are the typical solution for long-span applications, in combination with transverse bottom trusses for the casting cell. The trusses are spliced at midspan and connected to the inner web of the main girders to reduce the depth of the assembly. This type of connection prevents shifting the form shell outward over the main girder prior to launching, and the MSS is therefore opened by shifting the main girders outward on long pier brackets. The centre of gravity of the form shell is eccentric and inward during launching, and concrete counterweights are suspended outside the main girder to balance the load.

The bottom chords of the transverse trusses are connected to the bottom web-flange nodes of the main girders, and the top chords are connected to the inner web at mid-depth. Plate diaphragms with manholes and diagonal braces within the box cell transfer the load of the top chords to the outer web-flange nodes of the main girders. Mid-depth longitudinal slip-critical splices in the webs of the main girders are sometimes necessary for shipping. Plate diaphragms with manholes are also used at the application points of support reactions.

Screw jacks on the transverse trusses support the bottom form table of the casting cell for setting of the camber and crossfall. The top flange of the main girders supports props and turnbuckles for the rib frames of walls and side wings. Basically, the entire casting cell is an adjustable integrated platform supported on screw systems connected to the rigid assembly of box girders and transverse trusses (Figure 5.49). After the initial geometry setting, hydraulic lowering and lateral shifting of the main girders avoids any need for additional adjustment in the subsequent spans.

The span is released by retracting the support cylinders of the MSS, the central splices of bottom crossbeams and trusses are released, and the outer form shells are shifted outwards to create the central launch clearance. When the outer form shifts laterally on the main girders, shifting can be combined with hydraulic rotation of the central form sectors to vertical. The adjustable saddles that support the crossbeams on the main girders have sliding clamps that prevent uplift and overturning. Redundancy complicates setting of the camber and crossfall, but is necessary to control form overturning during launching. Long-stroke double-acting cylinders are used for lateral shifting of the bottom crossbeams, and the sliding clamps are equipped with low-friction surfaces.

Long front noses control overturning during launching. A C-frame that suspends the rear end of the main girders and rolls on the new span during launching may avoid launch tails and the rear set of pier brackets. When the outer form shifts laterally on the main girders, the girders may be permanently framed to the rear C-frame. These short MSSs are better suited to use on curved bridges: the weight of the C-frame enhances stability from overturning, and balancing bogies rolling under the bottom web-slab nodes of the deck may be applied to the C-frame to further shorten the launch noses. In the alongside MSSs, the outer form does not shift relative to the main girders, and the pier brackets have long side overhangs to shift the girders outwards for launching.

Figure 5.49. Integrated-platform MSS for heavy applications (Photo: Strukturas)

Compared to a twin-girder overhead MSS, the main advantage of an underslung MSS is the capability of carrying the outer form during launching. This simplifies the casting cycle, shortens the cycle time and offers other advantages when the area under the bridge cannot be disrupted or the piers are tall. Some machines are also able to reposition the pier brackets, which avoids any need for ground cranes. Compared to a single-girder overhead MSS, the casting cell is free from obstructions, and the cage can be prefabricated full length also in continuous spans.

The main limitation of most of these machines is the absence of portal cranes or winch trolleys that assist the operations of the casting cell. Light inexpensive hoists may be used to handle rebars when the cage is fabricated within the casting cell; the hoist is removed from the MSS after casting the first-phase U-span and repositioned after launching (Figure 5.50). Tower cranes anchored to the piers, rolling on the deck or carried by the MSS, are used to solve major load-handling requirements (Figure 5.51).

Underslung MSSs use multi-hook straddle carriers to deliver the prefabricated cage. The three-dimensional space frame that suspends the cage is as long and as wide as the typical span. The legs of the carrier are located at the ends of the cage and have motorised steering wheels. The casting cell is free from obstructions, and the front wheels of the carrier move forward along stiffened strips of outer form to the lowering location for the cage. Front telescopic legs are used to overcome the step between the deck and the outer form at the rear end of the casting cell, while the rear wheels remain on the deck. The carrier may be sheltered to cover the casting cell during span casting, and may be equipped with concrete distribution arms. Concrete may be delivered on the deck or pumped from the ground if the piers are not tall.

Figure 5.50. Light hoist for cage assembly in the casting cell (Photo: Strukturas)

Figure 5.51. Tower crane applied to an integrated-platform MSS (Photo: Strukturas)

The legs of the cage carrier are located as close as possible to the deck webs in order to minimise transverse bending in the side wings during cage transportation and to start the finishing work (which is mostly located at the edge of the wings) prior to deck completion. Cage delivery with straddle carriers is not compatible with the use of tower cranes on the deck.

The casting cell is anchored to the leading pier cap and the front deck cantilever to prevent displacements during filling. Form offset in curved spans is achieved by shifting the rib frames of walls and side wings laterally. The bottom form table is wider than necessary, and the wall shutters take support on it. The camber and crossfall are set using the support systems of cross-beams and bottom trusses.

Underslung MSSs require numerous walkways and working platforms. The form adjustment platforms are often extended to the launch noses for access to the next pier. Parapets are provided at the edges of the side wings, and a full-length walkway is suspended under the casting cell for operations on the central connection pins. The prestressing tendons are fabricated and tensioned from a stressing platform beyond the front bulkhead; the platform is also opened into two halves for launching, and provides access to the bottom walkway under the casting cell.

5.6.2   Support, launch and lock systems

Most underslung MSSs require three sets of pier supports and a rear suspension crossbeam. Two sets of supports are applied to the piers of the span to be cast, and the third set is removed from the preceding pier and applied to the next pier during span casting to receive the MSS during launching (Däbritz, 2011). In bridges over land, the pier supports are typically repositioned using ground cranes. A C-frame rolling on the deck may be used to suspend the rear end of the MSS during launching in order to avoid launch tails, the rear set of pier support, and the suspension crossbeam.

Four types of pier supports are available for an underslung MSS: pier brackets, W-frames on through girders, pier towers, and suspension frames supported onto the pier cap.

Pier brackets are the most common solution. They include braced crossbeams that support the main girder and provide lateral geometry adjustment, and diagonal props that support the cross-beams by transferring the load to the pier walls. When the pier is hollow, the props diverge to load the corners of the box section. Stressed bars connect the two brackets at the crossbeam level to resist transverse bending; a few bars are often provided also at the bottom of the diagonal props to enhance stability. Shear keys at the end of crossbeams and diagonal props are inserted into pier recesses for the transfer of vertical loads. Friction collars have been tested to minimise the number of holes needed in the piers, with contrasting results. Working platforms are applied to the crossbeams for the workers to reach the support saddles and clamping bars.

The pier brackets are typically applied and removed using ground cranes. When the piers are short, the diagonals may be propped from the bridge foundations. This solution avoids the need for recesses in the pier, simplifies reinforcement and reduces the erection loads applied to the pier. This solution is used frequently in the slender piers of LRT bridges.

Some MSSs carry two sets of self-repositioning brackets. Self-repositioning brackets avoid the need for ground cranes but complicate launching because the rear crossbeam suspends the

Figure 5.52. Self-repositioning of the rear pier brackets (Photo: NRS)

MSS during repositioning of the rear brackets (Figure 5.52). Some new-generation MSSs have been designed to reposition the rear brackets during cage fabrication.

Self-repositioning brackets have conventional geometry and are equipped with sliding restraints at all articulations for suspension. Hydraulic motors drive the brackets along the main girder and are disengaged after anchoring the brackets to the pier. Long-stroke double-acting cylinders shift the bracket transversely relative to the main girder to insert the shear keys into the pier recesses; the same cylinders are used to shift the main girder along the bracket to open the casting cell for launching. The brackets carry hydraulic systems powered by the PPU of the MSS.

The pier brackets support the MSS during launching with articulated saddles that permit longitudinal and transverse displacements and rotations in the horizontal and longitudinal plane. Roll assemblies on equalising beams are used for launching, in combination with rectangular rails welded to the bottom flanges of the main girders, noses and tails for guidance and the transfer of lateral loads. PTFE skids and edge guides are often used for transverse shifting in order to reduce the cost and complexity of the launch saddles. High-tonnage double-acting cylinders with mechanical locknuts lift the main girder from the launch saddles at the end of launching, and lower it back after application of prestressing to release the span.

W-frames on through girders are applied and removed using ground cranes (Figure 5.53). Light auxiliary frames assist the insertion of the through girders within the pier cell. Through girders are

Figure 5.53. Removal of a W-frame on through girders

used only on wide hollow piers; they are located as close as possible to the cell corners to increase the stability of the W-frames. The W-frames are assembled on the ground and placed on the through girders in one lift to minimise air working, even if this often requires two cranes for frame removal. The preceding set of pier supports is repositioned during span casting so as not to affect the cycle time.

The use of W-frames on through girders requires wide piers and narrow pier caps for control of overturning. One W-frame is used on either side of the pier to stabilise the through girders longitudinally and to halve the load applied to each frame. Geometry constraints may cause out-of-node eccentricity in the support reactions applied to the square truss of the MSS. Double-acting cylinders are placed on the through girders to lift the W-frames to the span-casting elevation and to lower them back to release the span. The support saddles are designed for the full load (span and MSS) and follow the launch gradient; they are typically based on PFTE skids for both launching and shifting.

W-frames on through girders are simpler and less expensive than pier brackets but require hollow piers, and the recesses in the pier walls weaken the pier and complicate reinforcement. When the piers are short, solid or narrow, crossbeams may be supported on towers from the bridge foundations. Short pier towers are stiff and inexpensive, the vertical loads are applied directly to the bridge foundations, the support cylinders may be applied at the base of the tower for easy access, and the pier braces the tower and resists lateral loads. The footings, however, must often be wider than is necessary to support the tower.

Suspension frames supported on the pier caps are rarely used with such heavy machines. Even though the suspension frames avoid recesses in the piers, they require stiff pier caps and ample vertical clearance between the deck soffit and the pier cap, and they are difficult to extract after casting the span.

In continuous bridges, a crossbeam suspends the rear end of the MSS from the front deck cantilever. The crossbeam shortens the casting span, controls the time-dependent stress redistribution of staged construction within the deck, and avoids settlement of the casting cell and steps in the soffit at the construction joint. The crossbeam is equipped with four adjustable support legs that apply the load over the deck webs. The legs include long-stroke double-acting support cylinders that load the hangers in one operation and cope with the gradient and crossfall in the deck.

Bundles of hanger bars cross the side wings of the deck and suspend the main girders of the MSS from the crossbeam (Däbritz, 2011). Ball-and-socket anchorages are used for the bars to minimise bending. The hangers are applied at the end of launching and removed after span release. Many hangers are necessary, as the rear crossbeam transfers 30–40% of the total load (span and MSS), and the bars are typically oversized to control low-cycle fatigue in multiple reuses. The use of load-tested bars is recommended.

The use of a rear C-frame permanently connected to the main girders simplifies launching and shortens the cycle time, but the main girders cannot be shifted outward and the outer form must therefore be designed for transverse shifting for the full launch clearance. The C-frame houses four double-acting cylinders that lift the MSS to the span-casting elevation in combination with the support cylinders of the front pier brackets. After application of prestressing, releasing the support cylinders lowers the main girders onto the launch bogies at the front pier brackets, and the launch bogies of the rear C-frame land on the rails on the deck.

Intermediate props from foundations are sometimes used to support the MSS in long-span applications or to increase the load capacity of an existing machine. Lateral wind on the unloaded MSS is a demanding design condition for the props and their bracing systems and tie-downs. When temporary piers are used to cast half-span segments, the MSS uses the same support systems at the temporary and permanent piers.

The two halves of light first-generation underslung MSSs for haunched slabs are launched independently with capstans. In heavier machines, redundant long-stroke double-acting cylinders housed within the launch saddles are used for launching. Hollow plungers pull oversized prestressing bars by means of anchor plates that are repositioned forward along the bars during launching (Rosignoli, 2000). Solid plungers act on perforated rails by means of lock pins. Paired cylinders are used for redundancy and to restrain the girder with one cylinder during repositioning of the other. Redundant launch cylinders are safer than capstans and provide smoother operations, but launching is slower (Rosignoli, 2002). In most machines, the two halves of the MSS are repositioned independently.

Double-acting cylinders are used to shift the launch saddles transversely along pier brackets and W-frames. When anchored into subsequent holes, the cylinders are paired for redundancy and to hold the saddle in place during repositioning of the anchor pins. The shift cylinders also act as transverse restraints during operations.

5.6.3 Performance and productivity

When the clearance beneath the deck is not an issue and the bridge has a constant and large plan radius, an underslung MSS is often the most cost-effective solution for span-by-span casting of medium-length bridges with constant 30–70 m spans. These machines are simpler and less expensive than overhead MSSs, access to the casting cell is unobstructed, site assembly and operations are relatively simple (Däbritz, 2011) and some new-generation MSSs are also able to reposition the pier brackets.

An underslung MSS can be easily reconditioned by modifying the rib frames and by adding or removing bottom crossbeams to adjust the length of the casting cell. Propping the MSS from foundations increases the load capacity, and a temporary pier placed at the third-quarter of continuous spans may double the casting span.

The casting cell is typically not sheltered. In projects of a size compatible with the investment, the reinforcement cage is prefabricated and the cage carrier may be equipped with full-length covering, concrete distribution arms and, in cold weather, steam curing plant. The outer form can be protected with thermal enclosures, and the camber and crossfall are easy to set from comfortable work locations.

When the piers are short and the area under the bridge is accessible, concrete may be delivered on the ground and distributed using two or three pumps to accelerate filling (CNC, 2007). A concrete pump at the rear end of the MSS may be enough for two-phase casting when the concrete is delivered on the deck. One pump on the deck and two pumps on the ground may be used for one-phase casting of long spans for fast filling and risk mitigation.

The typical cycle time for 40–50 m single-cell box girders with cage prefabrication and one-phase casting without steam curing is one span in 6 working days. When the cage is assembled in the casting cell, the cycle time for two-phase casting is 10 working days with minimised crews, one 10-hour shift, and three sets of pier supports. The cycle time increases by 1–2 days when self-positioning brackets are used.

When 80–120 m spans are cast with an MSS for 40–60 m spans, form adjustment to varying depth lengthens the cycle time to 3 weeks for the entire span (see Table 4.1). Balanced cantilever casting of a 120 m span with 5 m segments takes 12 weeks, and labour, prestressing and logistics costs are higher. On the other hand, temporary piers and foundations involve additional costs, the bridge must be low to the ground and built directionally, and several pairs of form travellers may be leased with the investment for an MSS for 60 m spans.

Repositioning the pier supports slows down reverse launching. Single-girder overhead MSSs are faster from this point of view, as the launch systems are on the deck. The two halves of an underslung MSS may be lifted on the deck with strand-jacking towers to be transferred along the deck. Tower–crossbeam assemblies of overhead MSSs may sometimes be used for lifting when the length of pier brackets or W-frames is sufficient to clear the side wings of the deck.

5.6.4 Structure-equipment interaction

An underslung MSS is normally used to cast an entire span. Solid, voided and ribbed slabs are cast in one phase, and box girders are cast in one or two phases. One-phase casting of box girders is not commonly done with underslung MSSs because of the low level of mechanisation typically associated with these machines.

Figure 5.54. Underslung MSS with PLC-controlled prestressing (Photo: BERD)

The main concerns with two-phase casting relate to the deflections of the MSS during casting of the top slab. Some new-generation underslung MSSs for long-span applications make use of PLC-controlled prestressing to minimise the deflections of the casting cell during two-phase casting (Pacheco *et al.*, 2007). The 900 ton, 120 m MSS for 2000 ton, 60 m spans shown in Figure 5.54 uses four pairs of pendular struts to deviate four 400 ton prestressing tendons at the thirds of the casting span. After releasing the span, hydraulic cylinders rotate the struts horizontal to avoid interference with pier brackets and the leading pier cap during repositioning of the MSS. A rear C-frame rolling on the new span during launching and two pairs of self-repositioning pier brackets make this machine fully self-launching on 113 m tall piers. The long launch nose allows access to the new pier for early preparatory work.

In less sophisticated machines, the suspension crossbeam is positioned on the front deck cantilever at the end of launching. The crossbeam suspends the main girders at the rear end of the casting cell and lifts the launch tails from the rear pier supports. Loading the front deck cantilever shortens the casting span, reduces the rear support reaction of the MSS due to the long front overhang in the casting cell, and reduces the time-dependent stress redistribution of staged construction within the deck. These effects are all advantageous, especially when post-tensioning is fully applied at short curing. However, releasing the span modifies the stresses in the continuous deck and the initial camber in the two leading spans.

Prior to application of prestressing, inner and outer forms are stripped to support the span on the bottom form table and minimise prestress losses. This operation can be avoided if the tendons are checked after span release and re-tensioned as needed.

Most underslung MSSs are supported only at the piers during launching. These machines are relatively heavy, and their weight may affect the design of piers and foundations. The longitudinal loads applied to the piers increase substantially when PTFE skids are used on W-frames on through girders due to the combined effect of high friction and of casting the span on inclined supports. For casting curved bridges, the MSS is shifted outward at the piers to compensate for the eccentricity of the midspan section of the casting cell. Additional load eccentricity may be necessary for traversing during launching. Staggered launching of the two halves of the MSS applies a transverse couple to the pier cap due to the differential in the support reactions, and should be avoided when W-frames on through girders are used to support the MSS due to the risk of uplift and overturning.

Compared to an overhead MSS, the centre of gravity of the wind drag area is below the deck, which reduces transverse bending in piers and foundations. The two halves of the MSS are more flexible laterally than a single-truss overhead MSS, and launching in the presence of wind should be carefully avoided.

REFERENCES

AASHTO (American Association of State Highway and Transportation Officials) (2008a) *Construction Handbook for Bridge Temporary Works*. AASHTO, Washington, DC, USA.

AASHTO (2008b) *Guide Design Specifications for Bridge Temporary Works*. AASHTO, Washington, DC, USA.

AASHTO (2012) *LRFD Bridge Construction Specifications*. AASHTO Washington, DC, USA.

ACI (American Concrete Institute) (2004) *Guide to Formwork for Concrete*. ACI, Farmington Hills, MI, USA.

ASBI (American Segmental Bridge Institute) (2008) *Construction Practices Handbook for Concrete Segmental and Cable-supported Bridges*. ASBI, Buda, TX, USA.

BSI (1996) BS 5975: 1996 Code of practice for falsework. BSI, London.

BSI (2008a) BS 5975: 2008 Code of practice for temporary works procedures and the permissible stress design of falsework. BSI, London.

BSI (2008b) BS EN 12812: 2004 Falsework. Performance requirements and general design. BSI, London.

CNC (Confederación Nacional de la Construcción) (2007) *Manual de Cimbras Autolanzables*. CNC, Madrid, Spain.

CSA (Canadian Standards Association) (1975) *Falsework for Construction Purposes*. Design. CSA, Toronto, Canada.

Däbritz M (2011) Movable scaffolding systems. *Structural Engineering International* **21(4)**: 413–418.

FIB (International Federation for Structural Concrete) (2009) *Formwork and Falsework for Heavy Construction. Guide to Good Practice*. FIB Bulletin No. 48. FIB, Lausanne, Switzerland.

Harridge S (2011) Launching gantries for building precast segmental balanced cantilever bridges. *Structural Engineering International* **21(4)**: 406–412.

Homsi EH (2012) Management of specialized erection equipment: selection and organization. *Structural Engineering International* **22(3)**: 349–358.

Pacheco P, Coelho H, Borges P and Guerra A (2011) Technical challenges of large movable scaffolding systems. *Structural Engineering International* **21(4)**: 450–455.

Pacheco P, Guerra A, Borges P and Coelho H (2007) A scaffolding system strengthened with organic prestressing – the first of a new generation of structures. *Structural Engineering International* **17(4)**: 314–321.

Povoas AA (2012) The utilization of movable scaffolding systems in large spans. *Structural Engineering International* **22**: 395–400.

Rosignoli M (1998) Misplacement of launching bearings in PC launched bridges. *ASCE Journal of Bridge Engineering* **3(4)**: 170–176.

Rosignoli M (2000) Thrust and guide devices for launched bridges. *ASCE Journal of Bridge Engineering* **5(1)**: 75–83.

Rosignoli M (2001) Deck segmentation and yard organization for launched bridges. *ACI Structural Journal* **23(2)**: 2–11.

Rosignoli M (2002) *Bridge Launching.* Thomas Telford, London.

Rosignoli M (2007) Robustness and stability of launching gantries and movable shuttering systems – lessons learned. *Structural Engineering International* **17(2)**: 133–140.

Rosignoli M (2010) Self-launching erection machines for precast concrete bridges. *PCI Journal* **2010(Winter)**: 36–57.

Rosignoli M (2011a) Bridge erection machines. In *Encyclopedia of Life Support Systems*, Chapter 6.37.40, 1–56. United Nations Educational, Scientific and Cultural Organization (UNESCO), Paris, France.

Rosignoli M (2011b) Industrialized construction of large scale HSR projects: the Modena bridges in Italy. *Structural Engineering International* **21(4)**: 392–398.

Rosignoli M (2012) Modena viaducts for Milan–Naples high-speed railway in Italy. *PCI Journal* **50(Fall)**: 50–61.

SAA (Standards Association of Australia) (1995) AS 3610-1995. Formwork for concrete. SAA, Sydney, Australia.

Bridge Construction Equipment
ISBN 978-0-7277-5808-8

ICE Publishing: All rights reserved
http://dx.doi.org/10.1680/bce.58088.167

publishing

Chapter 6
Forming carriages for the deck slab of composite bridges

6.1. Technology of composite bridges

In the last decades many factors have increased the competitiveness of composite bridges and extended their use to shorter spans. Progress in iron metallurgy has led to rolled steel plates with excellent and reliable mechanical properties. The development of design standards based on the strength of materials has led to a more accurate determination of structural safety. The progress of the fabrication techniques has replaced riveted joints with workshop-welded connections and a few splices bolted in the field. In recent years, the general increase in labour costs and the stagnant cost of steel has further accelerated the evolution of these techniques (Rosignoli, 2002).

A composite bridge offers several advantages. The high tensile and shear strength of steel plates combines well with the compressive strength of concrete. Rapidity of construction and the possibility of fabricating most of the structure in a workshop improve planning and risk management. The durability provided by different and renewable protective treatments and the possibility of modifying the structure over time enhance design flexibility, and the few structural elements the function of which can be clearly recognised enhance project aesthetics.

The composite sections are fit to simple spans, as the concrete slab resists axial compression and the steel girders resist axial tension and shear. In the negative bending regions of a continuous beam, axial compression at the bottom fibre requires thick flanges to control buckling, and axial tension in the concrete slab requires a great quantity of reinforcement to control cracking in areas directly exposed to traffic. Casting a bottom slab for double composite action is justified only for long spans, and in most cases the cross-section is open.

Redundancy of open composite sections is achieved in different ways. In the USA, redundancy is achieved using numerous poorly braced I-girders (AASHTO, 2012). Design standards in other countries recognise the capability of the lateral bracing, diaphragms and concrete slab to create multiple load paths in case of failure of a girder, and encourage the use of twin I-girders, which offer lower cost of steelwork, faster erection and simpler field splicing (BSI, 1991).

Different approaches to structural redundancy also affect the design of the concrete slab. Multiple I-girders are closely spaced and the slab is thin, has narrow side wings, and may be cast in place on corrugated metal or wooden forms or precast full depth. The concrete slab of a twin I-girder steel frame has long haunched wings and a wide central panel, the thickness and weight of the slab increase, and in-place casting and full-depth precasting both suggest the use of more mechanised methods.

The concrete slab is mostly cast in place. Forming is labour intensive, and left-in-place forms are frequently used, especially for the central slab panel between the webs. Materials include precast concrete acting compositely with steel beams or welded lattices, galvanised profiled steel sheeting, and reinforced plastic or fibre-cement sheeting, or similar (IABSE, 1997).

Left-in-place forms for highway bridges are mostly made of precast concrete planks reinforced with welded lattices or punched I-beams that resist the load exerted by fluid concrete and personnel. Precast planks may be considered as either structurally participating or non-participating with the second-phase concrete. In both cases the composite slab obtained by casting fresh concrete on precast planks is highly sensitive to the time-dependent behaviour of concrete.

When the composite action between precast planks and in-place concrete is relied upon, durability is a prime concern. Differential drying and thermal shrinkage affect durability, as humidity passing through full-depth cracks in the second-phase concrete gathers at the planks. If the concrete cover to plank reinforcement complies with design standard requirements above and below the bar grid, the plank is heavy and thick, and a thin in-place layer has often been associated with fatigue issues at the joints between planks. The typical solution is to reduce both concrete covers, which affects durability.

Structural continuity between adjoining planks may be achieved with lapping reinforcement projecting from units, embedded steel plates to be field-welded after plank assembly, or high-strength bolts. All these solutions may be affected by fatigue under repeated loading, are expensive, and involve a great number of field splices. For these reasons, the precast planks are often considered as structurally non-participating.

Differential shrinkage and partial composite action may also adversely affect durability when the planks are designed as non-participating. Some design standards prescribe that the requirements for cover to reinforcement and crack control applicable to in-place slabs be satisfied, ignoring the presence of the planks. Other standards require that the clear distance between the plank and the reinforcement in the second-phase slab exceeds the maximum nominal size of the aggregate, which limits the lever arm of reinforcement.

The weakest point of the non-participating planks, however, is their weight. This additional load does not increase the flexural capacity of the concrete slab, but does increase the weight of the steel grillage, and additional steel weight results from the constant thickness of the wide central slab strip between the webs. The plank dimensions restrain framing of the twin I-girders, and additional costs derive from transporting and lifting the planks into position.

When the piers are short and the area under the bridge is accessible, the planks are positioned using ground cranes. Tall piers or ground obstructions require the use of hydraulic trolleys that load multiple planks on their rear platform behind the abutment, roll along the girders of the steel grillage, and position the planks starting from the opposite abutment and proceeding backward.

In many instances the costs of left-in-place forms and in-place casting with forming carriages warrant consideration of alternatives based on prefabricated elements. Full-depth precast slab segments of short length and, when feasible, as wide as the entire deck may offer competitive solutions.

Full-depth slab precasting offers the same advantages as segmental precasting of prestressed concrete bridges, such as dimensional stability due to long curing times, accurate dimensions, and high quality resulting from mechanised casting processes. The main disadvantage of slab precasting is the small segment width deriving from transport requirements. This increases the number of construction joints in the slab, with related costs and durability issues.

Powerful ground cranes or crane trolleys rolling on the steel girders are necessary to handle wide, full-depth precast slab panels, and this further increases the final cost of the solution. For these reasons, full-depth segmental slab precasting is used rarely in non-cable-stayed major bridges.

The weak point of the precast deck panels is the great number of joints and shear connection pockets; leaking joints have been historically associated with this construction method. Special proprietary ultra-high-performance concrete (UHPC) mixes with a compressive strength of 120–200 MPa and a flexural strength of 15–40 MPa have been proposed for in-place casting of connections between high-performance concrete (HPC) precast deck panels. These materials have a high bond when cast against hardened concrete and a short bond development length with embedded reinforcement, which results in narrow in-place stitches that are stronger, tougher and more durable than the adjoining deck panels. The advantages include reduced joint size and complexity, improved durability and continuity, fast construction, and low maintenance (Rosignoli, 2002). In spite of the high adhesion and tensile strength of polymer grouts female-to-female grouted joints are typically post-tensioned.

An alternative construction method for the concrete slab of composite bridges with a constant and large plan radius involves segmental casting of a full-length continuous slab behind the abutment, combined with incremental launching of the slab over the steel girders (Figure 6.1). The casting cycle involves casting a 20–25 m deck segment on a casting bed behind the abutment, jacking the entire slab forward until clearing the casting bed, and splicing reinforcement and prestressing ducts of the new segment with the rebars and ducts protruding from the rear end of the launched slab.

Figure 6.1. Segmental casting and incremental launching of the concrete slab

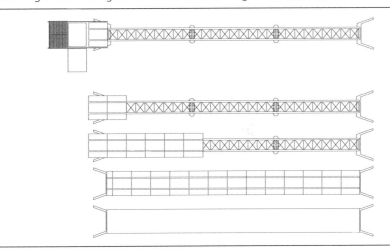

Upon launch completion, the continuous slab is connected to the steel girders using different methods: mechanical connectors, welding, filling of continuity pockets over the girders, and the like. The basic advantages are that the same formwork is used repeatedly, the concrete is cast and finished at a fixed and sheltered location, and the slab is structurally continuous.

The slab is launched with long-stroke double-acting hydraulic cylinders. During jacking, the rear segment is extracted from the casting cell on PTFE strips. The form table may be supported on the ground behind the abutment or may be suspended from the steel girders at any location. A fixed form table is less expensive than a forming carriage, the concrete slab is cast with a continuous process, and the logistics are substantially simplified. Slab launching also avoids interference with the area under the bridge, reduces shrinkage and creep cracking, facilitates the application of longitudinal prestressing, simplifies structural analysis, improves durability and reduces the risks to workers.

In the first applications of this construction method, the slab was launched directly over the flanges of the steel girders. When the slab reached its final position, shear connectors were welded to the flanges within shear pockets that were eventually filled with no-shrink concrete. This solution is affected by some limitations: field welding of shear connectors within small openings is expensive, filling continuity pockets distributed over the entire deck area is expensive and requires curing protection, localised connection forces tend to cause cracking over time, and the lateral support of the compression flange is localised at few points. Guiding the slab during launching is also difficult with this type of girder–slab connection.

Slab launching also caused new problems.

1   High sliding friction at inaccessible launch surfaces generates anomalous stresses in the steel girders and increases the cost of the thrust devices.
2   As the pre-cambered vertical profile of the steel girder is attained by joining straight units to form the shape, the support lines of the concrete slab are irregular. The flexural stiffness of the slab is significant at a local level, and support irregularities generate stress concentrations that may cause cracking. These phenomena are not really quantifiable, and the remedy is a generalised increase in slab reinforcement. Irregular loading of the steel girders also causes transverse bending in the top flanges, which may require thicker plates and vertical web stiffeners to protect the web–flange welds at the field splices.
3   Splice plates and bolted joints in the top flange prevent slab launching.

During launching, the concrete slab is supported on launch shoes that slide at the edges of the flanges. Load eccentricity or eccentric shoes twist the flanges, and this can lead to local yielding or damage to the web–flange welds. Eccentric positions of the launching shoes can be avoided with lateral guides, but alignment tolerances in the flange are practically unavoidable.

After some trials, continuous launch openings filled with non-shrink concrete at the end of launching were introduced in the concrete slab (Figure 6.2). Continuous openings avoid concentration of connection stresses and improve the lateral stability of the compression flange, but do not solve the other weak points of the original scheme, and actually create new problems. Transverse reinforcement connecting the slab strips must prevent instability of the side wings. As the vertical lever arm is small and the transverse forces are therefore high, buckling of the bottom compression bars may trigger overturning of the entire cantilever. Relative

Figure 6.2. Direct slab launching over the flange plates with continuous connections

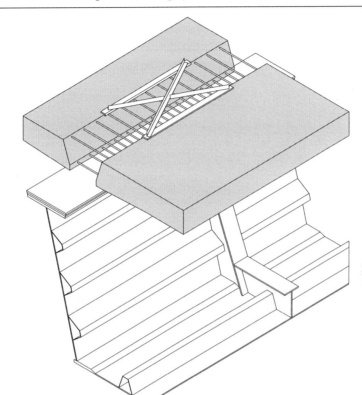

longitudinal movements between the slab strips caused by irregular frictional forces or stick-slip effects may also cause buckling of the compression bars, and it is therefore necessary to brace the slab strips.

Solutions based on vertical steel plates embedded in the concrete slab and emerging from the bottom surface have also been tested. At the end of launching, the plates are welded to the flange edges to achieve longitudinal continuity. In the embedded portion, the plates support headed stud connectors or are punched to attain concrete dowels. This solution avoids launch openings in the concrete slab and improves structural durability. It also distributes the connection forces uniformly and facilitates future replacement of the concrete slab by pulling it backwards after weld grinding. However, this solution requires stringent dimensional tolerances in the concrete slab and the steel girders and extensive overhead welding from mobile platforms, and the launch surfaces are not accessible. Lateral fillet-weld connections prevent transfer of transverse bending between the concrete slab and the steel girders, and may trigger fatigue issues.

Widening the launch openings permits workshop welding of headed stud connectors, but moves the slab support reactions to the tips of the flange. Stability of the cantilever slab strips also worsens due to longer compression bars in the launch openings and a shorter vertical lever arm. This solution still presents some of the weak points of the original scheme, and increases the flexural stresses in the top flange and in the web–flange welds.

Figure 6.3. Indirect slab launching

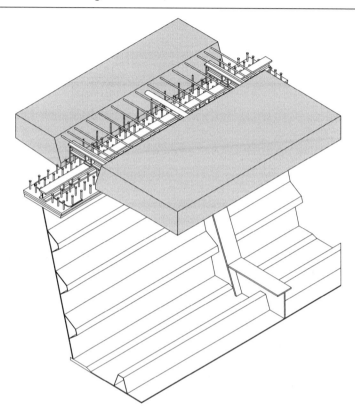

The weak points of the solutions examined so far may be avoided by launching the concrete slab on longitudinal rails supported on the flanges (Figure 6.3). The launch rails are welded to the top flanges with interposition of bearing plates. Short transverse I-beams connect the slab strips through the launch openings and are supported on launch shoes that control friction. During launching, the weight of the concrete slab is transferred to the steel girders through transverse I-beams, launch shoes, launch rails and bearing plates.

The flexural and shear stiffness of the transverse I-beams facilitates transfer of the slab support reactions. Their axial and flexural stiffness, in cooperation with transverse top reinforcement, avoids instability of the cantilever slab strips and limits their rotations. Flexural stiffness in the horizontal plane and the absence of differential launch forces between the slab strips avoid the need for temporary bracing between the strips.

The bearing plates of the launch rails compensate for flange surface irregularities (splice plates, bolts, pre-cambering discontinuities) and localise the loads applied to the steel girders, thus simplifying the design of the steelwork and any necessary local web stiffening.

The launch shoes reduce friction when launching uphill, and control friction when launching downhill. It is thus possible to launch the slab with low forces in the first case, and to avoid braking in the second case. The launch rails may be covered with materials that cooperate with the launch shoes in controlling friction, and, as the launch surface is immediately accessible, any unforeseen event can be easily solved.

On completion of slab launching, the launch openings are completed with reinforcement cages, and non-shrink concrete is poured to connect the slab to the steel girders without secondary stresses. The launch rails may be left in place or recovered. The launch shoes are recovered and the transverse I-beams are shimmed for final correction of the concrete slab elevation.

Launching the concrete slab over the steel girders offers several advantages over casting in place. The first advantage is avoiding the forming carriage. A self-launching forming carriage is not inexpensive, and short bridges often do not permit such an investment. In contrast, the investments needed for slab launching are limited to a form table supported on the ground, a system of launch rails and a pair of hydraulic cylinders that apply the thrust force, and these costs can be easily amortised in one project.

The use of a forming carriage may also be avoided with left-in-place forms, but precast planks are expensive and increase the weight of steelwork. In contrast, the launched slabs are entirely reacting, the composite cross-sectional efficiency increases, and the weight of steelwork decreases. Even neglecting these cost savings, avoiding the cost of precast planks, shipping and handling may cover the equipment cost for slab launching, even in short bridges.

Slab launching offers additional advantages. Structure durability improves because a monolithic slab is more stable than a two-layered composite slab, concrete cover may be thicker, and the bottom slab surface can be easily inspected at segment extraction from the casting cell and during the service life of the bridge. Imperfections and defects can be immediately corrected with minimum costs, and protective treatments can be applied at any time with the same ease (Rosignoli, 2002).

As the slab is cast in a fixed formwork, the top surface can be easily finished with a screed. Concrete is vibrated conventionally, and the screed is used to level and consolidate the top surface only. The screed expels the bleeding water from the top layer of concrete, thus locally reducing the water-to-cement (W/C) ratio. This increases the compactness and tensile strength of the concrete cover to top reinforcement, and reduces permeability. The screed also reduces the number of workers required for concrete finishing, and savings in labour costs soon cover the investment in the equipment.

Aesthetic advantages are not less significant. The launched slabs have a regular and ordered aspect, and launching permits the adoption of pleasant haunched shapes and rounded corners, at minimum cost. Surface finishing (sandblasting, high-pressure washing) is easier at form stripping, and deeper surface finishing (board-marking, engraved textures) is inexpensive because of the small form dimensions.

The reinforcement cage can be entirely prefabricated or handled in grids to improve labour rotation. Concrete pouring with buckets or conveyor belts is less expensive than pumping, the area below the bridge is unaffected, better work conditions and repetitive operations enhance the productivity of personnel, and the entire slab can be cast under a tower crane for easier handling of materials.

Launched slabs undergo most of the dimensional variations due to drying and thermal shrinkage before being connected to the steel girders. This results in low residual stresses, and the jacking action imparts a definitive compression to concrete. The use of non-shrink concrete for filling the launch openings also reduces the residual stresses in these slab areas, where the expansion action of concrete, contrasted by the slab strips, generates a diffused compression that prevents shrinkage contraction from creating cracks. Creep stresses are also low, due to the absence of longitudinal long-term stresses in the concrete slab and the dimensional stability resulting from long curing at joining.

Longitudinal prestressing of the concrete slab improves structure durability by avoiding full-depth cracks, and makes the slab contribution to the flexural capacity of the composite section constant along the deck, which simplifies the structural analysis and makes it more reliable. Longitudinal prestressing involves complex operations in a cast-in-place concrete slab while its introduction is easiest during construction of a launched slab. Prestress transfer to the steel girders is also reduced by the long curing of the concrete slab at joining.

HPC can be used without excessive worries, with additional economic advantages and without risks of thermal-shrinkage cracking. The use of HPC permits significant reductions in the slab thickness: 12.2 m wide precast slab panels made with 80 MPa concrete and with thicknesses ranging from 14 cm at midspan to 22 cm above the girders have already been used.

Slab launching also offers technological advantages. Most of construction occurs on the ground, in a sheltered location, in any weather conditions, and under a tower crane. The presence of workers and materials on the deck is minimised and there is no need to create access ways and working areas on the ground. The logistic are particularly cost-effective when the steel girder is also incrementally launched, as the same site facility can be used for both the steel girder and the concrete slab.

The concrete slab is 3–4 times heavier than the steel girder. During slab launching, there is a risk of both load eccentricity and load concentration, and the risks are not realistically quantifiable. Slab launching should always be carried out in close collaboration with the designer of the structure, and unexpected events should be recorded, investigated, and shared between all parties involved.

6.2.   Forming carriages

As an alternative to the use of full-depth precast deck panels, two-layered slabs or incremental slab launching, the concrete slab of a composite bridge may be cast in place segmentally by means of a forming carriage. In continuous bridges with 60–90 m spans, the length of the slab segments is typically one-third of the span, with construction joints at the span sixths and at midspan. Discontinuous casting sequences are used where two span segments are cast in 4–5 spans followed by the solutions of continuity at the piers to reduce the permanent tensile stresses in the negative-moment regions (Figure 6.4).

A forming carriage consists of a casting cell suspended from an overhead frame that rolls along the steel girder (Rosignoli, 2002). In some machines the overhead frame is supported on four hydraulic legs during launching and on multiple adjustable spikes during casting of the slab segment. Launch bogies and spikes are supported on rectangular rails welded to the flanges of the steel girder between lines of workshop-welded headed stud connectors (Figure 6.5). Rails

Figure 6.4. Segmental casting sequence for the concrete slab

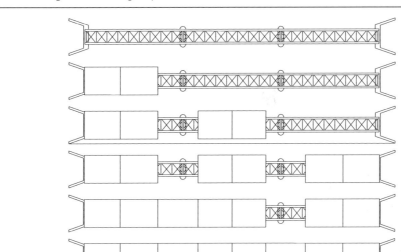

and spikes are embedded in the concrete and the spikes are extracted from the slab prior to repositioning the carriage. The front launch bogies are beyond the front bulkhead, and the rear bogies are behind the rear bulkhead or supported on the preceding slab segment.

Hybrid connection systems between the steel girder and the concrete slab have been used in many new-generation forming carriages. Headed stud connectors are combined with workshop-welded steel shapes that support the forming carriage during operations and launching (Figure 6.6). Crossfall shims are applied on top of the embedded steel shapes to support fixed launch bogies that set a horizontal rolling plane. The crossfall shims include a haunched bottom form to create a recess in the slab surface that is filled with non-shrink mortar after removal of the shim to create the concrete cover for the embedded shape. The forming carriages for fixed launch bogies are equipped with two full-length support frames that roll over the launch bogies during repositioning.

The overhead space frame of the carriage includes longitudinal trusses over the top flanges of the steel girder, transverse trusses at the end launch bogies and at the interior support legs, and lateral and diagonal bracing. Square tubes and bolted field splices are often used to stiffen the frame. The frame may support a covering and lateral enclosures to protect the casting cell in cold weather.

The casting cell includes a central form table and the shutters for the side wings. Hanger bars suspend the casting cell from the overhead frame during casting of the segment. During repositioning, the form table rolls on bogies applied to the lateral bracing of the steel girder. Segment geometry is repetitive, and the casting cell does not require major geometry adjustment.

When the concrete strength has been reached, the hanger bars are released and lifted, and the central form table is lowered onto the launch rolls and extracted. Forward extraction is used

Figure 6.5. Rectangular rails between lines of headed stud connectors for end launch bogies

Figure 6.6. Hybrid connection system supporting crossfall shims and fixed launch bogies (Photo: Doka)

for casting the next span segment, and backward repositioning for several spans is needed for casting the pier closures. The shutters for the side wings are suspended from the overhead frame and rotated to vertical prior to repositioning the carriage.

6.2.1   Loading, kinematics, typical features

The overhead frame is assembled on the ground at the most accessible pier and lifted onto the steel girder in one operation. The wing shutters are applied to the frame in a symmetrical sequence to minimise load imbalance, and the carriage is moved to the first casting location. Driving the carriage on multiple spans is a fast operation when the frame has hydraulic legs. If the carriage has full-length support frames for repositioning on fixed launch bogies, crossfall shims for 3–4 spans are necessary for the typical four-span casting sequence of the pier segments. The modules of the central form table are lifted on the launch rolls and connected to each other, and the form table is then moved under the carriage.

When equipped with hydraulic legs, the carriage is supported at the ends during repositioning and at multiple locations over the steel girder during segment casting. When equipped with full-length support frames, launch bogies on crossfall shims support the frame during launching and operations. The carriage is as light as possible for cost reasons and to minimise the load applied to the steel girder during operations. The carriage is stable longitudinally and does not require counterweights, but it is rarely anchored to the steel girder and overturning may therefore occur in the transverse plane.

The risk of overturning increases as the spacing of the steel girders reduces and the width of the side wings increases. Wide wings are often necessary with the U-girders for single-cell box girders to avoid two longitudinal field splices in the bottom flange plate (Rosignoli, 2002). Overturning is controlled by means of filling sequences for the casting cell that include casting the central strip between the flanges first, and filling the wings symmetrically afterwards. A U-girder is very flexible in torsion/distortion prior to achieving hollow-section behaviour, and symmetrical casting sequences are necessary also to avoid twisting the steel girder and risks of instability.

The segments are cast in one phase without cold joints; concrete is typically pumped from the ground (Figure 6.7). Concrete delivery on the deck requires multiple temporary bridges for the mixers to move over the solutions of continuity at the pier segments. On-deck delivery is often unavoidable when the bridge has tall piers on inaccessible terrains (Figure 6.8). Stiffer bracing or temporary steel bridges structurally connected to the steel girder are necessary at the solutions of continuity in the concrete slab in order to stabilise the steel girder.

6.2.2   Support, launch and lock systems

The concrete slab is cast directionally from abutment to abutment, although the casting sequence involves shuttling the forming carriage and central form table back and forth along the deck.

When the carriage is equipped with hydraulic legs, multiple adjustable spikes support the overhead frame during casting of the slab segments to minimise longitudinal bending in the frame and to spread the load applied to the steel girder. The spikes take support on the launch rails to avoid transverse bending in the flanges and interference with the stud connectors; the launch rails also facilitate load dispersal within the webs. Workshop-welded headed stud connectors are divided into two groups, and the launch rail is welded at the centre over the web. Motorised

Figure 6.7. One-phase casting of a 25 m span segment (Photo: Doka)

Figure 6.8. On-deck concrete delivery for a 30 m carriage for 80 m spans (Photo: Doka)

wheels drive narrow launch bogies on the rails, without interfering with the connectors; the flanges of the wheels are guided by the sides of the rail for transfer of lateral loads.

When fixed launch bogies are used in combination with embedded support shapes and crossfall shims, the support and launch systems of the carriage are above the slab and the geometry is less restrained. The launch bogies are equipped with lateral guides for the transfer of lateral loads, may be equipped with tie-downs for control of overturning, and also support the carriage during segment casting. Braced full-length support frames spread the casting loads along the carriage, but the load applied to the steel girder is more localised than with adjustable spikes. Redundant electric winches pulling haulage ropes anchored to the crossfall shims are typically used to move the carriage.

Hanger bars suspend the central form table and wing shutters from the overhead frame (Figure 6.9). Multiple support points for the shutters lighten the forms and carriage, but the large number of holes in the concrete slab increases the finishing costs. The hangers are applied during cage assembly; their presence complicates operations in the casting cell, and typically prevents cage prefabrication.

After releasing and lifting the hangers, the bottom form table is lowered onto the launch rolls and the wing shutters are opened in preparation for launching. The launch operations are different for the two types of carriage.

Figure 6.9. Internal view of a casting cell with multiple support spikes

1   When the carriage is equipped with hydraulic legs, long-stroke double-acting cylinders lodged within the legs lift the carriage, and the support spikes are extracted from the segment. During forward repositioning, the front legs roll on the steel girder and the rear legs roll on the concrete slab. Short rail segments supported on the slab include load-distribution plates and handles for the workers to pull them along the slab. When the rear bogie reaches the end of the rail segment, the closer support spike is shimmed with packs of steel plates, the hydraulic leg is retracted and the rail segment is pulled forward for a new cycle. During long-distance backward repositioning to cast the pier segments, both legs roll on rail segments supported on the concrete slab. When the legs reach the end of the segment, they are lowered onto the steel girder. PTFE plates are inserted within the packing shims, and the trolley is pushed backward on the low-friction support until the rear launch bogies cantilever out over the flanges.

2   Fixed launch bogies on embedded supports do not require lifting and lowering of the carriage, as the support frames are immediately ready for launching. Long-distance repositioning for casting the pier segments is faster and less labour intensive. On the other hand, embedded supports, crossfall shims, a great number of launch rolls, and finishing work for the concrete slab at so many support recesses involve additional costs.

6.2.3   Performance and productivity

A 20–30 m forming carriage with a PPU and hydraulic systems is not inexpensive. Its weight may be 0.20–0.25 ton per square meter of formed surface, and the costs of shipping, site assembly and decommissioning also depend on the weight of the machine. Although the contractor's depreciation strategy ultimately defines the cost share charged to the project, bridges shorter than 150–200 m are rarely compatible with such investments.

Productivity is one slab segment per week after the learning curve has been completed. Incremental launching of the steel girder is typically faster, and the slab casting rate governs construction duration (see Table 4.1). Incremental launching of the steel girder combined with in-place casting of the concrete slab with a forming carriage is one of the fastest construction methods for medium-span bridges (Figure 6.10).

Forward repositioning of the carriage and central form table takes a few hours, and geometry adjustment is very simple. Shuttling the carriage and central form back and forth along the deck, however, takes more time and disrupts the casting cycle. When the length of the bridge justifies the investment and the construction materials are delivered on the ground, a second carriage may be used for the pier segments (Figure 6.11). In this case, the pier segments are often shorter than the span segments. Fixed launch bogies on crossfall shims and embedded supports are the first-choice solution when two forming carriages are used on the deck.

6.2.4   Structure–equipment interaction

While the steel girder has stable elastic behaviour, the concrete slab is affected by time-dependent phenomena, and stress redistribution occurs within the structure over time.

In the absence of prestressing, cracking is practically unavoidable. The most critical tensile stresses are induced in the concrete slab during construction. Even when proper segmental casting sequences are adopted, the permanent loads resisted with composite action cause longitudinal tensile stresses in the support regions of the continuous beam. Live loads increase negative bending, but longitudinal tension is often present even in the midspan regions due to drying

Figure 6.10. Incremental launching of the steel girder combined with in-place casting of the concrete slab with a forming carriage

Figure 6.11. A 25 m carriage for span segments combined with a 15 m carriage for pier segments (Photo: Doka)

and thermal shrinkage, early creep of concrete, and the low compressive stresses resulting from composite action. Finally, longitudinal cracks over the entire bridge length are frequently found in the slab regions over the steel girders in the case of long side wings.

Slab cracking affects the concrete cover to the top reinforcement. A cracked slab is subjected to the adverse effects of water and de-icing salts, and this condition is particularly severe when the wearing surface is integrated with the slab. A protective asphalt layer is used in most cases, but perfect water-tightness is difficult to ensure, even when a membrane is properly installed between the asphalt and the concrete slab. Polyester concrete overlays directly applied to a roughened concrete slab are finding increasingly frequent application in the USA. Some design standards stipulate an increased concrete cover to the top reinforcement to provide for cracking and deck wear, but this is inconsistent with the practice of placing the reinforcement near the surface to control crack width. When the concrete slab is designed correctly, however, crack widths up to 0.2–0.3 mm are typically unable to cause corrosion of reinforcement.

As slab cracking is mainly due to flexural stresses and concrete shrinkage, several design standards provide instructions on both these aspects (BSI, 1991; CSA, 2000). The use of minimum reinforcement ratios in the slab regions where the tensile stresses exceed the tensile strength of concrete is often required. The minimum reinforcement ratio is often set to 1% of the gross cross-sectional slab area, this being increased to 1.3–1.5% in the support regions of the continuous beam. In these regions the tensile stress in longitudinal reinforcement is limited to 60–70% of the serviceability limit state stress limit to further control crack width. Additional provisions in the standards designate the quality and compactness of the concrete, the W/C ratio and the use of admixtures, as all these parameters influence both the tensile strength and the permeability of concrete.

To limit permanent tension and related cracking, segmental slab casting is often prescribed according to a discontinuous sequence in which the span segments are cast first, followed by the pier segments. This casting sequence reduces, or avoids, self-weight tensile stresses related to the casting sequence.

Complex step-by-step time-dependent analyses of the segmental casting sequence of the concrete slab are necessary because, when a new segment is cast in a span, cracking tends to occur at the end of the hardened segment in the previous span. The weight of the forming carriage and the fluid concrete applies a significant load to the steel girder, and camber design is further complicated by the creep differential between slab segments of different age. Time-dependent stress redistribution within the composite structure is analysed using an effective width of the concrete slab that is compatible with the expected shear-lag effects. Shell elements are used for the concrete slab in the most delicate cases.

Without preventive measures, the concrete slab tends to develop cracking over time. After hardening, there are at least five causes of cracking in service conditions: external loads, creep, drying shrinkage, external temperature variations and thermal shrinkage. Most of these phenomena are well known; thermal shrinkage of concrete, on the contrary, is a specific problem of composite bridges, is typically disregarded by the design standards, and may be a prime cause of permanent cracking.

During cement hydration, the temperature in the concrete slab increases by 15–30°C during the first 12–24 hours. This heating phase is followed by a cooling period of 150–180 hours. The development of the mechanical strength of the concrete is governed by cement hydration, and the elastic modulus of concrete is therefore higher in the cooling period than in the heating phase.

If the concrete slab and the steel girder act compositely from the moment the concrete sets, then this composite action prevents the expansion of concrete during the heating phase and its contraction during the cooling period. The restraint that the steel girder exerts on the concrete slab can be modelled by assuming constant but different values for the elastic modulus of concrete during the heating and cooling periods. Restrained thermal expansion during the heating phase generates compressive strains in the concrete that are recovered during cooling. The difference in the elastic modulus of concrete in these two phases produces permanent tensile stresses that are high when compared to the tensile strength of the young concrete.

The restraint action exerted by the steel girder can be represented by the ratio β of the cross-sectional areas of the steel girder, A_S, and of the concrete slab, A_C:

$$\beta = \frac{A_S}{A_C}$$

The current trend to reduce labour costs at the expense of increasing the quantity of steel, and with it the restraint action that the steel girder exerts on the concrete slab, increases the practical relevance of the effects of thermal shrinkage (Rosignoli, 2002).

Typical values of β for recent twin-girder bridges increase from 0.04 to 0.12 when the span increases from 30 to 80 m. Laboratory tests and numerical analyses show that the residual tensile stress σ_r in the concrete slab after hydration varies in the range 0.5–0.8 MPa for cross-sections with $\beta = 0.05$, 1.2–1.5 MPa for $\beta = 0.08$, and 1.6–2.0 MPa for $\beta = 0.12$.

These tensile stresses are significant when compared with the tensile strength of young concrete (2.0–2.5 MPa) and can lead to full-depth cracking of the concrete slab at an early age. They also reduce the available tensile strength of concrete, f_t, and an effective tensile strength, $f_{t,eff}$, should be used when considering the stiffness of the composite section:

$$f_{t,eff} = f_t - \sigma\rho$$

A simplified relationship between the residual tensile stress in the concrete slab and β can be expressed using the following equation:

$$\sigma_r = \frac{\alpha\beta^2 E_S^2 (E_{C.h} - E_{C.c})}{(\beta E_S + E_{C.h})(\beta E_S + E_{C.c})} \Delta T$$

where α is the coefficient of thermal expansion of concrete, $E_{C.h}$ and $E_{C.c}$ are the mean values for the elastic modulus of concrete during the heating and cooling periods, E_S is the elastic modulus of the steel girder, and ΔT is the maximum difference between ambient and concrete temperature during cement hydration.

A qualitative evaluation of the influence of β on the effects of thermal shrinkage in the concrete slab leads to the following conclusions. Hydration effects have limited influence on early cracking for $\beta \leq 0.05$. For $0.05 < \beta \leq 0.08$ thermal shrinkage reduces the effective tensile strength of concrete and produces a limited risk of early cracking. For $0.08 < \beta \leq 0.12$ hydration effects reduce the effective tensile strength more markedly, early cracking is probable, and actions for reducing the residual tensile stresses should be considered. For $\beta > 0.12$ hydration effects significantly reduce the effective tensile strength, high risks of early cracking arise, and actions for reducing the residual tensile stresses should be adopted.

Possible corrective actions are all addressed to controlling ΔT. They involve reductions in the cement content (provided that cement quality is reasonably constant) and in the W/C ratio, the use of low-heat cement, cooling of fresh concrete or during hydration by cooling pipes, and slight heating of the steel girder. Laboratory tests have shown that both the use of low-heat cement and concrete cooling can limit the residual tensile stress in the concrete slab to less than 1.0 MPa, even in the case where there is a high level of restraint ($\beta = 0.12$).

For technical and economic reasons, however, most of these options are impractical, and the classic solution consists of a careful design of passive reinforcement. In this case, full-depth cracks will open, and it will almost always be impossible to keep their width below 0.1 mm. Suitable detailing of bar diameter and distribution usually permits limiting the crack width to 0.15–0.20 mm, but crack width may increase over time because of progressive bond deterioration at the crack boundaries due to fatigue effects.

In the presence of diffuse full-depth cracks, the first protection of the concrete slab is the use of a waterproof membrane of good quality, firmly anchored to a sound concrete cover with low permeability, and perfectly sealed at the joints and at the lateral kerbs. A waterproof membrane controls the two main causes of most chemical and physical processes affecting the durability of structures: transport within pores and cracks, and water. Freeze–thaw cycles, the effects of de-icing salts, penetration of chemically aggressive agents, alkali–silica reactions, concrete carbonation and reinforcement corrosion all are related to these two causes.

However, the stability of the protective action of a waterproof membrane over time is a basic concern. Despite all precautions, sooner or later de-icing salts will cause chloride ions to penetrate into concrete. If the slab is cracked the rate of penetration is very high, and depassivation occurs quickly. In these circumstances, structure durability is controlled by corrosion development. Service life values of 20–40 years are sometimes mentioned, but this needs to be confirmed by further research.

The use of precast deck panels avoids most of the effects of thermal shrinkage because the hydration heat of the closure pours dissipates within the adjoining panels. The launched slabs offer the same advantages along with a much smaller number of construction joints.

REFERENCES

AASHTO (American Association of State Highway and Transportation Officials) (2012) *LRFD Bridge Design Specifications*. AASHTO, Washington, DC, USA.

BSI (1991) BS 5400: Part 3: 1982 Steel, concrete and composite bridges. Part 3. Code of practice for design of steel bridges. BSI, London.

CSA (Canadian Standards Association) (2000) CSA-S6-00. Canadian highway bridge design code. CSA, Toronto, Canada.

IABSE (International Association for Bridge and Structural Engineering) (1997) Composite construction: composite and innovative, International Conference, Innsbruck.

Rosignoli M (2002) *Bridge Launching*. Thomas Telford, London.

Bridge Construction Equipment
ISBN 978-0-7277-5808-8

ICE Publishing: All rights reserved
http://dx.doi.org/10.1680/bce.58088.185

Institution of Civil Engineers

publishing

Chapter 7
Self-launching equipment for balanced cantilever construction

7.1.   Technology of precast segmental balanced cantilever erection

Balanced cantilever construction is suited to precast segmental and cast in-place bridges. The deck is erected from each side of the pier in a balanced sequence to minimise load imbalance. This method is particularly advantageous on long spans and where access beneath the deck is difficult.

Balanced cantilever bridges typically have box section. Single-cell box girders may reach 18–20 m width, and transverse ribs, diagonal struts and combinations of ribs and struts have been successfully used to further widen the top slab. Ribbed slabs with a double-T-section have also been used in the past, although the moment of inertia of the cross-section is smaller and the centre of gravity is closer to the top slab, which complicates the transmission of the high negative moments that characterise this construction method. Prestressed concrete (PC) ribbed slabs comprising edge girders and crossbeams are often used for balanced cantilever erection of cable-stayed bridges because most of negative bending is resisted by the stay cables.

The deck may have constant or varying depth. Constant-depth segments are easy to cast, have been used on 30–90 m balanced cantilever spans, but are mostly competitive on 45–70 m spans where incremental launching and composite bridges are often less expensive and faster to erect. Constant-depth precast segmental box girders with trussed webs have also been erected as balanced cantilevers (Figure 7.1).

Varying-depth precast segmental box girders have been used on 80–180 m spans and are mostly competitive on 100–120 m spans (ASBI, 2008). The depth of the cross-section and the thickness of bottom slab and webs vary throughout the cantilever to calibrate the flexural and shear capacity to the demand of cantilever construction; the hammerhead segments, however, soon become too tall for ground transportation and too heavy for lifting, and spans longer than 120–130 m are typically cast in place.

Balanced cantilever erection of precast segmental bridges requires heavy lifters. The longer the bridge, the more powerful the erection equipment, and heavier and longer segments may be used. Segment erection using ground cranes, floating cranes or lifting frames on the deck permits free construction sequences, while the use of self-launching gantries is typically associated with linear construction from abutment to abutment. Linear construction requires pier availability at the right time, but the segments may be delivered on the deck, which accelerates construction and minimises the impact of ground constraints.

Precast hammerhead segments are mostly supported on bearings, and stability of the hammer during erection is an important consideration. The hammerhead segments should have the

Figure 7.1. Balanced cantilever erection of constant-depth precast segments using trussed webs

same weight as the other segments so as not to require special lifters. The hammerhead segments, however, include thick pier diaphragms that transfer deviation forces of external continuity tendons and self-weight shear from the webs to the bearings. The bottom slab is also thick due to the longitudinal compression applied by the cantilevers. In most cases, therefore, the hammerhead segments are thin and tall.

Thin segments facilitate placement as the front auxiliary leg of the gantry is also supported on the pier cap (Figure 7.2); however, thin, tall and heavy segments may be unstable during transportation and placement. The area beneath the bridge is rarely flat and the need for the hauler to climb slopes can destabilise the segments. Ground conditions worsen in wet weather, and the need for a solid road base from the precasting facility to the erection site can significantly increase the cost of site setup (Harridge, 2011).

Some hammerhead segments have been split horizontally to increase the width and reduce the height. This complicates prefabrication but simplifies delivery and placement. The bottom segment is placed on four jacks at the corners of the pier cap, the cap segment is placed onto it, and the front support crossbeam of the gantry is placed on the cap segment (Harridge, 2011). During these operations, the front auxiliary leg of the gantry is supported on brackets anchored to the front face of the pier.

Pier tables comprising three thin segments joined with longitudinal tendons or prestressing bars have also been used. The assembly is supported on jacks or shim packs while it is bedded on the bearings. If the bearings have dowels on the top plate, recesses are formed in the segments

Figure 7.2. Placement of the up-station starter segment of the hammerhead assembly (Photo: Comtec)

at the dowel locations and the gap is filled with mortar. If the top bearing plate has no dowels, the plate is grooved and the gap is filled with epoxy mortar (Hewson, 2003).

Jacks at the corners of the pier cap combined with vertical post-tensioning bars that anchor the hammerhead to the pier cap provide fast and accurate geometry adjustment and effective initial stabilisation of the hammer, but the limited flexural capacity of the connection (the bars are typically stressed to 50% of the tensile strength to be reused) is suitable for the load imbalance of only 1–2 cantilever segments on either side (ASBI, 2008). Tie-downs are often sufficient when an erection gantry stabilises the hammer or applies the segments simultaneously to both ends of the hammer to minimise load imbalance.

When the pier is narrow, steel brackets may be used to support the jacks and to anchor the tie-downs with a longer lever arm. Higher tie-down forces can be attained by using U-tendons embedded into the pier. These connection systems are flexible, complicate geometry control during cantilever erection, and cause discomfort to personnel on the hammer (Harridge, 2011).

Temporary pier towers from foundations, single tension/compression props, double compression props and auxiliary tension ties provide more flexural capacity, but are cost-effective only with short piers (ASBI, 2008). They are typically necessary when hammers supported on bearings are erected with bidirectional lifting frames because of the large load imbalance of staged construction.

187

As an alternative to precasting, the pier table may be cast in place at the end of pier erection, with a special casting cell or within precast shells. In-place casting reduces the design demand on transportation and erection equipment, which is often governed by the hammerhead segments. A thicker pier diaphragm also facilitates the development of heavy column reinforcement in the pier table in seismic regions.

Thin, non-reinforced, wet joints are used between the pier table and the starter segments to compensate for the geometry tolerances of in-place casting. The use of wet joints lengthens the cycle time of the span by a couple of days, as the erection equipment hangs the segments in place during curing. Geometry control is a basic requirement for cast in-place pier tables, especially with regard to the duct alignment of the cantilever tendons. Additional space for duct coupling across the closures can be provided by including small blockouts on either side (ASBI, 2008).

The use of precast shells is a hybrid between precasting and in-place casting. Two outer shells of the pier table are match-cast against the starter segments of the cantilevers and erected over rebar protruding from the pier. After cage completion, the pier diaphragm is cast using the outer shells as a portion of the formwork. Precast shells offer the stability of monolithic connections with the piers and the erection rapidity of match-cast joints with the starter segments (ASBI, 2008). The drawbacks include a large number of rebar couplers for the diaphragms and the need for accurate positioning of the shells to set a correct alignment for the cantilever.

The length of the segments increases along the cantilever to balance the weight. The length and weight of midspan segments are governed by handling and transportation requirements. Lengths up to 3.5 m are often transportable on public roads without excessive restrictions.

Cranes, gantries and lifting frames hoist the segment with a single central lift. Slotted holes were used for the slings in the first-generation lifting beams to hoist the segment at the crossfall required to align shear keys and gluing bars with the previously erected segment. Handling of segments is faster and more effective when pull cylinders or PLC-controlled synchronous lifting systems are applied to the slings. New-generation lifting beams are equipped with a tilt cylinder that rotates the segment in the transverse plane for alignment.

One or two cantilever tendons per web are anchored at every joint, depending on the maximum tendon size, top slab thickness, interference with web stirrups, and anchorage protection (ASBI, 2008). Detailing of tendons, anchorages and alignment keys in the top slab is standardised as much as possible. Top slab blisters are typically avoided for the cantilever tendons for reasons of weight and cost, but they are often unavoidable at the tip of the cantilever when the midspan closure is too narrow to apply the stressing jacks to the strand tails extending from the anchor heads. Top and bottom internal continuity tendons are always anchored in blisters at the web–slab nodes as there is no access to the segment face.

The segments become lighter towards the tip of the cantilever, and variations in the stressing load of the gluing bars are often necessary to ensure even pressing of the epoxy joints (ASBI, 2008). Bottom edge compression is maintained after tensioning the cantilever tendons to avoid tensile cracking in cured joints and poor adhesion and gaps in the uncured joints close to the tip of the cantilever due to insufficient weight of the end segments. Uneven pressing may cause varying joint thickness, which can affect the vertical alignment of the cantilever after several segments.

Small geometry inaccuracies accumulate when long deck sections are erected, and some correction to segment alignment may be necessary. This is achieved with shims of woven glass fibre embedded in the epoxy to increase the thickness of the joint. Although this technique achieves only a small angle break at any joint, the effect is amplified as subsequent segments are built on. Joint shimming is a last resort, because it causes stress concentration and prevents proper joint closure, which may cause voids, insufficient filling, penetration of epoxy into the ducts, and grout crossover (ASBI, 2008).

Permanent cantilever tendons are installed for most segments, and temporary gluing bars are therefore necessary only at the frontmost 3–4 segments of the cantilever during erection. When continuous coupled bars are used, bars, couplers, nuts and anchor plates are removed prior to applying the last segment of the cantilever, as it is often impossible to dismantle them after midspan closure.

Midspan closure pours are used in modern bridges to provide continuous connections between adjoining hammers. The new up-station cantilever is secured to the completed bridge with strong-backs bridging the closure gap. The strong-backs are anchored to the top slab of the first two segments on either side with shims and stressed bars housed in the lifting holes of the segments. The strong-backs are designed for the weight of forms and concrete for the closure pour, the residual load imbalance, the realignment forces for the cantilevers, the flexural effects of thermal gradients in the deck, and the friction resistance of the sliding bearings at the pier caps.

The closure pour is typically 0.5–1.5 m long. The weight of long closure segments may cause deflections and rotations of the closure joints as well as pier cap rotations in the leading hammer and in the two starter hammers of the bridge. For the precast segments, these deformations are compensated for by the casting curve.

The joint surfaces must be clean, free of laitance and roughened to expose coarse aggregate. AASHTO (2003) specifies 6 mm roughness, which is achieved by bush-hammering or chipping after sandblasting. Application of bond enhancer maybe specified to increase adhesion. The closure segment is cast directionally from one joint toward the other to avoid joint cracking in the bottom slab. When the segment is very long, two-phase casting with partial application of continuity prestressing prior to casting the top slab may be necessary.

Long viaducts may require expansion joints to alleviate the effects of temperature and time-dependent shortening. Different solutions are available for the expansion joints (ASBI, 2008) in combination with sliding bearings, monolithic connections at the piers, leaf piers that provide moment connections with minimal longitudinal shear stiffness.

1   Expansion-joint segments are used at the quarter of the span to transfer shear and to allow longitudinal movement and rotation. These joints rely on their location at the counterflexure point to minimise the angle breaks under live loading. The two halves of concrete seat-type hinges are blocked together during cantilever erection. Temporary top-slab tendons are installed through the hinge segment to erect the remainder of the cantilever. After midspan closure, the temporary tendons are removed and the hinge is released.

2   Midspan expansion joints are used to transfer shear and bending and to allow longitudinal movement. Continuity is achieved with two steel box girders that cross the joint within the box cell and which are anchored on either side. Upon completion of the first hammer,

the girders are lifted into position and inserted into the box cell. Upon completion of the adjacent hammer, the girders are pushed back through the expansion joint and secured in their final position. The weak points of this solution are the weight and complexity of the diaphragm segments, the difficult maintenance of bearings and steel girders, and the impossibility of replacing primary structural steel members located under a roadway expansion joint in case of leaking and corrosion.

3 Pier expansion joints are used to support the ends of continuous frames. Continuous frames require long end spans to avoid uplift at the end bearings under live loading. A pier table at the expansion joint is erected first. The two end segments are blocked and secured together to create a monolithic pier table, and the hammer is extended on either side with segments equipped with ducts for temporary cantilever tendons. After erection of the adjacent hammers and their connection to the expansion-joint hammer, the prestressing tendons through the joint are released and the blocking removed to release movement.

The most common erection methods for precast segmental balanced cantilever bridges are with ground cranes, floating cranes, lifting frames or light cranes on the deck, and self-launching gantries (Homsi, 2012). Bridge length and width, span length and sequence, plan alignment, pier geometry, access at ground level, and availability of second-hand equipment govern the choice between the different options. Availability of assembly areas, overhead obstructions such as power lines and overpasses, and the time for shipping, customs clearance, trucking, site assembly and commissioning of special erection equipment are additional factors.

- Ground cranes often give the simplest and most rapid erection procedures with the minimum of temporary works, but require good access to the deck throughout the bridge (Hewson, 2003). Crane erection imposes the lightest construction loads on the bridge, is compatible with tight radii and steep gradients, and the cranes are immediately available at the end of hammer erection. Rental cranes are readily available in many countries, and multiple hammers can be erected at once. The erection speed is typically 2–4 segments per day per crane. The main constraints on crane erection are access and tall piers, as balanced cantilever bridges are often selected in response to inaccessible terrains.
- Floating cranes are used frequently for low-level bridges over water. Access to the bridge must be maintained throughout the erection phase, and this can be expensive due to the cost of barges and tugboats. The erection schedule depends on weather and sea conditions, docking facilities are needed to load the barges, and personnel and material delivery are less efficient than with construction methods that offer better access conditions (Homsi, 2012). The cranes are sized to erect the hammerhead segments and to work under the tips of the cantilevers. Special care is required for placement of the midspan segments due to the handling requirements of working alongside the bridge. Midspan closure and the application of continuity prestressing are not on the critical path, and the cranes can be used to support multiple activities throughout the project.
- Commercial cranes are rarely used on the deck because they require counterweights for slewing and are therefore heavy. Light cranes have been used on the deck to handle the light segments of LRT bridges on short hammers, without major impacts on top-slab prestressing and load imbalance in pier cap stabilisation gear, piers and foundations. Loading the segments on the deck with a ground crane may reduce the need for counterweights and shorten the cycle time.
- Lifting frames are used on tall piers, long or curved spans, spans of different length, and spans over water where the lifters can handle heavier segments and barge delivery

minimises geometry and weight constraints. Lifting frames and derrick platforms are the standard solution for the erection of cable-stayed bridges; they are relatively inexpensive, but must be lifted on the pier table and removed from the deck prior to midspan closure, which disrupts the erection cycle. Hoisting segments over significant heights is time consuming and can affect the cycle time.

■ Self-launching gantries offer the fastest erection rates and the highest productivity of personnel; they also minimise ground disruption when the segments are delivered on the deck. Gantries are fit for operating over water or other such obstructions; however, they erect the hammers directionally from abutment to abutment and are delayed if problems occur at any pier or span. The gantries are able to reposition the pier supports, to erect the hammerhead segments, to stabilise the hammers during erection, and to position the stressing platforms for fabrication and tensioning of tendons (Harridge, 2011). The gantries are also used for access of personnel to the working areas, for delivery of materials, and as overhead service cranes for most operations.

7.2.   Lifting frames

The lifting frames are heavy lifters anchored to the deck to avoid the use of counterweights. These machines are simple to use and maintain and less expensive than commercial cranes of the same hoist capacity; they are also much lighter but, because of the absence of counterweights, they must be anchored to the deck at every working location (Figure 7.3). Lifting a commercial crane on the pier table requires powerful ground cranes, and the crane may require modifications to operate

Figure 7.3. Placement of the hammerhead segment using a lifting frame (Photo: Strukturas)

with the outriggers partially extended to load the deck over the webs. On the other hand, however, a commercial crane is a multi-purpose machine that can be resold or reused more easily than a lifting frame.

Some lifting frames have been designed to be assembled on the pier table with a few lifts of a tower crane anchored to the pier. Others are not much lighter than the segments they handle and require large cranes for relocation in one lift, but the frame can be repositioned on a new pier table in a few hours (Homsi, 2012). Heavy lifters for cable-stayed bridges are sometimes so heavy as to require strand jacking into position from pylon brackets. There are several categories of lifting frames.

- Pier-table lifting frames are used to strand jack into position entire PC spans delivered on barges or cast on the ground under the bridge. The pier tables are cast in place at the end of pier erection and secured to the pier caps. After lifting, the spans are connected to the pier tables with closure pours. These lifting frames have strand jacking platforms on either side. During lifting of the span, the front hoist cables of the leading frame are anchored to the pier foundations and stressed to minimise bending in the leading pier.
- Fixed lifting frames overhanging from the tip of the cantilever have limited load-handling capability and are used with PC and steel bridges when the segments are delivered on barges. Each segment must be delivered immediately under the final position, which may be demanding on shallow water or where there are land obstructions. Fixed lifting frames are also used to strand jack drop-in closure spans of steel portal frames.
- Self-launching derrick platforms are fixed lifting frames equipped with a rotating arm that can lift loads from behind or laterally. They have a large manoeuvring area but apply significant torque to the deck (Hewson, 2003). Derrick platforms are often used to erect composite cable-stayed bridges because edge girders, crossbeams and precast deck panels are handled individually, and the stay cables resist the applied torque.
- Mobile lifting frames pick up the segment on the ground close to the pier, lift it up to the soffit of the hammer, move it out beneath the cantilever, and are anchored to the tip of the hammer to cantilever out the segment and lift it up to the deck level.
- Bidirectional lifting frames are wheeled straddle carriers equipped with long cantilever noses on either side that support the runway for a winch trolley. These lifting frames load the segments from the ground or the deck, move them throughout the hammer, and erect them on either side (ASBI, 2008).

Lifting frames are typically used for balanced cantilever erection of precast segmental bridges with tight plan or vertical radii, for spans longer than 100–120 m that prevent the use of a self-launching gantry, or when the small number of spans does not justify the investment in a gantry. Lifting frames may also be used to load precast segments onto an underslung gantry for span-by-span erection; the segments are delivered on the ground or the deck, and the lifting frame is anchored to the leading end of the deck to feed the gantry.

Fixed lifting frames are frequently used to erect high-level cable-stayed decks over water (Gimsing and Georgakis, 2012), while low-level decks are erected more efficiently and with less load imbalance by using floating cranes. Heavy lifters are required for PC or steel deck segments, while lighter machines are used to handle members of composite grillages individually. Compared with the use of ground or floating cranes, the weak points of the lifting frames are the single-operation nature of these machines, their relative slowness, the limited load handling

Figure 7.4. Lifting of the hammerhead segment with a temporary erection frame (Photo: Comtec)

capability, and the disruption of the erection line when moving to the next pier after completion of the hammer.

Many lifting frames can be easily separated into a few major components. They are commissioned on the ground and placed on the pier table with a few lifts of a ground or floating crane. Temporary erection frames may be used on the pier cap to lift the hammerhead segments with the winch trolley of the lifting frame (Figure 7.4). After completion of the pier table, the erection frame is removed, the cart of the lifting frame is lifted onto the pier table, and the winch trolley is placed on the cart to complete the assembly.

Fixed lifting frames are anchored to the tips of the cantilevers and move out during deck erection, while mobile lifting frames transport the segments along the cantilevers. A fixed lifting frame is geometrically similar to an overhead form traveller for in-place casting; less stringent deflection requirements result in a higher payload-to-self-weight (P/W) ratio although lifting accessories reduce the nominal hoist efficiency of the crane. A lifting arm carries the top sheaves of the crane and is hinged to a base frame that is anchored to the deck over the webs. Wind bracing stiffens the base frame, post and lifting arm.

The hoist winches are installed at the rear end of the base frame to increase stability from overturning during repositioning. A long base frame is used to minimise the pull in the rear tie-downs and, by equilibrium, the load that the front supports of the frame apply to the deck. Cables with double-acting adjustment cylinders, reeved ropes or stand tendons connect the rear

Figure 7.5. Forward launching of the rail segments (Photo: SDI)

end of the base frame to main post and lifting arm for longitudinal movement of the payload at the end of lifting. Transverse movement capability is very limited. A self-launching underslung working platform is often used in the steel bridges to connect the new segment to the tip of the cantilever.

Fixed frames lift the segments from the ground or barges and have limited load-handling capability. The segment must be placed under the hoist for lifting, which is often easy to do with barges but often impossible in shallow water or where there are ground obstructions (Homsi, 2012). The fixed frames are among the least expensive lifters and have an excellent P/W ratio; they are assembled on the pier table using the same methods as used for the overhead form travellers, and are repositioned using self-launching rail segments anchored to the deck (Figure 7.5). The lifting frame is secured directly to the deck with rear tie-downs during operations, and during repositioning is secured indirectly through the anchorages of the launch rails.

Custom lifting frames have been designed for simultaneous erection of adjacent hammers (VSL, 2008). The common winch trolley has a long crane bridge spanning between the two hammers, and the hoist crab shuttles back and forth along the crane bridge to serve the two bridges (Figure 7.6).

A mobile lifting frame comprises a bottom cart that supports two longitudinal runway girders for a winch trolley. The winch trolley has wide side overhangs to hoist a spreader beam wider than the segment. The spreader beam is permanently connected to the winch trolley and is lifted at the ends to avoid interference between the hoist ropes and the side wings of the previously erected segments (Figure 7.7). Because of the demanding load geometry of crane bridge and spreader beam, mobile lifting frames are typically heavier than fixed frames.

Figure 7.6. Fixed lifting frame for simultaneous erection of adjacent hammers (Photo: VSL)

Figure 7.7. Mobile lifting frame (Photo: VSL)

The segment is picked up at the base of the pier and moved out beneath the cantilever. When it is not necessary to transport the segments along the cantilever, a mobile lifting frame may be operated at the tip of the cantilever like a fixed frame.

A mobile lifting frame is lighter than a commercially available crane with the same hoist capacity as it is devoid of counterweights and unable to slew and to tilt the boom. The counterweights required for a crane to balance the lifting couple often exceed one-half of the operational weight of the crane, and the weight difference is therefore significant. The light weight simplifies placing the lifter on the pier table at the beginning of cantilever erection and removing it from the hammer prior to midspan closure. It also diminishes load imbalance and top-slab prestressing, and the high P/W ratio results in high hoist efficiency. The absence of counterweights, however, requires anchoring of the frame to the deck before cantilevering out the segment.

During segment transportation, the winch trolley is blocked at the centre of the machine for longitudinal stability (Figure 7.8). After reaching the tip of the cantilever, the lifting frame is anchored to the deck, the stressing platform is detached from the hammer and moved out to the tip of the runway girders, the winch trolley is moved out to lift the segment into position, and the stressing platform is applied to the new segment. A second lifting frame performs the same operations on the opposite side of the hammer.

Most mobile lifting frames roll on rails anchored to the deck over the webs; some new-generation frames are equipped with tyres and steering wheels. These frames cannot be used on cable-stayed decks because the side overhangs of the winch trolley interfere with the cables. The rails can be full length on the hammer, or short rail segments can be launched with double-acting cylinders as

Figure 7.8. Segment transportation along the cantilever (Photo: Comtec)

Figure 7.9. Segment lifting with a bidirectional lifting frame and slewing lifting beam (Photo: Deal)

employed for fixed frames and form travellers. The mobile frames shuttle back and forth along the cantilever; short self-launching rails are slow to reposition, and fixed full-length rails are often preferred. Tyre frames offer unrivalled mobility. Compared to a fixed frame, a mobile lifting frame can pick up the segments at any location along the cantilever.

Bidirectional lifting frames can transport precast segments below or above the deck. The runway girders have long cantilevers on both sides and the winch trolley is equipped with a hydraulic slew ring to rotate the segment. The lifting frame picks up the segment at either end of the hammer (Figure 7.9); the segment may also be loaded on the hammer with a ground crane for double handling. The lifting frame is anchored to the deck, picks up the segment, lifts it up to the deck level, rotates it by 90° if necessary, and inserts the segment longitudinally into the cart body. The lifter is driven to the other end of the hammer and re-anchored to the deck, the stressing platform is removed from the hammer and moved out to the end of the runway girders, and the segment is extracted from the cart body, rotated by 90° and lowered into position, (Figure 7.10). The stressing platform is finally applied to the new segment.

These lifting frames are used when the segments can be delivered only on one side of the pier; they shuttle back and forth along the hammer, and steering straddle carriers are typically used for the

Figure 7.10. Segment rotation and placement (Photo: Deal)

cart body for faster operations. First-generation bidirectional frames were heavy and operated on rails anchored to the deck. Long overhangs are necessary in the runway girders on both sides of the cart to extract the segment from the cart prior to rotation and to suspend the stressing platform during this operation. Winch trolleys with side overhangs and wide lifting beams permit segment transportation below the cantilever.

A pair of fixed frames simplifies deck erection, but the segments must be delivered under the lifters. A pair of mobile frames facilitates segment delivery, but both sides of the pier must be accessible. A bidirectional frame requires a smaller pier table for assembly and provides unrivalled load handling capability, but the cycle time is longer and a ground crane is often necessary to load the segments on the hammer for double handling and faster erection.

7.2.1 Loading, kinematics, typical features

A fixed lifting frame is supported on the front segment of the cantilever, and tie-downs at the rear end prevent uplift and overturning during lifting of the segment. The hoist winch is applied to the rear end of the base frame for enhanced stability during operations and launching. The PPU may be separated from the lifter to be used for different operations during segment connection to the cantilever. High-tonnage double-acting cylinders with mechanical locknuts at the front supports and lighter rear support cylinders are used to set the base frame horizontal on bridges with grade and crossfall. The launch bogies are lifted from the rails during levelling and the rear tie-downs are stressed at the end of the operation.

After connecting the new segment to the cantilever, long-stroke double-acting launch cylinders push the launch rails forward over the new segment. The rails are anchored to the deck and the support cylinders are retracted to lower the launch bogies onto the rails. When the overturning couple of the unloaded machine prevails over the stabilising couple of PPU and rear hoist winch, tie-down bogies at the rear end of the base frame roll within the launch rails to prevent overturning. The cylinders used to push the rails forward are also used to pull the main frame along the rails to the next lifting location. The rear end of the base frame is secured directly to the deck during segment lifting, and during repositioning of the lifting frame is secured indirectly through the launch rails. The launch rails are located over the deck webs for direct load transfer.

The mobile frames typically roll on full-length rails and are devoid of tie-down bogies. The winch trolley is blocked at the centre of the cart during frame transfer to prevent instability. The support legs are equipped with vertical double-acting cylinders with mechanical locknuts to set the runway girders horizontal on decks with gradient and crossfall. Tie-downs are stressed to anchor the support legs to the deck prior to releasing the blocks of the winch trolley for segment placement. A PPU energises hydraulic pumps that supply drive motors, support cylinders, the hoist winch, shift cylinders for the crab and auxiliary systems. The PPU, hydraulic pumps and fluid tanks are installed on the rear side of the frame to balance the overturning moment and to improve stability.

The bidirectional lifting frames are light and sophisticated wheeled straddle carriers that provide fast movement along the hammer. A single PPU typically drives a hydraulic system to provide power to the hoist, drive, steering, braking, fan drive and charging pumps. Pumps and motors are controlled with proportional solenoid valves. Hydraulic lines and wiring are often installed internally to the frame structure for containment and protection from loads and sunlight.

An independently controlled steering cylinder is mounted on each wheel. Redundant internal position transducers provide cylinder position feedback to the steering computer. Some of these lifters offer up to four steering modes: four-wheel coordinated steering, two-wheel (front) automotive steering, two-wheel (rear) forklift steering, and four-wheel parallel (crab) steering. The steering computer synchronises all wheels during manoeuvring and virtually eliminates tyre scrub due to steering geometry errors. Enhanced manoeuvrability facilitates positioning of the support legs over the ducts of the tie-down bars for anchoring.

Double-acting support cylinders lift the cart from the deck and compensate for gradient and crossfall before positioning the segment. Integrated cylinder position sensors provide position feedback for closed-loop automatic cylinder synchronisation. Variable displacement pumps supply pressure and flow to operate the support cylinders. Pressures and temperatures in the hydraulic systems are monitored by control valves.

7.2.2 Support, launch and lock systems

The new-generation lifting frames are devoid of counterweights. Support cylinders set the lifter horizontal for segment placement and tie-downs control overturning. Because of their critical role in ensuring stability of the lifter, the tie-downs are typically oversized and redundant. Tie-down bars cross the top slab within ducts, are anchored to the bottom surface of the slab, and are stressed to provide frictional shear capacity and to prevent involuntary movements. The ducts are oversized for placement tolerance and to avoid shear transfer by contact. Tie-down bars must not be bent, earthed during welding, welded or heated. Damaged bars must be marked as such and disposed of immediately.

Launching a fixed lifting frame is a two-step process: first the rails are pushed forward for the entire new segment length and anchored to the deck, and then the frame is lowered on the rails and pushed forward over the rails. The lifting frame and rails are launched using the same set of long-stroke double-acting cylinders. The launch cylinders push the front bogies forward and are repositioned alternately during launching to avoid uncontrolled movements of the lifter. Tie-down bogies at the rear end of the frame roll within the launch rails to control uplift and overturning. The uplift force in the rear bogies is smaller than the tension in the tie-downs used during segment lifting, as this force is only related to longitudinal wind and the weight imbalance of the machine.

Some first-generation mobile frames were also supported on short self-launching rails. Compared to the fixed frames, however, the mobile frames travel along the cantilever for segment placement. The use of self-launching rails requires lifting the cart from the rails prior to pushing the rails forward. When repeated several times, these operations lengthen the cycle time and increase labour costs and the risk of missed rail anchoring due to complacency. The use of full-length rails shortens the cycle time and does not affect safety, as overturning during frame transfer can be effectively controlled with a proper location of the winch trolley within the lifter. The winch trolley is mechanically locked to the runway girders when not in use in order to avoid involuntary displacements.

The bidirectional lifting frames shuttle back and forth along the hammer. Only one lifter is used on the hammer for handling of two segments and two stressing platforms, and fast movements are indispensable. The runway girders are placed on a sophisticated straddle carrier and multiple steering alternatives facilitate final alignment of the lifter for installation of the tie-downs. Support cylinders lift the wheels from the deck during these operations, as in a standard wheeled crane.

7.2.3 Performance and productivity

Two fixed lifting frames for precast segmental box girders typically require a 10–12 m pier table for assembly; assembly duration depends on the design of the lifter and the field splices. The lifters for macro-segments of cable-stayed bridges are heavier and more complex; they are also typically longer to reduce the uplift force in the rear tie-downs and the load applied to the tip of the cantilever. Assembling heavy and long lifters on the pier table takes several weeks because the tower crane performs multiple operations and is not always available.

Two mobile lifting frames are also assembled at once. The lifters are prepared and tested on the ground and separated into large pieces for lifting. The lightest machines may be placed on the pier table with three lifts of the tower crane: one for the main cart, one for the top frame, and one for the winch trolley. After hammer completion, the lifters are moved back to the pier table and dismantled in a reversed sequence of operations. The same sequence is used for the bidirectional lifting frames.

Match-cast PC segments with epoxy joints are erected in a 1-day cycle after the learning curve has been completed. As the cycle time does not depend on segment length, longer segments delivered on barges are often used to accelerate erection. The composite grillages of cable-stayed decks take a longer time, as the edge girders, crossbeams and precast deck panels are handled individually, and fabrication of a pair of stay cables lengthens the cycle time. Weekly cycles are easy to achieve and maintain, and 4-day cycles inclusive of repositioning of the derrick platforms have been attained in bridges with simple deck geometry. Macro-segments of steel box girders take a longer time due to slow lifting and the duration of field splicing.

Hoisting segments over significant heights is time consuming and often affects the cycle time. Strand jacks and hydraulic winches with reeved hoisting ropes are the typical solutions for lifting. The power of the PPU governs the cycle time of strand jacks and the rotation speed and line pull of hydraulic winches. Heavy segments require many falls of rope and the hoist speed may slow down significantly.

7.2.4   Structure–equipment interaction

The interaction between fixed lifting frames and the deck being erected is mostly related to the weight of the lifters, which added to the hoist load increases the demand for longitudinal top-slab prestressing and the shear force in the webs.

Load imbalance of staged construction generates longitudinal bending in piers, foundations and temporary stabilisation gear at the pier caps. When the bridge is curved in plan, cantilever construction also causes transverse bending, and three-dimensional structural models are necessary for the analysis of staged construction. The effects of load imbalance are further amplified by the need to design piers and pier cap stabilisation gear for a condition of sudden accidental release of a precast segment, which is typically represented by a load equal to the weight of the segment but directed upward.

Load imbalance may be reduced by means of a non-symmetrical design of the pier table, which staggers the joints by half a segment in the two cantilevers. This solution is rarely used in precast segmental bridges as it doubles the number of types of segment and constrains the lifting sequence.

Mobile lifting frames pick up the segments close to the pier and move them out along the cantilevers, and load imbalance may be controlled by means of the relative position of the lifters. Bidirectional lifting frames are more demanding because only one lifter is used on the hammer and the weight of the lifter adds to the unbalanced load of the segment.

Self-launching gantries are often used to balance the hammers during erection, while the use of lifting frames requires stabilisation gear at the pier cap designed for the full load imbalance. Fixed and mobile lifting frames can be used on relatively long cantilevers, while the load imbalance of bidirectional frames limits their use to short spans and hammers framed to stiff piers. When the piers are short, pier towers, tension/compression props, double props or inclined ties may be used to stabilise the hammer.

Transverse bending in the deck cross-section is rarely a major issue because the rails of fixed and mobile frames and the cart wheels of bidirectional lifting frames are placed over the deck webs.

7.3.   Cable cranes

A cable crane is an aerial crane that uses a carrying rope to support a hoist trolley. Cable cranes are mainly used for dam construction. The carrying rope spans transversely over the dam and is anchored to the main body of the crane on one side and to a counter-carriage on the other side.

Main body and counter-carriage shift longitudinally along foundation beams anchored to rock to cover the entire dam excavation with the hoist trolley. Two or three cable cranes are used on large dams to meet the productivity of the batching plant and silo-buses.

Cable cranes are also used for casting arch bridges, where they assist in forming and concrete delivery for the arch, the spandrel columns and the deck. Because of the geometry of an arch bridge and the midspan sag in the carrying rope, the cable crane is supported on tall guyed masts at the deck abutments. The masts are pinned at the base and rotated by winches and reeved ropes in the transverse plane for the hoist trolley to cover the full width of the bridge. The carrying rope is anchored to the ground beyond the masts and deviated at the top. The masts are designed and built like crane booms, with welded connections and a minimal number of field splices.

Locked-coil ropes comprising hot-galvanised Z-wires wound around a helical strand core are used as carrying ropes for heavy loads in order to achieve a smooth outer surface that is resistant to abrasion. Smooth rope surfaces and self-lubricating inserts ensure permanent lubrication, high stability against lateral wind and absorption of vibrations. The outer surface is an ideal running surface for the PTFE wheels of the hoist trolley, and self-compacting Z-wires make the rope less sensitive to side pressure and protect the interior from corrosion. The locked-coil ropes are anchored with bearing sockets fabricated by brooming the end of the rope into the conical cavity of a steel cylinder and by filling the cavity with zinc–copper alloys with a melting temperature of 400–420°C. Pilot ropes are typically used to haul the carrying ropes into position.

One loop of haulage rope provides propulsion to the hoist trolley. The trolley is anchored to the ends of the haulage rope and shuttles back and forth along the carrying rope. The haulage rope is a continuously circulating cable moved by a capstan winch housed in the main body of the crane. When the trolley arrives at the end terminals of the carrying rope, the motor of the haulage winch stops and reverses direction. The haulage winch typically does not require high power due to the small changes in elevation along the carrying rope.

The hoist trolleys for heavy loads are articulated multi-wheel assemblies that cope with the carrying rope sag due to the weight of the trolley and the hoist load. The hoisting rope is fixed at the counter-carriage and stored on a hoisting winch in the main body of the crane. One or two loops of hoisting rope are used between the trolley and the hook to reduce the line pull; the loops have a trapezoidal geometry to diminish longitudinal load oscillations on braking of the trolley. A few reeves accelerate vertical load movements, but the hoisting rope has a large diameter and a powerful winch is necessary to provide the required line pull; the winch drum and sheaves also have large diameter.

Spring hangers suspend the haulage rope and hoisting rope from the carrying rope. The trolley opens the hangers and keeps them open during their movement throughout the trolley to release the hoisting rope for operations. The hangers support the haulage and hoisting rope; PTFE rolls are used to avoid rope abrasion, and thus to obtain longer service life and maintenance intervals.

7.4. Lifting platforms for suspension bridges

Cable cranes are also used for the erection of suspension bridges. Earth-anchored cables are used in major suspension bridges, and construction must therefore proceed sequentially. After completion of the pylons and anchor blocks, the catwalks are installed for wire spinning, cable compaction and application of cable clamps and hanger cables. Finally, lifting platforms are installed on the main cables to erect a streamlined box girder deck or a stiffening truss (Gimsing and Georgakis, 2012).

1   Steel pylons are erected using ground or floating cranes for the lower portion and climbing cranes following the pylon for the upper portion. RC pylons are cast in place with slip forms or jump forms. Slip forms offer faster erection but require continuous concrete supply, and jump forms are therefore the typical choice when concrete is delivered on barges. The pylons are slender longitudinally to follow the displacements of the cable system, and large wind-induced deflections and vibrations may therefore occur prior to fabrication of the cables. Wind-induced oscillations are more pronounced in free-standing steel pylons due to their lower mass and damping and larger flexibility, and may be controlled with temporary bracing cables or tuned-mass dampers (Gimsing and Georgakis, 2012).

2   The catwalks are supported on multiple strands located under the walkway surface. A pilot rope is pulled from one anchor block to the other with a tug boat and lifted from the seabed. When a helicopter is used, a lighter fibre rope is installed to pull the pilot rope in place. After installing the pilot rope, the catwalk strands are pulled across the span. The walkway surface is prefabricated with wooden crossbeams and wire mesh, reeled, lifted to the top of the pylons, and unreeled on the catwalk strands. Hand ropes supported by U-frames are added for stability, and lateral wire meshes are applied to safely enclose the working area. Light cross-bridges are provided across the span to allow movement of personnel between the two catwalks. Adjustable anchorages are used for the catwalk strands at splay saddles and pylon saddles to follow the increasing sag in the main cable during lifting of deck segments; frequent sag adjustment is also necessary for the catwalks early in the controlled-tension wire spinning for the main cables. When the navigation clearance allows it, convex cables connected to the catwalk strands with hanger cables are used to enhance wind stability, and additional stability is provided by bracing the catwalk to the main cable.

3   The main cables are fabricated by air spinning or with prefabricated parallel-wire strands (PPWSs); hot-galvanised 5–5.5 mm high-alloy steel wire is used in both methods. With the air spinning method, two spinning wheels suspended from an endless ring of haulage rope shuttle back and forth between the anchor blocks in opposite directions. Each spinning wheel positions two or four loops of wire at every pass. Counterweight towers are used between wire drums and spinning wheels to apply an initial tension to the wire by reeving it through counterweighted sheaves. With the high-tension method, the initial wire tension corresponds to the full self-weight stress, and the sag of each wire is adjusted individually. With the low-tension and the controlled-tension method (50% and 80% of the self-weight stress, respectively), bundles of 300–500 wires are temporarily supported on the completed part of the main cable and the sag is adjusted in one operation. Semi-circular strand shoes are used to anchor air-spun wire loops within the splay chambers of the anchor blocks. Full-length PPWS comprising 91 or 127 wires in a hexagonal pattern and with bearing end sockets are unreeled using haulage ropes on catwalk-supported rolls alongside the main cable to accelerate cable fabrication and to reduce labour demand; the first PPWS erected is used for sag adjustment of the following strands. At the end sockets, the wires of the PPWS are led through the holes of a locking plate and finished with button heads, and the conical cavity is filled with zinc–copper alloy or cold-casting resins with fillers. With both

air spinning and PPWS, temporary spacer plates are used to arrange the main cable in vertical or horizontal hexagonal patterns; vertical spacers are preferred with PPWS and low-tension air spinning. When all the wires are in place, the cable is compacted with a hydraulic collar into a circular shape. Temporary steel straps are used to keep the cable in place prior to final wrapping.

4   Vertical hanger cables are suspended from cable clamps comprising two half-collars tightened together to transfer tangential loads to the main cable by friction. The collars are retightened during deck erection, as the diameter of the main cable contracts due to Poisson's effect when axial tension is increased. Pre-stretched helical strands and locked-coil strands are used for the hangers; helical strands may be looped around grooved cable clamps, while locked-coil strands require fork sockets for top pinned connections. Bearing or fork sockets are used at the bottom connection in both cases. The hangers are delivered on barges and lifted into position or hauled along the catwalks on the support rolls of the PPWS. Inclined hangers aimed at providing some damping with the hysteresis of helical strands have shown early fatigue issues and their use has been abandoned.

5   When the deck is erected from the midspan toward the pylons, the segments are often applied symmetrically, although a centre of gravity eccentricity around 10% of the main span length may increase the flutter speed in the critical configuration, with 30–40% of the main span in place (Gimsing and Georgakis, 2012). Segments are simultaneously applied to the side spans to control longitudinal bending in the pylons, and the segments adjacent to the pylons are the last to be erected in order to minimise bending in the main cables. Initially, the field splices between the segments are left open so as not to counteract the extension of the main cables and the changes in geometry that follow the funicular self-weight curves of segmental erection; temporary ties have sometimes been applied to the top flange in the concave deck regions and to the bottom flange in the convex regions to enhance aerodynamic stability with no vertical flexural stiffness. The joints are welded as soon as the deck remains aligned in order to minimise self-weight bending at mean temperature in the deck and the pylons. The deck may also be erected from the pylons toward midspan and anchor blocks, and the two procedures may also be combined to tighten up the main cables with a few midspan segments.

6   Tensioned Z-shaped wire has been used in place of the traditional soft-annealed 3.5 mm galvanised wire for mechanical wrapping of the cable between the cable clamps. To increase corrosion protection, the completed cable is treated with zinc dust paste prior to wrapping. Air-tight elastomeric wrapping combined with dehumidification by dry air injection is expected to provide superior long-term corrosion protection.

The large load capacity of an earth-anchored cable system suggests erecting long and heavy deck macro-segments to minimise the number of lifts and field splices. Lifting struts anchored to the main cables have been used in several applications to handle full-width macro-segments up to 50 m long and 500 ton heavy, with multiple reeves of hoisting ropes running along the catwalks, deviated at the pylon top, and pulled by powerful winches anchored to a pylon crossbeam. With increasing span lengths, however, such long hoisting ropes become excessively flexible.

Self-propelled lifting platforms bridging between the main cables and equipped with strand-jacking systems provide direct load handling, and the additional load applied to the main cables is rarely a major problem (Figure 7.11). The lifting platforms carry the PPU, fluid tanks, hydraulic systems, control systems, two strand jacks and hydraulic collar sledges for movement along the main cables. Individual lifting platforms require supplementary stabilisation ropes to

Figure 7.11. Lifting platform for suspension bridges (Photo: OVM)

control tilting of the segment, while paired platforms may require a torsional hinge on one of the platforms to achieve uniform pull in the hoisting strands.

Although a lifting platform with two strand jacks may lift 1300–1400 ton and two paired platforms may therefore reach a huge capacity, suspending long deck segments with 3–4 hangers on either side complicates load adjustment and control of deflections. Full-width 20–50 m deck segments are therefore delivered on barges, or tugged into position by floating when the deck has a streamlined box girder and the inner diaphragms can be made watertight, and are lifted and suspended from two hangers on either side in one operation.

The ends of the lifting platforms are pinned to collar sledges that climb the main cables. The sledges spread the load along the cables and the end articulations allow the platform to swing to vertical whatever the local gradient in the main cable may be prior to and after picking up the segment.

The sledges operate like inverted friction launchers driven by long-stroke double-acting cylinders. The collars take support on the main cable prior to wrapping without interfering with the cable clamps for the hangers. Neoprene plates are used to protect the wires (Figure 7.12). Safety ropes prevent sliding in the final stages of deck erection, when the lifting platforms operate close to the pylons.

7.5. Launching gantries

The launching gantries for balanced cantilever erection of precast segmental bridges are modular machines designed to be assembled behind an abutment and to erect the hammers in a sequential

Figure 7.12. Self-launching collar sledge (Photo: OVM)

manner from abutment to abutment. The gantries achieve fast erection rates and minimise ground disruption when the segments are delivered on the deck. Most gantries also stabilise the hammers during erection, which simplifies the stabilisation gear at the pier caps, reduces longitudinal bending in piers and foundations, facilitates aligning the cantilevers for midspan closure, and offers significant cost savings in long bridges comprising a great number of piers (Harridge, 2011).

The launching gantries erect rectilinear or slightly curved spans crossing rivers, steep valleys and other obstructions that prevent the use of ground cranes. The spans are typically shorter than 120 m. The height of the piers is critical for erection with ground or floating cranes, but almost irrelevant for gantry erection when the segments are delivered on the deck. Long balanced cantilever spans have varying depth, which suggests the use of overhead gantries.

The new-generation gantries are sophisticated machines that are generally customised for the bridge project, and they therefore represent a major investment. The number of spans must justify the investment, and their distribution should avoid transportation of the gantry over long distances. Convertible gantries for balanced cantilever erection of long spans and span-by-span erection of shorter approach spans are efficient machines but may overload the shortest spans with their weight. Large variations in the span length and tight plan radii suggest segment erection using lifting frames.

The first-generation gantries had their length minimised to reduce cost and weight. They are 4–5 segments longer than the longest hammer. This length permits the gantry to sit on the

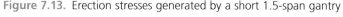

Figure 7.13. Erection stresses generated by a short 1.5-span gantry

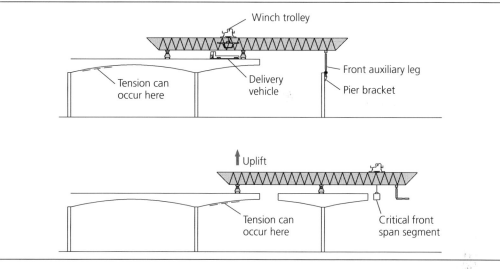

hammerhead segment and to place the rear support frame two segments back from the tip of the front deck cantilever. This gap is necessary to avoid edge tension in the epoxy joints, especially when erecting segments of the down-station cantilever beyond the leading pier, and allows space for the delivery vehicle on the bridge (Harridge, 2011).

Short 1.5-span gantries have been used intensively in the past. The weak point of these machines is that the delivery vehicle has to move up to the tip of the deck cantilever. The rear support reaction of the gantry, added to the weight of trailer and segment, often generates tension at the bottom fibre of the previous span that requires temporary prestressing (Figure 7.13). Tension can also arise at the bottom fibre of the front cantilever during placement of segments beyond the leading pier due to the uplift load applied by the gantry (Harridge, 2011).

The gantries for balanced cantilever erection are designed for negative bending, and varying-depth trusses have been used extensively in the past. Some first-generation single-truss gantries are equipped with a deviation tower at the central support and symmetrical stay cables that support the front cantilever. When the segments are delivered on the deck, the deviation tower is an A-frame crossed by the truss, and its support legs allow the segments to pass through when rotated 90° to the bridge. A tall portal frame is used at the rear support for segment delivery throughout the frame and to traverse the gantry in curved bridges (Figure 7.14). The stressing platform for the stay cables is typically located at the top of the tower.

When the A-frame is supported on a crossbeam for enhanced geometry adjustment, the hydraulic legs of the crossbeam cope with the deck gradient and crossfall, and the A-frame is devoid of support cylinders. Tubular trusses are often used for the A-frame, a single plane of stay cables is generally preferred, and a fan layout simplifies pull adjustment from the top stressing platform. A few cables may be anchored to the bottom chords to provide torsional restraint to the truss. Triangular trusses are preferred in these gantries for simpler structural nodes at the cable anchorages (VSL, 2008). A few cable-stayed gantries have been designed with braced I-girders

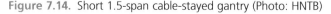

Figure 7.14. Short 1.5-span cable-stayed gantry (Photo: HNTB)

in place of a triangular truss, but the cost of cable anchorages soon offsets the labour savings of robotised welding, and use of this solution has been abandoned.

The underslung winch trolleys of these gantries span between the bottom chords of the truss and the field splices are detailed for the wheels to pass through. These gantries are able to place the hammerhead segments, but these operations are on the critical path of launching, which is slow, complex and labour intensive. If the hammerhead segments can be placed using ground cranes, the cycle time is shorter and the gantry can be lightened further.

Short 1.5-span cable-stayed gantries for balanced cantilever erection of 100–120 m spans are geometrically compatible with span-by-span erection of 50–60 m spans. The deviation tower is placed at the end of the completed deck and the front cable-stayed cantilever takes support at the next pier and suspends an entire span of segments (see Figure 3.8). The PPU is applied to the rear end of the gantry along with counterweights and tie-downs to balance the pull in the stay cables. The gantry is supported at the bridge piers, and the long rear balancing span provides ample room for segment loading on the deck.

New-generation convertible gantries, and even machines designed exclusively for balanced cantilever erection, rarely use stay cables due to the complexity of operations and the high labour costs associated with the first-generation cable-stayed gantries. Nevertheless, a few powerful stay cables may still be a cost-effective solution when the bridge includes a few longer spans and the remainder can be erected without activating the cables or with just a few of them activated (Figure 7.15).

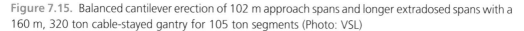

Figure 7.15. Balanced cantilever erection of 102 m approach spans and longer extradosed spans with a 160 m, 320 ton cable-stayed gantry for 105 ton segments (Photo: VSL)

Single-truss gantries are very versatile for these applications, as the support frames are narrow and fit well between the pylons of extradosed bridges (Figure 7.16). Cable-stayed gantries are also cost-effective when the deck is wide and the spans are short, as negative bending during segment erection far exceeds positive bending during launching, and the gantry may therefore be launched without releasing the stay cables.

The new-generation gantries for balanced cantilever erection have a modular twin-truss overhead configuration. High-grade steel trusses are lighter than braced I-girders and facilitate mobility of workers, and high flexural efficiency and field splices designed for fast assembly compensate for higher fabrication costs due to hand welding. The length of the trusses is about two times the typical span length in order to move the rear support of the gantry close to the pier; in spite of their length, however, these gantries are often lighter than the first-generation 1.5-span gantries due to the use of modern high-grade steels. The central truss modules are often compatible with span-by-span erection of 40–50 m spans to facilitate reuse of the gantry and to erect the deck sections between the abutments and the first and last hammer without props from foundations.

The trusses are tall to resist negative bending from the front cantilever and to position tall hammerhead segments delivered on the deck. The segment is lifted close to the winch trolley to move over the support crossbeams during forward movement, and the total depth of segment, lifting accessories and lower sheaves must be smaller than the combined depth of launch saddles and trusses. Three-hinge portal cranes spanning between the top chords can handle precast segments taller than the trusses, but are heavier and more expensive than flat-frame winch trolleys and are therefore rarely used.

Figure 7.16. Cable-stayed gantry operating within the pylons of an extradosed bridge (Photo: VSL)

Narrow rectangular trusses with two top chords, two bottom chords, lateral bracing between the chords, and braced diagonals and verticals were used intensively in the 1970s. Diagonals and verticals converge into nodes at the chords and at the mid-depth of the truss. Longitudinal braces connect the mid-depth nodes to control local buckling and to further lighten the members. Braced chords were soon replaced with full-width flange plates stiffened at the edges to reduce the need for hand welding, to increase the moment of inertia, and to provide a full-length walkway for access to the winch trolleys (Harridge, 2011).

Some rectangular trusses were so tall, narrow and potentially unstable that horizontal outer trusses were applied to the flanges to control lateral sway. Varying-width outer braces comprising a mid-depth chord and diagonals connected to the main chords of the truss were also tested for their ability to resist wind loading on the front cantilever. Rectangular trusses are still popular in emerging countries, where the low labour cost facilitates depreciation of the assembly costs and the modular nature of design permits shipping most of the gantry in 40 ft containers. Despite their structural efficiency, with the generalised increase in labour costs and the less and less frequent recourse to custom design, these gantries have lost most of their appeal.

Constant-depth triangular trusses comprising one top chord and two braced bottom chords are the first-choice solution for new-generation gantries in industrialised countries. Triangular trusses use the inclination of the diagonals in the transverse plane to control lateral sway of the top chord under positive bending due to segment handling in the rear span. The bottom chords are more prone to lateral sway because of the primary role of negative bending in balanced

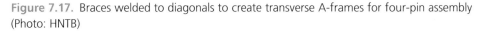

Figure 7.17. Braces welded to diagonals to create transverse A-frames for four-pin assembly (Photo: HNTB)

cantilever erection, and lateral bracing between the chords cooperates with the inclined diagonals in controlling out-of-plane buckling.

Triangular trusses are more expensive to fabricate than rectangular trusses due to their complex geometry, but they have a much smaller number of field splices. Stiffened I-shapes are used for the chords, and I-shapes or round or square pipes with central end plates are used for the diagonals. Verticals are sometimes used to prop the bottom chord at the centre of the panel to control local bending over the launch bogies. Lateral bracing between the bottom chords includes one or two planes of X-frames, connections designed to minimise displacement-induced fatigue, and sufficient flexural stiffness to resist vibration stresses. Braces may be workshop welded to the diagonals to create transverse A-frames that require only four end pins for site assembly (Figure 7.17).

The full two-span gantries have two support crossbeams and two auxiliary legs. The front cross-beam is anchored to the hammerhead segment, the rear crossbeam is anchored to the rear pier table, the front auxiliary leg supports the gantry during placement of the hammerhead segment and repositioning of the front crossbeam, and the rear auxiliary leg supports the gantry during repositioning of the rear crossbeam. The length of the trusses is designed to cover the hammer and front deck cantilever, and to provide a rear overhang behind the rear crossbeam to load precast segments delivered on the deck.

The shorter the gantry, the closer the rear crossbeam is to the tip of the front deck cantilever. Full two-span gantries are supported over the rear pier table, and cost savings in reinforcement, post-tensioning and associated labour throughout the bridge soon offset the cost of longer trusses. The cost increase from a short 1.5-span gantry to a full 2-span gantry is much less than a proportion of the length because the most expensive components of these machines are the winch trolleys and support crossbeams, which are necessary anyways. Placing the hammerhead segments, repositioning the support crossbeams, and launching the gantry are also simpler, faster and less labour intensive.

One or two winch trolleys are used to handle the segments. The winch trolleys of the first-generation gantries house two electric winches: the hoist winch handles the segments and the translation winch drives the trolley along the gantry by means of a capstan. The capstan is anchored to the ends of the gantry and kept in tension by lever counterweights. Two winch trolleys require two capstans for independent movements.

Nowadays electric winches are used rarely. Hydraulic winches with reeved hoisting ropes are used to pick up segments on the ground, and gimbal-mounted pull cylinders and hoist bars with lock pins are valid alternatives when the segments are delivered on the deck. As the trolley has a PPU and hydraulic systems on board anyway, hydraulic motors replace translation winches and capstans. Braked motorised wheels provide safe and redundant operations when the bridge has a longitudinal gradient, and the runway of the winch trolley is therefore an inclined plane.

The hoist is mounted on a crab that rolls or slides laterally along the crane bridge. Shifting the hoist point within the winch trolley provides geometry adjustment capability during segment handling, and is simpler than traversing the entire gantry. Capstans, motorised wheels or double-acting cylinders drive the hoist crab along the crane bridge. The dead end of the hoisting rope may be anchored to a pull cylinder for low-speed movements of the lifting beam. Lateral double-acting cylinders or hydraulic motors are used to rotate pull cylinders and hoist bars within the crab for segment rotation. The new-generation winch trolleys are sophisticated cranes that include multiple hydraulic systems for segment manipulation.

Most winch trolleys carry a PPU that supplies the hydraulic systems and electric controls. When a winch is used to hoist the segments, the PPU also supplies electric power to the hydraulic slew ring and the tilt cylinder on the lifting beam. In the short 1.5-span gantries the PPU is sometimes placed at the rear end of the trusses so as to lighten the winch trolley and to control overturning during launching, and the winch trolley is supplied with bus-bars or flexible cables and suspenders.

The winch trolleys are among the most expensive components of a gantry, and some gantries therefore have just one trolley. Long bridges suggest an investment in a second winch trolley to shorten the cycle time; two trolleys may also apply the segments simultaneously to minimise load imbalance (Harridge, 2011). In most cases, however, the gantry is used to stabilise the hammer during erection, and simultaneous application of segments is not necessary.

7.5.1   Loading, kinematics, typical features

Twin-truss overhead gantries are easily adaptable in both the length and the spacing of the trusses. They are able to cope with variations in span length and deck geometry, and, as they are located above the deck, they are unaffected by ground constraints.

Tight plan radii are a major counter-indication as the landing pier is offset from the chord of the previous span, and this complicates launching on long spans. Vertical curvature is rarely an issue, as the support legs of the crossbeams can be shimmed with packs of steel plates when the stroke of the support cylinders is insufficient for geometry adjustment.

Most of these gantries are tall. Portal cranes are rarely used, and handling tall segments with flat-frame winch trolleys requires tall trusses. The depth of the support saddles helps in creating vertical clearance between the lifting beam and the support crossbeams, and in some gantries the support saddles have been shimmed to lift the trusses further. Tall trusses, however, are the typical solution, as the flexural efficiency of the cross-section also increases.

The single-truss cable-stayed gantries are much taller, as the segments are handled beneath the truss and the A-frame adds substantial depth. Stay cables and top stressing platform increase the wind drag area, and the height of these machines may pose problems of lateral stability during launching, requires oversized tie-downs for operations, and may overload tall and narrow piers. The height of the gantry may also be a limitation when passing beneath power lines.

The twin-truss gantries operate in different ways depending on how the segments are delivered. If the segments are delivered on the deck, the winch trolley picks them up at the rear end of the gantry, lifts them close to the crane bridge, and moves them out within the clearance between the trusses (Figure 7.18). After passing the hammer stabilisation props and support crossbeam, the segments are lowered beneath the trusses, rotated by 90° and positioned (Figure 7.19).

Figure 7.18. Segment transfer over the mobile stabilisation prop of the hammer (Photo: HNTB)

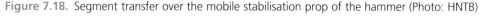

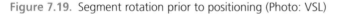

Figure 7.19. Segment rotation prior to positioning (Photo: VSL)

When the gantry carries two winch trolleys, a longer rear overhang is needed in the trusses to store the rear winch trolley when the front trolley picks up the segment. After clearing the lifting area behind the rear crossbeam, the rear trolley is moved forward to pick up the second segment. The winch trolleys operate independently and can place the segments simultaneously at the opposite tips of the hammer (Rosignoli, 2010).

If the segments are delivered on the ground or on barges, one or two winch trolleys reach down and lift them up to the deck level. Longitudinal movements of the trolley along the gantry and transverse movements of the hoist crab within the trolley provide more flexibility in positioning haulers or barges under the hammer than the use of fixed lifting frames.

When the front auxiliary leg does not slide along the gantry, the trusses also have a front overhang to move the front winch trolley over the pier cap for placement of the hammerhead segment (Figure 7.20). The front overhang may be equipped with a service crane for lifting of bearings, pier brackets and support jacks from the ground. The front overhang is designed for the full load of winch trolley and the segment. The rear overhang of trusses carrying two winch trolleys

Figure 7.20. Front overhang beyond the fixed auxiliary leg in an 860 ton gantry for 100 m spans (Photo: HNTB)

should also be designed for the full load in order to avoid collapse of the overhang in case of uncontrolled backward running of a loaded winch trolley (Rosignoli, 2011).

The lifting beam is permanently connected to the hoisting winch and is applied directly to the segments (without spreader beams) with stressed bars that cross the top slab (Figure 7.21). The anchor bars are typically located on the inner side of the web–slab node for easy access to the bottom anchorages at segment release. Outer bars and bars anchored beneath the bottom slab are rarely used due to the difficulty of access for removal. The bars are overdesigned and rarely stressed beyond 50% of the tensile strength for 20–30 reuses (unless damaged) prior to disposal. The stressing jacks may be permanently applied to the lifting beam for faster operations, and ball-and-socket anchorages are used to cope with geometry tolerances. Prestressing creates frictional shear capacity that avoids bar bending and shearing in case of segment displacements. Shims may be necessary to set segment crossfall, although most lifting beams provide hydraulic tilt.

The geometry of the lifting beam allows for a variety of shapes and sizes of the segments. Articulations and hydraulic systems provide horizontal rotation and transverse tilt. The hydraulic systems are supplied by the PPU of the winch trolley through a retractable power cable or by flying connections at the end of lifting (ASBI, 2008). Combined operations of the winch trolley and lifting beam are typically radio controlled.

The top-slab prestressing tendons are fabricated and tensioned from two stressing platforms applied to the tips of the hammer. Long tendons are typically stressed from one anchorage and completed from the opposite anchorage. The initial load at the dead anchor provides information on the friction losses to be used in calibrating the stressing. When the stressing platforms are not applied to the segments prior to lifting, they are stored on the hammer and applied during

Figure 7.21. Lifting beam with hydraulic slew ring and tilt cylinder (Photo: HNTB)

segment gluing. Multiple tendons are grouted after the application of new segments in order to move these operations out of the critical path.

7.5.2 Support, launch and lock systems

The support systems of a full two-span gantry include two crossbeams and two auxiliary legs. The crossbeams are supported on the hammerhead segment of the new hammer and over the rear pier table and carry rails for traversing the gantry. The crossbeams are anchored to the deck with tie-downs that prevent uplift, apply the load over the deck webs, and have side overhangs to create the clearance between the trusses needed for handling of segments and repositioning of the cross-beams. Long crossbeams and crossfall shims are necessary when two adjacent decks are erected simultaneously by shifting the gantry from one bridge to the other (Figure 7.22).

Simultaneous erection of two adjacent decks is faster and less labour intensive than erecting one deck first and the second deck after reverse launching. The gantry can handle a new pair of segments during the application of prestressing to the adjacent hammer, and reverse launching for gantry repositioning is also avoided (Meyer, 2011). However, the crossbeams are more expensive and double stabilisation gear is necessary at the pier caps with one line of bearings, because the gantry cannot be used to stabilise the hammers during erection. When the deck is continuous with the piers, the pier table may be erected using ground cranes or be cast in place to further accelerate gantry operations on the critical path.

One or both crossbeams restrain the gantry longitudinally during operations; double restraint enhances robustness and redundancy (Rosignoli, 2007) but locks in thermal stresses. Four adjustable support legs are used on every crossbeam to cope with the deck gradient and crossfall

Figure 7.22. A 108 m, 278 ton gantry with two winch trolleys for simultaneous erection of adjacent decks with 45 m spans and 60 ton segments. Shimmed support saddles provide additional clearance for segment transfer above the support crossbeams (Photo: VSL)

(Figure 7.23). Double-acting cylinders with mechanical locknuts are positioned over the deck webs and spaced longitudinally to control overturning under the longitudinal loads applied by the support saddles. Stacks of steel plates are used for packing when the stroke of the support cylinders is insufficient; packing is cheap but labour intensive (Harridge, 2011).

Tie-down bars resist uplift forces. Transverse uplift may arise because of traversing operations or the large wind drag area, and longitudinal uplift may arise due to thermal loads applied by the trusses; launch loads and the longitudinal projection of the support reactions due to inclined launch bogies. The tie-downs are prestressed to create frictional shear capacity so as to prevent involuntary displacements of the crossbeams and shearing of the tie-down bars (ASBI, 2008). The prestressing force should be calculated disregarding the compression generated by the self-weight and payload in order to provide a reliable anchorage to the gantry in case of accidental release of a precast segment. More than one crossbeam has overturned longitudinally in response to accidental segment release due to insufficient tie-down bars.

Box girders or braced I-girders are used for the crossbeams. Box girders are faster to assemble, often more difficult to paint and inspect, and typically more robust in case of accidents such as longitudinal overturning of the crossbeam. The crossbeams support the trusses with articulated saddles designed for launching and traversing. The typical saddle comprises two transverse bottom bogies or PTFE skids that shift along the crossbeam, and two longitudinal top bogies that support the bottom chords of the truss.

Figure 7.23. Crossfall adjustment in a 65 ton suspended crossbeam during gantry assembly (Photo: HNTB)

Cast-iron rolls and equalising beams are used for the top launch bogies. The launch bogies tilt longitudinally to cope with flexural rotations in the trusses and the launch gradient, and are assembled on a saddle that rotates in the horizontal plane for geometry adjustment and traversing. When the launch bogies are equipped with counter-rollers and bidirectional restraints are applied to all the articulations, the crossbeam may be suspended from the trusses to be repositioned with motorised wheels. In most gantries, however, the crossbeams are repositioned using the winch trolleys.

Rectangular rails welded to truss chords and crossbeams facilitate dispersal of the support reactions and provide guidance and transfer of lateral loads by acting against the flanges of the rolls. When the bottom bogies are replaced by PTFE skids, the flanges of the crossbeam are loaded for most of their width and are designed for transverse bending, and lateral guides transfer the load to the flange edges. Both crossbeams are equipped with transverse double-acting cylinders that also provide lateral restraint during gantry operations. Launching always occurs on bogies to accelerate the operations and minimise friction. Paired long-stroke double-acting cylinders acting within perforated plates provide the launch force in most cases. The two trusses are connected at the ends and cannot be launched individually.

Two solutions are used for the auxiliary support legs of full two-span gantries: independent legs that are shifted along each truss, and articulated pendular legs connected to the trusses at fixed locations. Independent legs have minor vertical adjustment capability and are used when the pier tables are erected using ground cranes or are cast in place. The pendular legs comprise a braced planar frame pinned to the bottom chords of both trusses and equipped with articulated extensions; as the frame is pinned to the top and bottom, it is unable to carry moment in the longitudinal plane. Double-acting longitudinal cylinders rotate the frame in the horizontal plane to the local plan radius, and hinges are used at the top to set the frame vertical, irrespective of the truss gradient. Vertical double-acting cylinders at the bottom of the frame are used to take support on stepped pier caps, to adjust the frame length in bridges with vertical curvature, and to recover the deflection of the cantilever truss during launching. Deflection recovery and adjustment of vertical geometry require long-stroke cylinders on such long spans.

Pendular extensions are used to modify the frame length beyond the geometry adjustment capability of the support cylinders. Short frame configurations are used during assembly behind the abutment and to support the gantry on the deck for maintenance of the crossbeams and during reverse launching. Double-acting cylinders drive rotation of the pendular extensions. Mechanical locks should be used when the legs are fully extended for reliable control of out-of-plane buckling. Work platforms are necessary at the articulations.

Multiple pendular extensions are necessary in varying-depth decks with spans of different lengths, as the depth of the hammerhead segments varies along the bridge. When fully extended, long articulated legs are among the most delicate components of these gantries and are prone to specific forms of instability. Most auxiliary legs provide some extent of rotational end restraint to the main trusses, which facilitates control of out-of-plane buckling.

The short 1.5-span single-truss cable-stayed gantries require different types of launch supports because the central A-frame is integral within the truss and cannot be used for support during launching. Friction launchers or adjustable towers with paired launch cylinders support the truss during launching and are dismantled at the end of repositioning to feed the segments from the deck throughout the rear support frame. When the segments are delivered on the ground, the launch supports may be stored at the rear end of the gantry to diminish the uplift load applied to the front deck cantilever during handling of the down-station segments of the hammer.

Launch capstans are rarely used due to the weight of these machines and the risks of launching with tension ropes. Double-acting cylinders are used to shift the support saddles transversely along the crossbeams. When anchored into sequential holes in the crossbeam, the shift cylinders are paired to hold the saddle in place with one cylinder during repositioning of the other. The two launch saddles may be connected to simplify geometry control and traversing operations.

When the deck is supported on one line of bearings, the hammer may be stabilised with shim packs or support jacks at the corners of the pier cap and prestressing bars or U-tendons embedded in the pier. Steel brackets attached to the pier cap may be used on slender piers, and props from foundations may be used when the piers are short (ASBI, 2008). These solutions generate longitudinal bending in piers and foundations, and often require the use of a second winch trolley to minimise load imbalance by applying segments simultaneously at both hammer tips. Investment in a second winch trolley can offset the labour savings from minimised stabilisation gear at the pier cap, but the second crane shortens the cycle time and may be resold with the gantry. The break-even point mainly depends on the number of spans to be erected.

Figure 7.24. Simultaneous segment application with two winch trolleys (Photo: HNTB)

In most cases a full two-span gantry is used to stabilise the hammers during erection, with adjustable props connecting the deck to the trusses. This avoids the cost of stabilisation gear at the piers, and load imbalance is resisted through the axial load at the rear pier and does not cause bending in the piers or foundations. The stabilisation props are applied to the rear cantilever of the hammer so as not to further load the front cantilever of the gantry. If the gantry has only one winch trolley, the segment of the front cantilever is placed before the corresponding segment of the rear cantilever in order to minimise positive bending in the gantry.

The first two or three segments of the cantilever are stabilised with support jacks and tie-downs at the pier cap (Figure 7.24). The props are then applied and moved backward as erection proceeds to keep the load in the props to a manageable level (Harridge, 2011). Two fixed sets of adjustable props or a mobile stabilising arm are necessary (Figure 7.25). A mobile arm is lighter than two pairs of fixed props but more expensive and complex to operate. Both types of props apply transverse loads to the gantry due to horizontal hammer rotations under lateral wind. Sliding bearings at the pier are locked during hammer erection and released prior to stressing of continuity tendons.

The typical erection sequence of a full two-span gantry is as follows.

1   The gantry is launched forward until the front end of the trusses reaches the landing pier. During this first phase of launching, the rear crossbeam is moved forward with the gantry supported at the front crossbeam and the rear auxiliary leg.

Figure 7.25. A 23 ton mobile stabilising arm (Photo: HNTB)

2 The front auxiliary leg is configured for the required length and takes support on the front edge of the pier cap. When the pier is narrow, steel brackets are applied to the front face of the pier to support the auxiliary leg.

3 The front winch-trolley places the hammerhead segment and holds it in place during anchoring to the pier cap. Then the winch-trolley places the rear starter segment on temporary support jacks. During these operations the gantry is supported at the front auxiliary leg and at both crossbeams.

4 The front winch-trolley picks up the front crossbeam and moves it onto the new pier table. During this operation the gantry is supported at the front auxiliary leg and at the rear crossbeam.

5 The front auxiliary leg is released and the gantry is launched forward until the front cantilever is slightly longer than the hammer to erect. During launching the rear auxiliary leg takes support on the deck to move the rear crossbeam forward to the rear pier table. Multiple traversing operations are necessary for launching along curves. The gantry is fully self-launching and no ground cranes are necessary.

6 The winch-trolley picks up the segments, moves them into position and holds them there until they are joined to the tip of the hammer with temporary prestressing bars while the epoxy glue in the joints cures. New segments are placed on either side of the pier in a balanced sequence. The segments can be applied simultaneously to each tip of the hammer or can be erected with a staggered sequence that requires only one winch-trolley. The stabilisation props of the hammer are progressively moved backward to control load imbalance.

7 When the hammer is complete, the rear cantilever is aligned and locked to the front cantilever of the completed deck and a midspan closure segment is cast in place.

Bottom-slab continuity tendons are installed across the midspan segments to resist positive moment and complete the connection. The stabilisation props for the hammer are finally removed.

7.5.3 Performance and productivity

Full two-span gantries are the first-choice solution in rectilinear bridges with 100–120 m spans and tall piers over inaccessible sites. Short 1.5-span gantries are rarely used nowadays. The number of spans governs depreciation of the setup costs of the precasting facility, the investment in transportation means and the gantry, and the additional costs of shipping and site assembly.

Both types of gantry are fully self-launching and able to erect the hammerhead segments. Compared to the use of lifting frames, no need for ground cranes, on-deck segment delivery and no disruption of the erection cycle to remove the lifters from the deck and to reposition them on the new pier table are major advantages, as balanced cantilever bridges are typically chosen in response to inaccessible sites and tall piers. The lifting frames, on the other hand, allow simultaneous erection of multiple hammers and can operate on curved bridges.

Launching gantries have been successfully used on spans ranging from less than 50 m to about 120 m. Balanced cantilever construction of short spans is slower than span-by-span erection, but does not require a gantry able to suspend an entire span of segments (Meyer, 2011). Long-span bridges often have shorter approach spans that are also erected as balanced cantilevers if a convertible gantry is not available.

The number of segments that can be erected in a day depends on several factors, and bridge design has perhaps the greatest influence (Harridge, 2011). If the top-slab cantilever tendons can be anchored every two segments, the time required to fabricate and stress the tendons halves and productivity increases. The use of two winch trolleys enables work to proceed at both ends of the hammer, and the segment delivery time and the lifting time from the ground or barges also influence the time schedule and productivity.

An average productivity of 2 + 2 segments per day is easy to achieve and maintain. Gantries equipped with two winch trolleys may erect 3 + 3 and even 4 + 4 segments every day, even though the hammer deflects many times during erection and surveying the growing cantilevers to ensure proper geometry by joint shimming often becomes critical at such erection rates. Surveying and geometry control are as important during segment erection as during casting, and, because of the length of the cantilevers, small orientation errors can have significant effects.

Erection is faster when the segments are delivered on the deck and proper delivery rates can be maintained. Lifting segments from the ground or barges is time consuming over significant heights, and often affects the cycle time. Hydraulic winches with reeved hoisting ropes are the typical solutions for lifting. The power supplied by the PPU governs the rotation speed and line pull of the hydraulic winches, heavy segments require many falls of rope, and the hoist speed may slow down significantly. Extremely efficient post-tensioning operations are also required (ASBI, 2008), for two reasons:

1   At least two, and occasionally four or six, tendons are fabricated and stressed in every cycle. These are critical path activities, as the next pair of segments cannot be installed until the tendons have been completed.

2   The tendons are fabricated and tensioned from stressing platforms at the tips of the hammer. The stressing platforms are removed and repositioned in every cycle, and also these activities are on the critical path. Integrating the stressing platforms with segment placement can therefore have a positive impact on productivity.

Erecting two adjacent hammers simultaneously by shifting the gantry from one alignment to another enhances productivity and reduces the criticality of operations. The gantry releases the segments as soon as the gluing bars are stressed, and can erect a new pair of segments on the other hammer during tendon fabrication. Erecting $2 + 2$ segments on two hammers offers the same productivity as $4 + 4$ segments on one hammer, but with better control of operations. Double pier cap stabilisation gear is necessary, as the gantry cannot be used to stabilise the hammers.

When the gantry stabilises the hammer during erection, operations are more complicated and productivity may reduce slightly. After the learning curve has been completed, a productivity of one 100 m hammer per week can be achieved regularly, and can sometimes be exceeded if there are no expansion joints in the span. The gantry is typically repositioned to the next span in one day.

Cantilever and continuity tendons are both grouted after completion of the hammer to avoid the risk of grout crossing over into empty ducts. If the allowed period for leaving a stressed tendon ungrouted is exceeded, corrosion inhibitors can be used to delay grouting further.

7.5.4   Structure–equipment interaction

During balanced cantilever erection, different load conditions are applied to a structure, and these change continuously. Erection with a launching gantry is often the most complex in this regard, as hammer erection and repositioning of the gantry include multiple steps, and the load of the gantry is applied at different locations. In addition to longitudinal bending and shear, transverse bending and torsion can be significant during gantry traversing on tight curves.

Balanced cantilever erection of 100–120 m spans requires long and heavy gantries, and significant loads are applied to the bridge. Short 1.5-span gantries apply the rear support reaction close to the tip of the front deck cantilever, which often requires temporary prestressing in the deck and modifies the camber after launching the gantry to the new span. Segment delivery on the deck also generates high localised loads.

Stability during launching is a prime consideration in determining the length of the trusses. Because of the lower stability of short 1.5-span gantries, the PPU and fuel tanks are applied to the rear end of the gantry, which increases the load applied to the deck and complicates power distribution. Temporary post-tensioning has been used to strengthen the deck sections affected by gantry loading. The temporary tendons are relieved after repositioning the gantry and reused in the subsequent hammers. Although labour intensive, temporary prestressing has been used successfully in short bridges that did not justify the investment in a full two-span gantry (Harridge, 2011).

The two-span gantries take support at the pier tables, and their interaction with the deck is less pronounced. However, the trusses are longer, the wind drag area is larger, and transverse bending in the piers and foundations is more marked. The weight of the gantry is typically

Figure 7.26. A 60 ton winch trolley for 200 ton segments equipped with overhang runways for two auxiliary 5 ton hoist blocks that assist deck post-tensioning (Photo: HNTB)

controlled by using high-grade steel; modern steels were not available for the first-generation gantries, and their weight is often a limitation to their reuse.

Repositioning the rear crossbeam requires taking support on the deck with the rear auxiliary leg. The load applied during this operation is rarely prohibitive, and its location can be adjusted. Multiple repositioning of the crossbeams is necessary to traverse the gantry when launching along curves; curved vertical profiles may also require multiple repositioning depending on the available stroke in the support cylinders. Numerous anchor holes are necessary in the top slab at every support location. Missing or misaligned anchor holes require the drilling of cores into the top slab, because operations without anchorages can be critical for gantry stability.

These gantries are very flexible. This is rarely a problem during segment erection as the segments are handled individually and deck geometry is determined by the segments as they were cast. When the gantry is used to stabilise the hammer, however, the flexibility of the gantry may cause movements in the hammer. During midspan closure these movements may misalign the adjoining cantilevers, and strong-backs and stressed bars are necessary at the closure pours to prevent displacements.

Operating winch trolleys may also cause displacements in the hammer. Winch trolleys carrying the PPU, fluid tanks and multiple hydraulic systems may be particularly heavy (Figure 7.26). During midspan closure the winch trolleys handle the strand coils for the continuity tendons, and the load deflections of the gantry can cause cracking in the closure pours if the strong-backs are removed too early.

The weight of the gantry may cause differential settlement in the foundations, especially when the piers are short, the deck is lightweight and the weight of the gantry is a major component of the foundation load. This may occur when a full two-span gantry for highway bridges is used to erect lighter LRT spans. As the deck is erected directionally, differential settlement of foundations may cause locked-in stresses in the bridge.

Lateral wind loading on the gantry should be considered in the design of tall piers. The taller the piers, the higher the design wind load, the bigger the advertising banners the contractors will want to put on their machines. Gantries and piers should be designed for wind loading applied to most of the solid area between the chords, which nullifies one of the advantages of trusses versus braced I-girders.

7.6. Technology of in-place balanced cantilever casting

In-place casting of balanced cantilever bridges with pairs of form travellers is the standard solution for curved bridges and PC spans longer than 70–120 m, depending on bridge geometry. Spans up to 250 m have been reached, although spans longer than 210 m are rare. The length of the segments is 2–3 m at the root of the cantilever and progressively increases to 5 m for reasons of weight, load imbalance and load capacity of the travellers. Segments longer than 5 m are rarely used because form flexibility may cause cracking and geometry defects, and the travellers are more expensive.

A standard form traveller is designed for a maximum segment length of 5 m. The weight of the steel frame ranges from 25 ton to more than 100 ton depending on segment weight. The total weight of the traveller, including the casting cell, is typically in the range 70–80 ton for single-cell box girders, and may exceed 130 ton in wide bi-cellular box girders. Segment weight varies in the range 75–300 ton.

In-place balanced cantilever casting offers limited potential of acceleration and the time-schedule dictates the number of pairs of travellers to be used simultaneously. When the adjoining cantilevers face each other at midspan, one traveller is lowered on the ground and the other is used to cast the closure pour. Bottom slab tendons are installed to make the deck continuous, and the second traveller is lowered to the ground and repositioned on the next pier table.

Hammerheads connected monolithically to the piers were used in the early applications of balanced cantilever casting in combination with midspan expansion joints designed for transfer of shear and release of axial displacements and rotations. In addition to complex closure operations with diaphragms and bearing seats, midspan hinges caused large time-dependent deflections in the cantilevers and angle breaks at the midspan joints due to moment discontinuity. Nowadays, midspan expansion joints are used only in very long bridges to separate long continuous frames comprising several spans.

The expansion joints are designed with two lines of sliding bearings within the box cell of each cantilever. Through-steel box girders are restrained by four lines of bearings at the top and the bottom to transfer bending and shear throughout the joint and to allow longitudinal displacement. Midspan expansion joints complicate the closure operations with complex diaphragms and the need for inserting the steel girders into the box cell after removal of the first traveller, and the continuous frames of the bridge are therefore made as long as is practicable (ASBI, 2008). Longitudinal bending from thermal and time-dependent effects may be controlled with flexible leaf piers in combination with sliding bearings at the end piers and abutments.

The weight of segments and form travellers is a large source of load imbalance during segmental construction. Load imbalance generates longitudinal bending in foundations, piers and stabilisation gear at the pier caps and may be reduced with an asymmetrical design of the pier table that staggers the construction joints by half a segment in the two cantilevers. The same result may be achieved by casting one half segment starter on one side of a symmetrical pier table. In-place casting provides more geometry flexibility than segmental precasting, even if the pairs of staggered segments must be cast sequentially, and this may increase the cycle time. When the deck is supported on sliding bearings, releasing the pier cap stabilisation gear after achieving midspan continuity avoids residual bending in the piers; however, the piers are designed for friction at the sliding bearings in service conditions, and the cost saving is rarely the factor that drives the choice.

The pier table is cast at the end of pier erection with a special casting cell. Assembly of form travellers dictates the optimum length of the pier table. In early applications the pier table was designed to accommodate a pair of travellers, and lengths of 8–10 m were not infrequent. With new-generation overhead travellers, fast splitting assembly requires 6.5–7 m pier tables, the use of temporary braces or side assembly require 5 m pier tables, and the hammerhead may even be as wide as the pier cap when a suspension-girder MSS is used for in-place balanced cantilever casting of long deck segments.

Props from foundations or frames on pier brackets support the casting cell of the pier table. A tower crane anchored to the pier assists the operations. Segment geometry is complex, the working space is limited and the segment is typically divided into numerous casting phases. Reinforcement is also complex, numerous top slab ducts and some vertical tendons have to be installed, and construction durations of 2–4 months are not infrequent for the pier table.

The first-generation overhead travellers in the early 1950s were long, to enhance the stabilising action of rear counterweights and to reduce the vertical loads generated by the overturning couple. In the new-generation travellers, rear tie-down bogies roll within launch rails anchored to the deck to prevent overturning during repositioning, stressed tie-down bars balance the traveller during casting and avoid the use of counterweights, and the travellers are lighter and shorter. The travellers are supported on short self-launching rails, and repositioning is a two-step process: first the rails are pushed forward and anchored to the deck over the new segment, and then the traveller is lowered onto the rails and pulled forward.

An overhead traveller has a number of longitudinal frames corresponding to the number of webs in the deck. The frames are supported on the front deck segment over the webs and are anchored to the second segment with tie-downs and support jacks that prevent vertical displacements. Hanger bars suspend the casting cell from two transverse trusses or crossbeams overhanging from the main frames (Figure 7.27). The rear truss connects the frames at the front support of the traveller, and the front crossbeam is placed at the tip of the front overhangs. After tensioning the top slab tendons of the new pair of segments, the forms are stripped and the travellers are released and moved forward to the next casting location.

The casting cell is adjustable to varying deck geometry and road alignment. Working platforms are incorporated around the casting cell. A stressing platform is suspended beyond the front bulkhead for fabrication and tensioning of the top slab tendons. The bottom form table is permanently suspended from the overhead frame of the traveller. The outer forms for side wings and webs are

Figure 7.27. Three-dimensional geometry of an overhead traveller (Drawing: Doka)

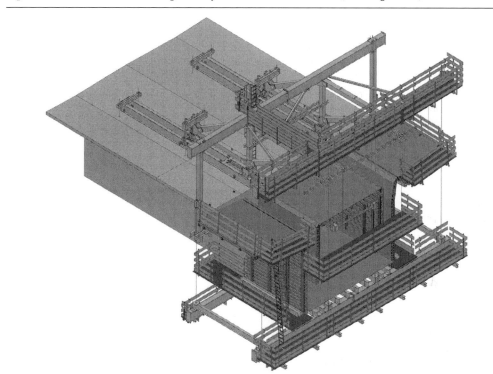

moved forward during repositioning of the traveller to close the casting cell as soon as possible. After fabrication of the rebar cage for the bottom slab and webs, the inner tunnel form is pulled forward along rails suspended from the front deck segment and the front crossbeam of the traveller for fabrication of the top slab cage.

Two form travellers are assembled on the pier table at once. It typically takes 2 weeks to assemble the overhead frames of the travellers and another 2 weeks to assemble and suspend the casting cells. Casting the starter segment takes another 2–3 weeks. The subsequent segments are typically cast in a 5-day cycle after the learning curve has passed. Concrete with early high strength is used to tension the top slab tendons as soon as possible.

Underslung travellers are used rarely in varying-depth PC decks, while they are the first-choice solution for in-place casting of cable-stayed decks to minimise interference with the cables. In underslung travellers for PC box girders, a full-width crossbeam supported on the front deck segment suspends the bottom frame of the traveller with hanger bars to provide geometry adjustment in varying-depth decks. In the travellers for constant-depth arch ribs, a C-frame suspends one longitudinal main frame on either side of the rib, and the two frames suspend the bottom platform of the casting cell. In the travellers for cable-stayed box girders with a central plane of cables, the C-frame is replaced by two C-hangers supported over the deck webs to avoid interference with the cables and to minimise transverse bending in the cross-section. In the underslung travellers for ribbed slabs comprising edge girders, floorbeams and two planes of stay cables, vertical hangers take support on the outer corner of the edge girders and suspend the bottom frame of the traveller.

In all types of underslung traveller, the longitudinal frames have a front cantilever and a rear stabilisation tail. The front cantilever supports the casting cell, and balancing bogies at the rear end of the frame roll along the deck soffit to prevent overturning.

With both types of traveller the segment is cast in one step (bottom slab first, then webs, and then the top slab), proceeding from the front bulkhead backwards to avoid settlement at the construction joint. The inner web shutters are equipped with vibration windows; additional windows are necessary close to the front bulkhead when precast anchor blocks are used for the top slab tendons. Concrete with early high strength is used to shorten the cycle time in combination with the use of precast anchor blocks.

In-place casting with form travellers is a relatively slow process. After the learning curve for a typical 100 m span has passed, casting a 10 m pier table may take 10 weeks, assembling the travellers on the pier table may take 3 weeks, casting nine pairs of 5 m segments may take another 9 weeks, and casting the closure pour at midspan may take a final week. Casting a 100 m span with two crews and 20 m of casting cell (10 m for the pier table and two 5 m travellers) may therefore take 23 weeks.

The cycle time can be shortened, without major additional cost, by assigning the casting cell for the pier table to a third crew so that the two activities can proceed in parallel. If the pier tables are cast before the travellers become available, the cycle time on the critical path reduces to 13 weeks (see Table 4.1). However, two tower cranes, two sets of stair towers, three crews and two supervisors are necessary.

Leasing a second pair of travellers reduces the cycle time of cantilever construction to 7 weeks. The critical path is now governed by the cycle time of the casting cell for the pier tables, and five crews and three supervisors are necessary. Depending on the length of bridge and typical span and the complexity of the pier tables, a break-even point can be found between the number of pier table casting cells and form travellers to minimise construction duration. This requires, however, several tens of meters of casting cell and numerous supervisors, crews, tower cranes and stair towers.

Another weak point of in-place balanced cantilever casting is that logistics follow production (Homsi, 2012). All materials are delivered on the ground and lifted on the deck, which requires powerful tower cranes. The crews must also reach the working area through stair towers anchored to the piers. Tall piers suggest the use of elevators to save the workers' energy for the real work, which requires additional investment.

An advantage of balanced cantilever casting is that the hammers can be built as soon as the piers become available. Long spans can also be built using relatively inexpensive equipment, as the deck supports its self-weight and the form travellers, even if this typically results in heavy superstructures and high prestressing costs. The load imbalance of staged construction generates longitudinal bending in the piers and foundations; transverse bending and torsion are also not negligible in curved spans. The effects of load imbalance are amplified by the need to design piers and pier cap stabilisation gear for a condition of sudden accidental release of a loaded form traveller, and can be mitigated with asymmetrical pier tables and staggered construction joints.

Balanced cantilever casting with form travellers has few alternatives on the longest spans, while faster alternatives based on linear construction from abutment to abutment exist for 80–130 m spans. Suspension-girder MSSs are used for linear balanced cantilever casting. A 1.5-span girder is supported on the leading pier table and the front cantilever of the completed deck. The girder stabilises the hammer during construction and suspends two long casting cells that shift from the pier table toward midspan to cast the two cantilevers. After midspan closure, the girder is launched forward to the new span, and the casting cells are rolled along the girder to the new pier table and set for the new starter segments.

Linear balanced cantilever casting offers several advantages. Even if the logistics still follow production, materials can be delivered on the deck and partial cage prefabrication saves time and removes activities from the critical path. Access to the working areas is simpler, production is more continuous, and no construction equipment is assembled on the pier tables and dismantled upon midspan closure. Deck prestressing is less expensive because of the lesser influence of staged construction, and the only real limitation is pier availability at the right time. The pros and cons of the different alternatives are compared in Table 4.1.

7.7.   Overhead form travellers

A typical overhead traveller for single-cell box girders consists of two longitudinal frames connected by a front crossbeam and a rear truss at the centre of the traveller. The front crossbeam suspends the bottom platform of the casting cell and the rails for repositioning of the outer shutters of webs and side wings and the inner tunnel form. The rear truss suspends the bottom platform of the casting cell, and often also the outer forms, and it also stiffens the traveller. As the two cantilevers of the hammer must be balanced, the form travellers always work in pairs.

The main frames of the first-generation travellers were long and heavy to enhance the stabilising action of the rear counterweights. Multiple transverse trusses were used to suspend a matching number of bottom crossbeams in the casting cell, which further increased the weight of the traveller and the demand for top slab prestressing in the deck.

New-generation travellers are much lighter and have no counterweights. Stressed tie-down bars at the rear ends of the frames control overturning during casting, in combination with support jacks that prevent accidental downwards movements and provide lateral shear capacity by friction. Because of their critical role in assuring stability to the traveller, redundant tie-downs cross the top slab and are anchored at its bottom surface. The ducts are oversized to ensure geometry tolerances and to prevent shear transfer to the bars by contact.

The main frames are located over the deck webs for direct load transfer; bi-cellular box girders therefore use three frames, and twin- or three-cell box girders use four frames. Each frame comprises a vertical post over the front support, a rear triangular frame to the rear tie-downs, and a front triangular cantilever that supports the front crossbeam (ASBI, 2008). The rear truss connects the posts of the main frames and cantilevers out to suspend the casting cell beyond the edges of the deck wings. The truss is connected to the upper portion of the posts to facilitate access to the casting cell from the previous segment.

The front crossbeam suspends the bottom frame of the casting cell, the main working platform at the bottom slab level, the stressing platform at the top slab level, and a number of spreader beams corresponding to the number of form sets for top slab and webs. The inner tunnel form for the top slab and webs rolls longitudinally on a rail frame suspended from the previous deck segment and supported on a front spreader beam. The outer forms for the side wings and webs may also roll on rail frames suspended under the side wings, or may be carried directly by the traveller; in either case the outer forms include bottom working platforms for form adjustment and tying, and top working platforms for concrete finishing and fabrication of transverse post-tensioning.

Spreader beams at the stressing platform level are also used to suspend the bottom frame of the casting cell. Soffit elevation is adjusted by shortening the lower hanger bars with hollow plunger jacks from the anchorages at the spreader beam. Angle breaks greater than 0.003 radians are typically corrected over multiple segments.

Wind bracing connects the rear truss, front crossbeam and main frames, and stiffens the traveller. Monorail hoist blocks assist delivery of materials and the operations of the casting cell, and further stiffen the traveller. The main frames, rear truss and wind bracing are modular assemblies of members comprising rigid welded nodes and conical pins for fast assembly of field splices (Figure 7.28). Square tubes without end plates are used for most members to avoid welding

Figure 7.28. Conical pins at the connections facilitate site assembly

and to simplify inspection. Large out-of-node eccentricity is accounted for in the design. The modular nature of design facilitates reconfiguration of the traveller to different deck dimensions and segment length. The hanger bars of the casting cell are adjusted during operations to fit segment geometry.

The frame of the traveller is as stiff as is practicable to control the deflections of the casting cell. Trusses are often used instead of the front crossbeams in long-span applications due to the weight of the starter segments. The need for stiffness is further pronounced in the travellers for wide single-cell box girders, due to the length of the side overhangs of the overhead frame. A practical limit for the deflection at the tip of the traveller is 25 mm, and the formwork beams are often designed for a maximum deflection of 1/400 of their length. Multiple post-tensioning bars in through ducts are used to anchor the bottom frame to the soffit of the previous deck segment to avoid joint settlement during casting of the segment.

Pinned field splices and the small number of main frames make wind bracing poorly effective in providing redundancy. In a typical two-frame traveller, failure of the tie-downs of one frame would double the load on the other frame. Load redistribution would require oversized frames and bracing, which would ultimately increase the weight of the traveller and make launching more difficult. Robustness is therefore ensured with oversized tie-downs, and possibly with two pairs of tie-downs on each frame for redundancy. Redundant design of tie-downs prevents the risk of accidental release of a loaded traveller, and this load condition is often assessed for the self-weight of the traveller only (loss of the traveller during repositioning).

The self-launching frame of an overhead traveller may weigh between 25 ton and more than 100 ton depending on the segment weight and dimensions. New-generation travellers are light, less expensive than the first-generation travellers, and much easier to assemble and recondition. Light travellers minimise top slab prestressing, and fast assembly is a must because many parts of the traveller are reassembled on each new pier table.

The casting cell consists of a set of modular wooden forms (Figure 7.29). The inner tunnel form is adjustable to cater for a varying-depth section with thicker webs and bottom slab at the starter segment; it typically comprises an upper portion designed for midspan deck geometry and a modular set of bottom extensions. The width of the bottom form table is also adjustable when the box girder has inclined webs. Inner and outer forms are pulled forward on rail frames suspended from the top slab of the previous segment and supported on the front spreader beams of the traveller; the rail frames are repositioned during launching of the traveller. In many new-generation travellers the overhead frame carries the outer form during launching (Figure 7.30).

Two levels of working platforms are incorporated around the traveller and beyond the front bulkhead. A suspended rear trailing platform facilitates access to the construction joint in the bottom slab for finishing. The working platforms are particularly complex when multiple operations are performed in the casting cell, such as assembling inner diagonal props for the central top slab strip or outer props for the side wings.

7.7.1 Loading, kinematics, typical features

Robustness and redundancy are difficult to achieve in a pure cantilever system designed for maximum weight saving. A heavier traveller would not necessarily be safer during casting and

Figure 7.29. Modular wooden forms (Photo: Doka)

would be riskier during launching. The typical approach is, therefore, to increase reliability rather than strength.

Tie-downs at the rear end of the frames prevent uplift during segment casting, and support jacks avoid accidental downward and lateral displacements by providing vertical constraint and friction shear capacity. A load factor of not less than 2 is recommended for the design of tie-downs and form hangers. Two pairs of tie-downs should be used on each frame for redundancy. The pairs of bars are anchored into short tie beams: if one bar fails, the pair is lost and the second pair resists the uplift force.

High-alloy post-tensioning bars may fail suddenly and without warning due to inclusions and metallurgical defects or low-cycle fatigue, and the dynamic effects can therefore be significant. The following steps are recommended to ensure robustness and redundancy of an overhead form traveller.

1 Use two pairs of tie-downs in every frame to ensure anchorage symmetry in case of bar failure. Design each pair of bars for the full design load.
2 Use only one size of bar in the traveller to avoid confusion regarding anchor hardware.
3 Keep the design stress in the tie-downs sufficiently low to resist load redistribution and dynamic effects.

Figure 7.30. Front and rear hangers suspend the outer form from the overhead frame during launching (Photo: Doka)

4 Limit the number of reuses of tie-down bars and anchor nuts in relation to the preload level.
5 Do not allow couplers in the tie-downs.
6 When the bars are cut in bundles, check the absence of notches, partial cuts or other such damage.
7 Prevent the use of damaged bars and anchorages. Mark damaged bars and dispose of them immediately.
8 Avoid bar bending and shear with ball-and-socket anchorages and no contact in the holes through the top slab.
9 Protect the bars from physical damage and avoid welding around tie-downs, earthing during welding or heating.
10 Preload the tie-downs to avoid decompression of the rear supports under full design load. Residual compression is necessary to transfer lateral loads by friction and to avoid displacements of the traveller.
11 Use similar care for the rail tie-downs. Most accidents with the form travellers occur during repositioning, when the load path passes through the rails, and rail geometry may cause significant prying action in case of bar failure.

Double-acting cylinders with mechanical locknuts are used at the front support to set the traveller horizontal on decks with gradient and crossfall. After tensioning the top slab tendons and

233

Figure 7.31. Support cylinder lifting twin articulated launch bogies from the rail

repositioning the rails, the cylinders are retracted to lower the launch bogies on the rails. The cylinders are designed not to interfere with rail repositioning (Figure 7.31). After anchoring the rails to the new deck segment, tie-down bogies rolling within the rails are used at the rear end of the traveller to prevent overturning during repositioning. The rear end of the traveller is secured directly to the deck during casting, and during launching is secured indirectly through the rails. Main frames and launch rails are located over the deck webs for direct load transfer.

The bottom frame of the casting cell is suspended from the traveller and moves forward during launching. The outer forms may also be repositioned with the traveller or may be rolled forward on rail frames at the end of launching. After cage fabrication for the bottom slab and webs, the inner tunnel form is pulled forward and adjusted to the new segment geometry, and the cage of the top slab is fabricated to complete segment reinforcement.

New anchorages for the form rail frames and new ducts for tie-downs and anchor bars of launch rails and casting cell are embedded in the new segment. Operating overhead form travellers requires a great number of embedded items to be positioned with great accuracy in every deck segment.

7.7.2 Support, launch and lock systems
Pairs of identical form travellers are used on the hammers for load balance and to accelerate assembly. The length of the pier table dictates the geometry of the launch rails and the sequence

Figure 7.32. Full-length pier table of a three-cell box girder with corrugated-plate webs (Photo: Doka)

of operations to assemble the travellers. A 10–12 m pier table is typically necessary to accommodate two independent travellers (Figure 7.32). Three techniques may be used with shorter pier tables: fast-split assembly, temporary bracing, and side assembly.

With the fast-split technique, the rear cantilevers of one traveller are modified to overlap with the frames of the other traveller. The bottom chords and diagonals of the special traveller include U-forks to avoid conflicts with the standard traveller (Figure 7.33). The launch bogies of the two travellers are designed to use the same rails. After anchoring the rails to the pier table, the standard traveller (the right-hand traveller in Figure 7.34) is assembled without tie-down bogies and anchored to the deck with tie-down bars, and then the special traveller is assembled and anchored with a different set of tie-down bars.

After casting the two starter segments and tensioning the top slab tendons, the special traveller pushes the rails over the left starter segment, and a second set of rails is assembled under the standard traveller. Rear tie-down bogies are applied to the special traveller to reposition it over the left starter segment with standard operations. This permits having access to the rear end of the standard traveller to assemble the tie-down bogies, release the tie-down bars and launch it over the right starter segment. The fast-split technique is used for pier tables ranging in length from 6.5 m to 12 m.

On shorter pier tables, the rear cantilevers of the two travellers may be replaced with temporary cross braces to cast the two starter segments with a common overhead frame (Figure 7.35). This technique works on pier tables as short as 4–5 m and is frequently used also in the cable-stayed

Figure 7.33. Rear U-forks on the special traveller (left) for fast-split assembly (Photo: NRS)

Figure 7.34. Fast-split assembly on a short pier table (Photo: NRS)

Figure 7.35. Temporary bracing replaces the rear frames on a short pier table (Photo: Strukturas)

bridges. After casting the starter segments, the launch rails are extended over the new segments to support auxiliary front bogies that stabilise the travellers during removal of the temporary braces. One traveller is moved forward to create the room for assembly of the rear frames, and the auxiliary front bogies are removed to complete launching with the rear tie-down bogies (Figure 7.36). The same operations are repeated on the other traveller.

A third possibility is assembling the form travellers individually. Side assembly also works with 4–5 m pier tables, but it lengthens the cycle time of the hammer. A first traveller is assembled on the pier table and the starter segment is cast (Figure 7.37). After tensioning short top slab tendons anchored within the pier table, the traveller is positioned for casting the second segment, and this creates sufficient room to assemble the second traveller on the pier table.

Repositioning a form traveller is a two-step process: first the launch rails are pushed forward for the length of the new segment and anchored to the deck, and then the traveller is lowered on the rails and pushed forward. The length of the launch rails is twice the longest segment. The rails provide a smooth surface for rolling the traveller, provide guidance and lateral load transfer, and minimise the effects of surface irregularities in the top slab.

The rails and traveller are repositioned using the same set of long-stroke double-acting launch cylinders (Figure 7.38). The cylinders push the front bogies forward when launching uphill, and are repositioned alternately during launching to avoid uncontrolled backward movements of the traveller. Neither workers nor extra load should be allowed on the traveller during launching.

Figure 7.36. Extension of the launch rails and application of the rear frames (Photo: Doka)

Figure 7.37. Side assembly of the first traveller (Photo: Doka)

Figure 7.38. Long-stroke double-acting launch cylinder and movable contrast plates (Photo: Doka)

Pantograph clamps are rarely used to anchor the launch cylinders to the rails because these devices work only under active thrust and offer no safety in downhill launching. Whatever the hammer gradient may be, one of the two travellers will necessarily be launched downhill. Pins and perforated rails are a low-cost solution to safely anchor double-acting launch cylinders, and the traveller can also be pulled backward if necessary.

Two tie-down bogies roll within the launch rail at the rear end of the main frame for redundant control of overturning during launching. After removing the support jack, a vertical pull cylinder anchored to the rail with a sliding clamp lowers the frame to release the tie-down bars. Releasing the cylinder transfers the load to the tie-down bogies. The load in tie-down bogies and pull cylinder is smaller than the load in the tie-down bars used during segment casting, as it is related only to the weight of the traveller.

7.7.3   Performance and productivity

Fast-split assembly and temporary bracing typically require 3–4 weeks to assemble a pair of travellers. Casting the starter segments takes 1–2 weeks after the learning curve has passed. With fast-split assembly, the starter segments are therefore cast in 4–6 weeks. Temporary bracing requires another week for removal of the braces and completion of the travellers; this technique is hence 1 week slower. Side assembly takes 2 weeks for every traveller, but the second traveller cannot be assembled prior to launching the first one; also in this case, therefore, the two starter segments are typically cast in 6–7 weeks.

The cantilever segments are typically cast in a 5-day cycle after the learning curve has passed. The operations are distributed over the 5 days as follows.

1 Stress the transverse tendons in the top slab. Strip outer and inner forms and remove the anchorages of the casting cell. Push the launch rails forward and anchor the rails to the top slab. Complete stressing of the cantilever tendons. Retract the front support cylinders and then remove the rear support jacks. Actuate the rear pull cylinders to engage the safety clamps, remove the rear tie-down bars, and release the pull cylinders to transfer the uplift load to the tie-down bogies.
2 Reposition the traveller over the new segment. Extend the front support cylinders, set the overhead frame horizontal with the rear pull cylinders, insert the rear support jacks, apply and stress the rear tie-down bars, and tighten the locknuts of the support cylinders. If the overhead frame does not carry the outer forms during launching, roll the outer forms forward to close the casting cell. Level and fix the outer forms in position. Install the prefabricated bulkhead.
3 Fabricate the steel cage for the bottom slab and webs; the cage is typically tied within the casting cell. Place tendon ducts in the bottom slab and the anchorages of continuity tendons ending in the segment. Form the blisters of continuity tendons.
4 Roll the inner tunnel form forward. Adjust the geometry and install ties between inner and outer forms for the webs. Fabricate the steel cage for the top slab. Place tendon ducts, the anchorages of the cantilever tendons ending in the segment, and the embedded items needed to operate the traveller on the new segment.
5 Survey the final geometry. Cast the segment in one phase using a crane and bucket. Use chutes through the front bulkhead or openings in the top slab form to cast the bottom slab.

The segments are cast starting from the front bulkhead and proceeding backwards to minimise settlement at the construction joint. The bottom web–slab nodes are cast first, followed by central strip of the bottom slab, webs and top slab. The inner form of the webs is equipped with windows for vibration of 0.50–0.60 m symmetrical lifts; additional windows close to the front bulkhead are necessary when precast anchor blocks are used for the top slab tendons. Longitudinal and transverse top slab tendons are fabricated during segment curing after removing front and lateral bulkheads.

The constant-depth closure segments at the abutments are typically cast on ground falsework in two or three phases.

7.7.4 Structure–equipment interaction
The interaction between form travellers and the bridge being erected is mostly related to the weight of the travellers, which increases the demand for top slab prestressing and the shear force in the webs. Removal of the travellers at hammer completion modifies the deflections of the cantilevers and releases shear and bending. The travellers of the adjoining hammer are removed after achieving midspan continuity, and the new hammer is therefore free on one side and connected to the completed deck on the other side. Removing the traveller weight modifies permanent stresses and camber in the bridge, and additional stress modifications result from releasing the stabilisation gear at the pier cap.

The project time-schedule also affects some structural details. When two hammers are erected simultaneously, the clearance at the midspan closure must permit unrestricted operations of the travellers (Figure 7.39). When the closure pour is too narrow to thread stressing jacks on the strand tails extending from the anchor heads, top slab blisters are cast within the end segments

Figure 7.39. Midspan closure (Photo: Doka)

of the hammer. A few top slab continuity tendons anchored in blisters are used in the midspan region to resist negative bending, and this requires some adaptations in the casting cell geometry at the top slab level.

Long and wide deck segments may be affected by geometry issues and cracking due to the flexibility of forms and overhead frame. Form flexibility is more pronounced in the heavy starter segments of long cantilevers and in wide decks.

7.8. Underslung form travellers

In the underslung form travellers for PC box girders, one or two full-width crossbeams or C-frames supported on the front deck segment suspend a two-segment bottom frame beneath the bottom slab of the box girder. The bottom frame has a front cantilever that supports the casting cell and a rear stabilisation cantilever. Balancing bogies at the end of the rear cantilever roll under the bottom web–slab nodes to prevent overturning during repositioning of the traveller.

Bottom crossbeams or transverse trusses connect the main trusses under the casting cell and stiffen the traveller (Figure 7.40). Additional longitudinal trusses are often necessary to support the thick bottom slab of the starter segments of long cantilevers. The outer forms and front stressing platform are supported on the bottom frame and provide minimal stiffness to the assembly. The inner tunnel form rolls on suspended rail frames as in an overhead traveller.

Underslung travellers are more complex to operate and inspect than are overhead travellers. When the pier table is 10–12 m long, an underslung traveller is fast to assemble, as the launch rails, crossbeam and a rear lifting crossbeam are placed on the deck in four lifts, and the bottom frame is strand jacked into position in one operation.

Figure 7.40. Underslung form travellers (Photo: Strukturas)

Underslung travellers are well suited for use with constant-depth decks, as the casting cell makes up part of the traveller framing. These travellers are, therefore, frequently used for cable-stayed decks and cable-supported arch ribs, and less frequently for varying-depth PC box girders.

Unobstructed access to the casting cell is a major advantage in cable-supported decks and arch ribs and also allows cage prefabrication. In prestressed composite decks, the corrugated plate web panels can be easily lowered into the casting cell to be welded to the previous segment. However, the casting cell operations require the use of a service crane on the deck, which increases the load imbalance.

In balanced cantilever decks comprising PC slabs and trussed webs, long underslung travellers supporting two staggered casting cells may be used to cast two slab segments from node to node of the truss panel (Figure 7.41). Handling heavy diagonals for the webs requires auxiliary service cranes on the deck.

In most PC box girders, the steel cage is assembled within the casting cell, and the overhead travellers offer the advantage of easier sheltering and insulation. Monorail hoist blocks suspended from the overhead frame are less expensive and much lighter than service cranes operating on the deck to assist the operations of the casting cell.

7.8.1   Loading, kinematics, typical features

One or two adjacent crossbeams suspend the bottom frame of the traveller from the front deck segment. A C-frame with telescopic form hangers may replace the crossbeam for geometry adjustment, and provides a flexural connection with the bottom frame in varying-depth decks. In most cases, however, the casting cell is suspended with hanger bars. The crossbeam takes

Figure 7.41. Underslung travellers with staggered casting cells (Photo: VSL)

support on the deck with launch bogies that include double-acting support cylinders with mechanical locknuts. The cylinders lift the crossbeam from the launch rails to cast the new deck segment and to reposition the rails forward over the new segment. After tensioning the top slab tendons and anchoring the launch rails to the new segment, the support cylinders are released to lower the front bogies on the rails for launching.

Balancing bogies at the rear end of the main frames roll beneath the bottom slab of the previous deck segment to control overturning during launching. The rear bogies incorporate balancing cylinders with mechanical locknuts for geometry adjustment of the casting cell in cooperation with the support cylinders of the front crossbeam and to resist the overturning couple of segment casting.

The casting cell consists of a set of modular forms. The bottom form table and web shutters are adjustable to cater for a varying-depth section. The bottom form table, outer forms for webs and side wings, front bulkhead and stressing platform make up part of the traveller and move forward with it during repositioning.

Side working platforms over the bottom frame facilitate form adjustment. The working platforms are extended to the rear cantilevers to provide access to the balancing bogies. Conventional working platforms are applied to the casting cell at the top slab level.

After fabricating the steel cage for the bottom slab and webs, the inner tunnel form is pulled forward and adjusted to the new segment geometry. The inner form rolls on a rail frame suspended from the top slab or supported on the bottom slab of the previous segment and the bottom form table of the traveller. The steel cage may be prefabricated, although cage delivery on the hammer requires a wheeled crane on the deck.

7.8.2   Support, launch and lock systems

Repositioning an underslung traveller is a two-step process. First the launch rails are pushed forward for the segment length and anchored to the deck. Then the front support cylinders and rear balancing cylinders are retracted to activate the launch bogies, and the traveller is moved over the new segment. The rails and traveller are pushed forward using the same set of cylinders. The launch cylinders are repositioned alternately during launching to avoid uncontrolled movements of the traveller.

Paired long-stroke double-acting cylinders provide redundancy during repositioning. Pantograph clamps or contrast plates are sometimes used to anchor the cylinders when the deck has no gradient. When the deck has a gradient and one of the travellers is therefore launched downhill, pins and perforated rails are used for enhanced safety and to pull the traveller back if necessary.

Launching an underslung traveller is more complex than launching an overhead traveller and requires the presence of personnel on the traveller. The rear balancing bogies have a fundamental role in the control of overturning; load limitation to manageable levels requires long rear cantilevers in the main trusses, which further increases the weight and flexibility of the traveller. The rear balancing bogies are equipped with rubber wheels, and irregularities in the bottom deck surface make motion irregular. Motion irregularities are amplified when the bottom frame is suspended from hanger bars devoid of flexural stiffness.

However, dismantling the travellers after midspan closure is simpler. The overhead travellers lower the casting cell on the ground, but then the entire self-launching frame must be moved back to the pier table to be disassembled with the tower crane. An underslung traveller lowers the casting cell on the ground by strand jacking, and then the crossbeam and launch rails are lowered in three lifts.

7.8.3 Performance and productivity

When the pier table is 10–12 m long, two underslung travellers can be assembled at the same time. After anchoring the launch rails to the pier table, the crossbeam is placed on the rails, and main frame and casting cell are strand jacked in one operation and suspended from the crossbeam. Fast-split assembly and side assembly rarely work on shorter pier tables because the rear balancing frames conflict with the pier. Assembly with temporary bracing requires working under the deck, which is complex, risky and labour intensive.

The casting cell is permanently applied to the bottom frame. The first assembly takes 3–4 weeks, while the following reassemblies are generally faster. Casting the starter segments takes another 2–3 weeks prior to getting past the learning curve. The cantilever segments are cast in a 5-day cycle in a sequence of operations similar to that followed when using an overhead traveller. Underslung travellers require a small number of embedded items, which may be an advantage where there are congested steel cages.

7.8.4 Structure–equipment interaction

The interaction between form traveller and the hammer, pier and foundations is similar to the interaction seen with an overhead traveller. When the hammer is stabilised with pier brackets, the rear balancing frame of the traveller may conflict with the stabilisation gear during casting of the starter segment. Cage prefabrication requires a service crane on the hammer, which further increases longitudinal load imbalance and top slab prestressing.

Underslung travellers are suspended beyond the edge of the side wings, and are therefore more deformable than overhead travellers, where the main frames are located over the webs. Long and wide deck segments may be affected by geometry and cracking problems due to the flexibility of the bottom frame, hangers and crossbeam. Additional support lines may be provided outside the webs by hanger bars suspended from the crossbeam and crossing the top slab. Hangers within the box cell conflict with the operations of the inner tunnel form.

7.9. Form travellers for concrete arches

Underslung travellers suspended from full-width crossbeams are used for in-place casting of cable-supported arch ribs. The crossbeam is supported on the front rib segment and suspends the bottom frame of the casting cell by means of a triangular frame on either side of the rib (Figure 7.42). Balancing bogies at the rear end of the bottom frame roll beneath the rib soffit at the previous segment to control overturning during launching. Rubber wheels are used in the balancing bogies to avoid damaging the young concrete.

Hydraulic cylinders with mechanical locknuts release the front rollers and rear launch bogies prior to casting the segment. After anchoring the crossbeam to the arch rib, the segment is cast in one

Figure 7.42. Two 18.7 m, 122 ton underslung travellers for 7.9 m, 267 ton arch rib segments (Photo: NRS)

Figure 7.43. Underslung travellers for solid rib segments (Photo: Strukturas)

phase and the top shutter is closed with the progress of filling to facilitate concrete vibration in the bottom slab and webs. The casting cell is stiffer than the casting cell of an underslung traveller for PC box girders as the arch ribs are devoid of side wings. Shorter crossbeams and triangular suspension frames also transfer the load to the rib webs more directly.

The casting cell does not require major geometry adjustment. The length, width and depth of the rib segments are typically constant, and the parabolic profile of the arch is achieved by offsetting a circular casting cell. Casting operations are even simpler in solid arch ribs (Figure 7.43). The cables that support the cantilever rib are applied after repositioning the traveller to avoid interference with the crossbeam and to increase the age of concrete at stressing of the stay cables.

Working platforms are incorporated around the traveller. The working platforms often need modifications over the progress of construction due to gradient reduction. The inner tunnel form of the box rib is stripped from the previous segment and rolled forward within the cage of bottom slab and webs for the next segment.

7.9.1 Loading, kinematics, typical features
The bottom frame of the casting cell includes two edge trusses connected by crossbeams that support the form table. The top crossbeam suspends the edge trusses with a triangular frame on either side of the rib for direct load transfer and simple connections. Field splices with

single large-diameter pins facilitate assembly and inspection, are insensitive to vibrations, and are frequently used in all types of form traveller.

Front cylinders at the main crossbeam and rear cylinders at the balancing bogies provide geometry adjustment for the casting cell. The front cylinders press the rear end of the bottom form table against the rib to avoid settlement and steps at the construction joint, and the rear cylinders set the elevation of the front end of the casting cell by rotating the traveller about the crossbeam.

7.9.2   Support, launch and lock systems

Launching is a two-step process: first the rails are pushed forward for the entire segment length and anchored to the arch rib, and then the crossbeam is lowered onto the rails and pushed forward. PTFE launch skids are often used at the crossbeam due to the high design load. The rails and traveller are repositioned using the same set of long-stroke double-acting cylinders. Paired launch cylinders are used for redundancy, and strand brakes incorporated within the launch rails further enhance safety and robustness in the launches at the springing of the arch, actions which are still within the crew's learning curve period.

Launching is complicated by the gradient of the first rib segments. Launch rails and contrast systems for the thrust cylinders are designed for the weight of the traveller. Adjustable brackets anchored to the top slab of the arch rib are used to lock the crossbeam in place and to transfer the longitudinal loads of segment casting. The brackets may be permanently connected to the crossbeam for fast repositioning during launching.

7.9.3   Performance and productivity

The travellers are assembled at the springing of the arch. The rear balancing section of the bottom frame is placed on the rib foundation and kept in place with props and diagonal ties. The casting cell is temporarily assembled on the rear section of the bottom frame to cast the starter segment (Figure 7.44). The front section of the bottom frame is applied to the rear section, and the outer form is stripped and shifted forward to the final position on the traveller.

Launch rails and anchor brackets are applied to the top slab of the starter segment, the front support cylinders are applied to the brackets, and the crossbeam is placed on the cylinders and connected to the bottom frame with a triangular frame on either side of the rib. Working platforms are applied to the traveller, and the temporary props and ties are removed. The traveller is thus ready for casting the second rib segment and self-launching.

The casting cell makes up part of the traveller and it typically takes 4–6 weeks to assemble the entire machine. Casting the starter segment takes 2–3 weeks. The following rib segments may be cast in a weekly cycle after the learning curve has passed, even if the operations of the cable crane are often the bottleneck of productivity. Compared to an underslung traveller for PC box girders, minor geometry adjustment and the absence of post-tensioning simplify the casting cycle.

The new pair of stabilisation cables is applied to the arch rib after launching the traveller. The steel cage may be prefabricated and delivered with the cable crane, but in most cases it is assembled within the casting cell. Concrete distribution arms applied to climbing platforms are fed via pipes supported on the rib. Concrete feeding with buckets via cable crane requires powerful hoists and is typically slower.

Figure 7.44. Casting of the starter segment

The segments are cast in one phase proceeding from the construction joint upwards. Concrete with early high strength is used to launch the traveller and to tension the stabilisation cables as soon as possible.

7.9.4 Structure–equipment interaction

The interaction between the form traveller and the arch rib is mostly related to the weight of the traveller, which increases the load in the stabilisation cables and the flexural and shear demand in the rib. The travellers are removed from the arch prior to midspan closure to facilitate the final geometry adjustment. Midspan jacking is rarely necessary, as relieving the stabilisation cables after midspan closure generates initial compression in the ribs.

Long rib segments may be affected by geometry and cracking issues due to the flexibility of the forms and suspension frames. In a few heavy travellers the tip of the casting cell has been suspended by stabilisation cables, although this complicates the operations and lengthens the cycle time. Front suspension is also poorly effective in the arch crown area due to cable inclination.

7.10. Form travellers for cable-stayed and extradosed bridges

Numerous cable-stayed bridges have been designed with two planes of stay cables supporting a composite deck grillage comprising steel edge girders, crossbeams, stringers and a PC deck slab (IABSE, 1997). Precast deck slab panels are made continuous with the steel grillage by means of in-place stitches over the top flanges; this increases deck stiffness and strength, and involves lower short- and long-term costs than using an orthotropic steel deck. Plate girders are typically used for the edge girders of single-deck bridges, and deep trusses are used for double-deck bridges because of clearance and ventilation requirements. The stay cables are anchored over the edge girders or in tubular anchor brackets bolted to the outer face of the webs.

A composite grillage is much lighter than a PC deck, and this generates cost savings in stay cables, pylons and foundations. The deck is erected using light self-launching derrick platforms, which handle the modules of the edge girders, the crossbeams and the deck slab individually. Lighter erection equipment and less load imbalance reduce the flexural stresses of staged construction in pylons and foundations. The drawbacks are the cost of the steel grillage, the maintenance costs over time, the need for an efficient temporary restraint between the deck and the pylons during balanced cantilever construction, and the need for a relatively wide deck to ensure lateral stability of the long cantilevers.

A triangular self-launching derrick platform supports a vertical mast at the front 90° corner (Figure 7.45). Diagonal braces connect the mast to the other two corners of the base frame to transfer lateral loads. The triangular base frame facilitates assembly of two opposite derricks on a short pier table. A slewing ring at the base of the mast supports a lifting arm that can be lifted or lowered to modify the load radius. The PPU and hoisting winch are applied at the rear corner of the base frame to increase stability against overturning.

The slewing arm can lift laterally or from behind, and the torque applied to the deck is resisted by the stay cables. Edge girders, stringers, crossbeams and deck slab panels are handled individually,

Figure 7.45. Self-launching derrick platforms (Photo: VSL)

and the design load of these cranes is therefore relatively low. The edge girders are cantilevered out from one cable anchor point to the next one without temporary supports; crossbeams and stringers are applied during fabrication and tensioning of the new pair of stay cables; and the precast deck slab panels are finally applied to stiffen the frame (Gimsing and Georgakis, 2012). Continuity between the deck panels and the steel grillage is achieved by filling solutions of continuity over the top flanges with non-shrink concrete. Structural continuity between adjoining panels is achieved using lapping reinforcement projecting from the units, prestressing tendons or coupled prestressing bars. The transverse joints are gradually subjected to axial compression as the cantilever length increases due to the longitudinal component of the pull in the stay cables.

The interaction between the self-launching derrick platforms and the composite deck is complex and often nonlinear as the deck is flexible. The lifting platforms are anchored to the concrete slab for better control of load dispersal and buckling, and are long in order to reduce the load imbalance due to the hoist couple.

Several cable-stayed bridges have also been designed with PC ribbed slabs comprising solid edge girders and steel or PC crossbeams. The use of multiple stay cables, in both harp and fan arrangements, with closely spaced anchor points reduces the demand for flexural stiffness in the deck. A ribbed slab is easier to cast than a box girder, the stay cables are anchored at the bottom of recess pipes embedded in the edge girders, which directly resist the longitudinal component of the pull in the stay cables, and maintenance is less expensive than for a composite grillage. However, a PC ribbed slab is heavier than a composite grillage, the load imbalance of staged construction is larger, and two planes of stay cables are still necessary because of the poor torsional constant of the open section.

The torsional stiffness and strength of a wide box girder integral within a central pylon are typically sufficient to support the deck with a central single plane of stay cables. Many cable-stayed bridges carrying separated highways have been designed with a single plane of cables to simplify the pylon, to reduce the number of anchorages, to improve the aerodynamic stability of the deck with a streamlined profile of the cross-section, and to reduce the drag coefficient.

Cable-stayed box girders may be made of steel, PC or composite construction. Streamlined macro-segmental steel box girders erected using floating cranes or lifting frames are used for the longest spans due to the higher strength-to-density ratio of steel. PC is the typical choice for shorter spans due to the lower cost of materials and less maintenance needs over time. Prestressed composite box girders with two concrete slabs and steel corrugated-plate webs are earning popularity for the extradosed bridges in the 100–200 m span range because of the weight saving, the high flexural efficiency of the cross-section, and the capability of corrugated-plate webs not to interact with longitudinal post-tensioning (Rosignoli, 2002).

Compared to a PC ribbed slab, the constant of torsion of a PC box girder is 2–3 orders of magnitude higher, the moment of inertia is one order of magnitude higher, and the cross-sectional area is similar. A box girder is, therefore, more sensitive to imposed deflections, but is less unstable under the compression loads applied by the stay cables, and it has approximately the same self-weight.

Modern stay cables are made of individually protected, parallel seven-wire strands housed within butt-welded steel pipes or HDPE pipes connected by fusion. Protection includes an epoxy coating or a corrosion-inhibiting coating and individual polyethylene sheathing applied to the regular or galvanised strand. PPWSs are more compact than strand stay cables and are sometimes used to reduce wind drag in long-span bridges. In either case, the recess pipes are welded or bolted to the anchorage bearing plates and installed prior to casting the segment.

Single-cell PC box girders supported by a central single plane of stay cables are often too wide and heavy for segmental precasting with delivery on barges and erection with floating cranes (low-level decks) or lifting frames (high-level decks). Segmental precasting has been used successfully in very long bridges when the cable-stayed deck could be divided into twin box girders, and the same cross-section could be used also for the approaches (ASBI, 2008). In most cases, however, single-cell PC box girders are cast in-place using form travellers.

Because of the marked flexural stiffness and strength of a box girder, overhead travellers may often be repositioned prior to fabrication of the stay cables in order to avoid interference (Figure 7.46). This is almost unavoidable for the first segments of the cantilever, as the cables are almost vertical. In some travellers the side overhangs of the rear truss have been braced to the main posts and the central truss panel has been replaced with a top tie to minimise interference with the stay cables.

Figure 7.46. Overhead traveller for the central plane of stay cables (Photo: Doka)

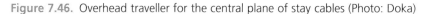

Figure 7.47. Overhead traveller for extradosed bridges (Photo: Doka)

Overhead travellers facilitate sheltering and insulation of the casting cell but complicate cage prefabrication. Monorail hoist blocks suspended from the overhead frame are used to facilitate insertion of cage panels and steel struts into the casting cell, as well as to handle the strand coils and stressing jacks. Two form tables rolling on rail frames are used to form the side wings with outer struts, and two inner sets of forms for the top slab and web are necessary in a single-cell box girder with diagonal struts within the cell. The front crossbeam of these travellers therefore suspends six spreader beams.

Overhead travellers are often used for in-place casting of extradosed bridges because the cables have a harp arrangement with constant inclination and the large moment of inertia of the cross-section allows cable fabrication after advancing the traveller (Figure 7.47). New-generation extradosed bridges often use steel corrugated-plate webs, and the traveller may be easily equipped with monorail hoist blocks to handle the web panels within the casting cell and to keep them in place during welding to the preceding panels. The form tables for the concrete slabs do not interfere with the steel webs, and the web panels may therefore be longer than the slab segments so as to shorten the cycle time and to reduce the number of field welds.

The underslung travellers for single-cell box girders with a central single plane of stay cables are suspended from two C-hangers. Each hanger rolls on a launch rail anchored to the deck over the web and suspends a longitudinal truss under the bottom web–slab node. The central clearance between the C-hangers avoids interference with the stay cables. Balancing bogies at the rear end of the trusses roll beneath the deck soffit to control overturning during launching.

Figure 7.48. Underslung traveller for a cable-stayed ribbed slab (Photo: Strukturas)

Transverse trusses or crossbeams connect the front cantilevers of the main trusses to support the bottom form table of the casting cell and to stiffen the traveller. The cable-stayed box girders typically have constant depth and the bottom crossbeams do not interfere with the geometry adjustment systems of the integrated platform.

A ribbed slab is simpler to cast than a box girder, and the use of an underslung traveller further simplifies the operations and minimises the interference with the stay cables (Figure 7.48). In an underslung traveller for ribbed slabs, two vertical hangers rolling on the outer edge of the deck suspend modular rectangular trusses. The length of the trusses is about 2.5 times that of the typical deck segment (Figure 7.49). Rear balancing bogies roll beneath the soffit of the edge girders to control overturning during launching. Bottom crossbeams connect the front cantilevers of the trusses to support the casting cell and to stiffen the traveller.

The travellers for cable-stayed box girders and ribbed slabs are heavier than the travellers for non-cable-supported balanced cantilever bridges due to the wider deck and the longer segments based on cable spacing. Modern cable-stayed bridges use closely spaced stay cables to facilitate erection and to lighten the deck, and the casting cell of the traveller may be 9–12 m long. Stiff modular truss assemblies and three-dimensional space frames are necessary for the bottom frame of the traveller to control form deflections. The shutters do not require major geometry adjustment, as the cable-stayed bridges are typically rectilinear and the segments have repetitive geometry and constant depth and length.

Working platforms are incorporated around the traveller. A stressing platform beyond the front bulkhead facilitates fabrication and tensioning of longitudinal slab tendons. Working platforms on the main trusses facilitate access to the rear balancing bogies for geometry adjustment and to

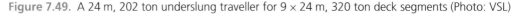

Figure 7.49. A 24 m, 202 ton underslung traveller for 9 × 24 m, 320 ton deck segments (Photo: VSL)

operate the balancing cylinders. The working platforms are also used for access to the anchorages of the stay cables. Top lateral working platforms are also necessary when the deck is post-tensioned transversely.

7.10.1 Loading, kinematics, typical features

The form travellers for cable-stayed bridges are conceptually similar to the travellers for non-cable-supported box girders or arch ribs. Launch devices and operations are also similar. The travellers are designed for heavier loads, depending on the length and width of the segments, and controlling the deflections of the casting cell is a major issue because of the flexibility of such long frames and crossbeams.

The typical erection cycle of a segmental PC cable-stayed bridge consists of alternately casting the segments and fabricating and stressing the stay cables. The peak erection stresses are typically reached in the deck (negative longitudinal bending) and the stay cables (pull in the frontmost cable) during pouring of segments, when the weight of the segment adds to the weight of the form traveller. Shear lag effects are also more critical during construction than during service.

The flexural stiffness of the box girders spreads the load of the new segments more evenly among the stay cables, reduces negative bending in the deck, and allows erection of short unsupported front cantilevers. Casting short segments facilitates control of longitudinal load imbalance on such long spans, the spacing of stay cables may be divided into two or three segments, and re-stressing the cables between pours of short segments further enhances control of negative bending. The segments of flexible ribbed slabs are often as long as the full spacing of the stay cables, but deck bending, stay cable overloading and longitudinal load imbalance must be reduced.

Load imbalance during construction may be resisted with moment-resisting pylons, temporary diagonal cables anchored to pylon foundations to reduce bending in the critical pylon sections at the deck level, temporary vertical cables anchored to special foundations to also control buffeting, and temporary stabilising cables connecting the top of the pylon to outside anchors.

The balanced cantilevers of cable-stayed bridges can reach exceptional lengths that make asymmetrical wind loading more likely. Aerodynamic stability during construction is critical, and horizontal and vertical wind loads during construction are often determined by means of wind tunnel testing. Stay cable vibrations and wind/rain-induced vibrations may also occur during construction when the vibration periods of the stay cables and hammer are similar.

7.10.2 Support, launch and lock systems

Repositioning an overhead traveller for cable-stayed box girders is a two-step process: first the launch rails are pushed forward and anchored to the new deck segment, and then the support cylinders of the traveller are retracted to activate the front launch bogies, the rear tie-downs are released to transfer the load to the tie-down bogies, and the traveller is pushed forward.

The rails and traveller are launched using the same set of cylinders. Paired long-stroke double-acting cylinders provide redundancy during launching, and restrain the traveller during repositioning of the adjacent cylinder to prevent involuntary movements. Through pins are used to anchor the cylinders to perforated launch rails and to pull the traveller back if necessary.

In the new-generation travellers, the outer form shell is suspended directly from the overhead frame and repositioned during launching. The form table for the central top slab panel is divided into two strips to form the central anchor beam for the stay cables without conflicts with the diagonal struts. After fabrication of the reinforcement cage for the bottom slab and webs, the inner forms are rolled forward on two rail frames supported on the front spreader beams of the traveller.

Launching an underslung traveller requires the presence of personnel on the traveller and is further complicated by the weight of these machines. The rear balancing bogies have a basic role in the control of overturning; for heavy applications, cast-iron rolls on equalising beams are used in combination with short rail segments applied to the soffit of the edge girders.

7.10.3 Performance and productivity

A cable-stayed bridge is a self-anchored system. The load capacity of the cables depends on transferring the horizontal component of the pull in the cables through the deck, and therefore the bridge must be erected sequentially (Gimsing and Georgakis, 2012). During balanced cantilever erection, one or two cables are installed on either side of the pylon every time an anchor point is reached, and productivity of deck erection and cable fabrication are tightly interrelated. Bridges with a harp configuration of the stay cables may be faster to erect, as only the lower portion of the pylons must be completed prior to installation of the first stay cables; in these bridges, however, the productivity of slip forms or jump forms for the RC pylons may also affect the erection rate.

Steel pylons are erected using ground or floating cranes for the lower portion and with climbing cranes for the upper portion. Derricks with self-climbing brackets are an alternative to conventional climbing tower cranes. Powerful hoisting winches may be used at ground level to lighten the crane and to hoist faster with a lower reeving ratio.

RC pylons are cast using slip forms or jump forms. Slip forming offers faster erection rates but requires continuous concrete supply, and is therefore used on land or at water locations that can be reached with temporary feeding lines from the coast. Jump forms are the typical solution when concrete is delivered by barges, even though erecting tall pylons with 4–5 m lifts in a weekly cycle takes many months. Both casting methods require temporary props between the inclined legs of A- or diamond-shaped pylons to control transverse bending during leg extension. Trusses anchored to the pylon legs are used to support permanent post-tensioned crossbeams. Multi-level working platforms are often applied above the casting cell for fabrication of multi-segment reinforcement cages and application of anchor hardware for the stay cables.

Compared to an earth-anchored suspension bridge, deck and cable erection can start prior to completion of both pylons, and the two hammers can be erected independently. As a drawback, however, balanced cantilever erection is typically much slower than lifting macro-segments for suspension from the main cables.

The pier table of a cable-stayed composite grillage is relatively simple to erect. The steel frame is assembled on the ground, strand jacked into position, and suspended from the pylon with the first set of stay cables. In low-level bridges the steel frame may also be erected, using ground cranes, on four temporary piers. Precast deck panels are assembled on the steel frame and made continuous in order to stabilise and strengthen the assembly, and self-launching derrick platforms are applied to the pier table to erect the cantilevers. The presence of a support crossbeam between the pylons further simplifies the operations.

Casting the pier table of a PC ribbed slab or box girder is more complex and requires a special casting cell. The casting cell may be supported on ground falsework (low-level decks) or brackets applied to the pylon (high-level decks), or may be suspended from the pylon by means of temporary stay cables. The design of the casting cell depends on the type of connection between the deck and the pylon (suspended or supported ribbed slab, monolithic connection of the box girder to a single pylon) and on pylon geometry (lateral pylons with or without crossbeam, central pylon). Cantilever casting cells supported on brackets at the central pylon and suspended casting cells are particularly complex and flexible.

The pier table is one of the most difficult parts of construction, and its length is minimised in relation with the traveller assembly. When overhead travellers are used, the outer form of the casting cell for the pier table is rarely reused in the travellers. The travellers are assembled using the same techniques as applied for non-cable-supported balanced cantilever bridges; assembly is complicated by the width of the travellers and the small dimensions of the pier table. Assembly is more complex when the pier table is suspended from the pylon, due to the presence of the first stay cables.

The casting cells of underslung travellers could be supported on ground falsework to cast the pier table; in most cases, however, the pier table is cast using a special casting cell, and the travellers are assembled on the ground, or on temporary trusses when over water, and strand-jacked into

Figure 7.50. Strand-jacking of two underslung travellers assembled on support trusses (Photo: VSL)

position from brackets anchored to the pylons (Figure 7.50). Temporary bracing connects the two casting cells during the construction of the starter segments. Launch rails and vertical hangers are assembled on the starter segments during fabrication of the new stay cables, and the casting cells are separated during the first launch to complete the travellers and make them self-supporting, (Figure 7.51).

The cycle time for a pair of segments depends on the cross-section of the deck, the length and width of the segments, and the technique used to fabricate the stay cables. The PC ribbed slabs are simpler to cast than the box girders, as no inner forms and diagonal struts are required, but two stay cables are installed instead of one. The stay cables are fabricated on the deck and lifted in one operation, or the strands are individually pushed into stay pipes and kept in position by the guide strand of the cable.

7.10.4 Structure–equipment interaction

The weight of the travellers governs the structure–equipment interaction. The cable-stayed decks are wide and the segments are long due to the cable spacing. Long and wide segments require stiff casting cells, and the weight of the travellers often exceeds one-half of the segment weight. The stay cables are often designed to balance the self-weight shear in the deck against the vertical component of the pull in the stay cables, to control longitudinal bending, and the weight of the traveller may necessitate overdesign of the stay cables and top slab tendons. Longitudinal load imbalance of staged construction induces high flexural demand in the pylons and foundations

Figure 7.51. Completion of underslung travellers during the first launch (Photo: VSL)

that cannot be reduced by means of staggered construction joints because symmetrical stay cables dictate the segment geometry.

The local effects of the loads applied by the traveller and stay cables are more demanding in the ribbed slabs due to the poor moment of inertia of the cross-section; in addition, loading is applied to the edges of the deck and load dispersal is less favourable. Overhead and underslung travellers are supported at four points: the rear balancing load is directed upward, and the load that the front supports apply to the deck is therefore larger than the total weight of the traveller and the segment. Lengthening the rear frames reduces the load imbalance but adds weight and worsens the general load imbalance of the staged construction in the structure.

The edge girders of the ribbed slabs require a few days of curing, and the longitudinal moment generated by the traveller and stay cables may exceed the cracking moment in both positive and negative bending. Several solutions have been tested to avoid cracking in the edge girders.

1   The segments may be cast in two phases: edge girders first, and the crossbeams and top slab after partial tensioning of the new pair of stay cables. Supporting the edge girders with the new cables reduces the longitudinal bending generated by the traveller.
2   Precast anchor blocks may be used for the stay cables so that the new cables can be fabricated and partially tensioned prior to casting the rest of the segment in one phase. Props connecting the anchor blocks to the previous segment resist the longitudinal component of the pull in the stay cables. This solution complicates tensioning of the stay cables and longitudinal post-tensioning of the edge girders.
3   The new stay cables may be used to support the traveller during casting of the segment. After curing, the cables are transferred to the segment to release the traveller.

4   The front bulkhead of the traveller may be suspended from the pylon by means of temporary stay cables to also reduce the flexibility of the casting cell.

The mass and wind drag area of the form travellers may trigger or amplify wind-induced vibrations of the hammer prior to midspan closure. Torsional stiffness of the pylon and the foundations and the in-plane flexural stiffness of the deck are the main controlling factors of transverse vibrations and lateral stability. Props from foundations used in the lateral spans to control longitudinal load imbalance should not increase the first vertical frequency beyond 50–60% of the first torsional frequency at each step of staged construction to avoid risks of divergent aeroelastic instability and flutter.

Geometry control is an essential part of construction monitoring, and is a basic tool for checking the stresses in the deck and the stay cables. The segments are erected according to a casting curve that depends on the theoretical geometry, construction loads, elastic modulus and time-dependent properties of the concrete, the apparent elastic modulus of the stay cables, and the construction time schedule to ensure correct final alignment of the deck and pylons (ASBI, 2008).

The movements of the deck and pylons are measured during lifting or casting of segments and during stressing of stay cables to assess the actual stiffness of the structure and to correct the casting curve if necessary. Bending in the cantilevers is checked by comparing actual and theoretical elevations at each joint and lateral deflections of the top of the pylon for a given erection phase.

The stay cables are typically made of individually protected seven-wire strands. The strands are installed and stressed one at a time using calibrated monostrand jacks. Isotensioning stressing sequences are based on monitoring the force in a pilot strand using a load cell, and stressing every strand to that force. Elongation is measured during stressing to check the correlation between the stay cable force and design assumptions about the elastic modulus of the cable and deck and the stiffness of the deck and pylon.

The pull in the stay cables is also checked by surveying the elevation of the same survey marker over a full erection cycle (lifting or casting of segments and tensioning or re-tensioning of stay cables). Loss of elevation may indicate insufficient pull in the cables or excessive segment weight. The deflections are measured early in the morning to reduce the effects of thermal gradients within the deck and between the concrete deck and the stay cables. Positive thermal gradients create a downward deflection of the cantilever during the day, and when the gradient dissipates, upward deflections result into a loss of pull in the stay cables. Thermal gradients measured using thermocouples may be used to correct the casting curve.

The deck deflects up and down during erection, and the amplitude of the deflection range increases with the length of the cantilever. If the deck is stiffer than assumed, the deflection range will be narrower, and modifying the pull in the stay cables to bring the deck to the correct elevation at each cable stressing could overload the structure. The erection of stiff box girders is, therefore, governed by the pull in the stay cables.

Except for the starter segments close to the pylon, the elevation of flexible ribbed slabs and composite grillages varies greatly with small variations in the stay cable forces, and the stresses in the deck do not depend greatly on the deck geometry. With flexible decks, the control of deck elevation often has priority over the control of the pull in the stay cables, and the stay cables are stressed early in the morning when the deck geometry is not affected by thermal gradients (ASBI, 2008).

In both cases, segment alignment is adjusted by means of corrections to the casting cell geometry. Geometry discrepancies due to heavier segments or inaccurate pull of the stay cables are corrected by re-stressing the stay cables. Usually several stay cables are re-stressed to control bending in the deck and the pylon. Some additional capacity of the stay cables is often provided during design to account for construction variations.

In-place casting of an extradosed box girder is similar to the construction of a non-cable-supported balanced cantilever bridge, because the cables do not carry the weight of the deck directly. Geometry is controlled through the casting curve, and discrepancies cannot be wholly corrected by re-stressing the cables due to the stiffness of the deck. The cables are therefore stressed once, like post-tensioning cables, and grouted after stressing.

7.11. Suspension-girder MSSs

As an alternative to the use of deck-supported form travellers, a PC balanced cantilever bridge can be cast using a suspension-girder MSS supported at the leading pier and on the front deck cantilever (Figure 7.52). To cast the cantilevers two 10–12 m casting cells suspended from one or two girders shift symmetrically from the pier table to midspan. After achieving midspan

Figure 7.52. A 158 m, 1600 ton suspension-girder MSS for balanced cantilever casting of 120 m spans with 11.3 m, 680 ton two-phase segments (Photo: ThyssenKrupp)

Figure 7.53. Launching of the suspension girder (Photo: ThyssenKrupp)

continuity, the girder is launched forward to the next span, and the casting cells are moved to the new leading pier and set close to each other to restart the cycle (Matyassy and Palossy, 2006).

The use of a suspension-girder MSSs is suitable for spans up to 100–120 m and rectilinear or slightly curved bridges of length sufficient to amortise the investment. The length of the MSS is about 1.3 times the maximum span and 1.5 times the typical span, in order to move the rear support of the girder as close as possible to the rear pier. The length of the girder is rarely a problem, as an overhead girder does not interfere with the piers. The height of the machine above the deck may cause clearance issues when passing under power lines.

Square trusses, braced I-girders, box girders or hybrid assemblies are used for the suspension girder. Diaphragms and lateral braces distribute the torsion applied by the casting cells and support systems, and constant-depth girders permit unrestricted movement of casting cells rolling along the top-flanges.

Single-girder MSS are lighter and easier to launch, while twin-girder MSS provide enhanced torsional stability. The overall geometry is similar to a short 1.5-m-span gantry for balanced cantilever erection of precast segments, but the girders are heavier because of the load applied by such long casting cells. Overturning during launching is rarely an issue as the casting cells may be moved to the rear end of the girders to act as counterweights (Figure 7.53).

The casting cell comprises a top platform that sits on the main girder, or spans between the girders, and suspends two lateral space frames that support the outer form. The space frames are hinged to the top platform and can be opened and closed hydraulically. After application of top slab prestressing, the casting cell is lowered by releasing vertical cylinders at the sliding saddles that support the top platform on the main girders. The entire MSS may also be lowered by releasing hydraulic cylinders at the main supports when the girder is not used to stabilise the hammer. The casting cell is separated into two halves by releasing connection pins at the centre of the bottom frame. The casting cell is opened, shifted to the next casting location, reclosed and lifted into position. Repositioning the two casting cells does not require ground cranes and takes a few hours.

Ribbed slabs with double-T-section are rarely used on 100–120 m spans. The box girders are cast in one or two phases. The inner tunnel form for one-phase casting remains within the segment during repositioning of the casting cell and is extracted into the steel cage of the bottom slab and the webs of the new segment, like the process when using a form traveller. Two-phase casting with horizontal joints at the top web–slab nodes requires modular shutters for the webs and a sliding form table for the top slab. Two-phase casting facilitates stripping of the inner shutters and avoids the horizontal cracks that sometimes affect one-phase casting due to the settlement of fresh concrete in tall webs. Despite their cost, these MSS offer several advantages.

1   The suspension girder permits access to the working locations from the completed deck. Mobility of workers is simpler and safer. Concrete is delivered on the deck, pumped to the casting cells through pipes supported on the girder, and distributed using hydraulic arms. Reinforcement and light auxiliary prestressing materials are also delivered on the deck and moved into the casting cells by winch trolleys.
2   If the deck is supported on one line of bearings, the girder stabilises the hammer during construction and avoids stabilisation gear at the pier cap. Hammer stabilisation is also useful when the deck is continuous with tall piers, to reduce the longitudinal bending of staged construction in the piers and foundations.
3   Casting cells suspended at multiple points from an overhead girder are so stiff that the deck segments may be 10–12 m long and wider than 20 m. Midspan closures and the end deck segments at the abutments are also cast using the MSS.
4   Less top slab post-tensioning is required throughout the bridge as no erection loads are applied to the cantilevers. One-half of the anchorages and prestressing operations can also be saved because of the longer segments.
5   Transferring the casting cells to the new pier is a matter of hours instead of the 3–4 weeks required to dismantle and reassemble a pair of form travellers.
6   The use of tower cranes, stair towers and elevators is limited to the duration of pier erection.
7   The saving in labour costs may be substantial.

The main limitations of these machines are the initial investment, the cost of shipping and site assembly, and the need to erect the hammers directionally from one abutment toward the other. Bridges with tight plan or vertical radii are another counterindication.

The suspension-girder MSSs can be easily reconfigured for balanced cantilever erection of macro-segments, and vice versa. The casting cells are replaced by lifting platforms for strand-jacking of macro-segments cast on the ground, or delivered on barges and the rest of the machine does not require modifications. Both configurations of machine are compatible with shorter spans and variations in deck geometry. In the MSS configuration, these machines are also unaffected by ground constraints.

Figure 7.54. Framing systems of the casting cell (Photo: ThyssenKrupp)

7.11.1 Loading, kinematics, typical features

The outer form comprises side wings, adjustable web shutters and a bottom form table supported on a stiff platform with screw jacks for vertical adjustment. The bottom platform includes transverse trusses or space frames suspended from truss hangers. Longitudinal bracing stiffens the assembly and resists longitudinal loads and torsion (Figure 7.54). The bottom platform supports the outer form, provides anchorage for form adjustment, and anchors the tie-downs that prevent floating of the inner tunnel form during one-phase casting. The casting cell includes multiple levels of working platforms and suspends catwalks for access to the central splice in the bottom platform.

The form hangers are equipped with long-stroke double-acting cylinders that rotate the bottom portion of the outer frame in the transverse plane. After lowering the casting cell and releasing the central splice in the bottom platform, the outer frame is opened to move the casting cell to the next casting position. Hydraulic operations accelerate repositioning and leave the auxiliary winch trolleys available for other operations. The central splices in the bottom platform are opened and reclosed at every cycle, and frontal connections with one transverse pin at the bottom flange and two longitudinal shear bolts at the top flange are often preferred because of the rapidity of operations.

Balanced cantilever bridges with 100–120 m spans have a varying-depth box section. The outer web shutters are tall and require stiff rib frames to control form deflections during filling and to provide stability during launching. When the outer framing system is so stiff, the inner shutters may be anchored to the outer ribs, and are therefore very simple. This simplifies operations, as most of the stiffness is provided by a self-repositioning frame.

The outer frame is suspended from a top platform that carries the concrete distribution arm and a covering for the casting cell. The top platform is supported on adjustable bogies that roll along the main girders. Support cylinders with mechanical locknuts are housed within the support bogies to provide geometry adjustment for setting the camber and crossfall. Wind loading on such tall casting cells generates significant torsion, and the support saddles may require sliding clamps to avoid uplift prior to filling.

The form hangers and outer frames may be designed for the full load or may be integrated with hanger bars crossing the casting cell and suspended from the top platform. Hanger bars prevent operations of the winch trolleys, increase the labour demand and complicate slab finishing, but facilitate two-phase casting as the deflections of the casting cell are minimised. The inner and outer forms are anchored to the previous segment prior to filling the casting cell in order to prevent joint settlement and displacements. The concrete level is raised symmetrically during filling to minimise torsion.

The auxiliary winch trolleys roll on runway beams overhanging from the main girders to assist the operations of the casting cells and delivery of materials from the completed deck. The winch trolleys also handle the stressing platforms for the top slab tendons, and assist in the repositioning of the launch saddles.

The steel cage is fabricated within the casting cell. In the absence of clearance requirements for cage delivery, the main girders are kept close to the deck to simplify the launch supports and stabilisation props for the hammer. The casting cells may be easily insulated, and these MSSs have found several cold-weather applications where the only limitation is the availability of fresh concrete.

Short pier tables are cast in place at the end of pier erection. A tower crane and stair tower are already available, assembling a narrow casting cell on the pier cap is not very labour intensive, and several complex activities can be removed from the critical path to shorten construction duration and to mitigate their risks. The support systems of the MSSs and the bottom frames of the casting cells are also much simpler. The pier tables are cast on the permanent bearings and stabilised by tie-downs and support jacks at the corners of the pier cap; they are slightly longer than the pier cap to simplify setting the casting cells of the MSSs for the starter segments.

The MSS is supported on the leading pier table and the front deck cantilever during operations and launching. The support saddles are anchored to the top slab over the webs, and no cross-beams are necessary in rectilinear bridges. Avoiding crossbeams and support towers balances the labour costs of in-place casting of the pier tables, and launching is also faster and simpler.

7.11.2 Support, launch and lock systems
Suspension-girder MSSs are used in rectilinear or slightly curved bridges. Traversing the MSS is not necessary and the deck segments are cast with minimal lateral offset, which avoids the need for support crossbeams or minimises their length.

When the pier tables are not cast prior to the arrival of the MSS, the casting cells are set close to each other to cast two full-length pier-table segments on the permanent bearings. This requires a tower–crossbeam assembly at the leading pier and a front auxiliary leg for launching, while the rear supports of the MSS are still anchored to the deck over the webs. Support towers designed for such loads complicate the inner forming of the pier tables and lengthen the cycle time. Launching is also complicated by the need to extract segments of tall braced towers from the pier tables.

Launch saddles designed for such loads are often based on PTFE skids, which are thinner and less expensive than articulated bogies. PTFE skids or bogies support the main girders under the webs and cope with flexural rotations and the launch gradient. The field splices in the girder are designed with frontal connections and stressed bars for a flush launch surface. Minor lateral adjustment may be provided by shifting the launch supports laterally along the bottom anchor frame of the launch saddles.

Friction launchers or paired long-stroke double-acting cylinders anchored into perforated rails provide the launch force and restrain the MSS during repositioning and operations. Synchronised launchers minimise the loads applied to the piers and the distortion of the casting cells due to staggered launching. The winch trolleys reposition the support saddles during launching, without need for ground cranes.

7.11.3 Performance and productivity

When short pier tables are cast in place at the end of pier erection, a suspension-girder MSS can cast a deck segment on both sides of the pier in weekly cycle (see Table 4.1). The segments may be very wide and 10–12 m long, while the form travellers rarely exceed 5 m; productivity is more than double, and a higher level of mechanisation provides additional labour savings.

The camber and crossfall are easily set, the casting cells may be sheltered and insulated for continuous production during bad weather, and steam curing is also easy to apply if required. Construction materials are delivered on the deck and handled using winch trolleys. Concrete is also delivered on the deck and pumped into distribution arms carried by the MSS.

Casting a short pier table is simpler and faster than casting a 5–12 m pier table for the assembly of two form travellers; the pier table is also simpler, as the MSS stabilises the hammer during construction and less stabilisation gear is therefore necessary at the pier cap. Major labour savings also result from no disruption of the casting cycle for repositioning of the casting cells on the new pier.

However, there are also weak points to this option. In addition to the high initial investment, the site assembly and final decommissioning of such heavy machines are complex and take time, labour and cranes. The main girders are composed of a great number of modules, which facilitates shipping and reuse in future projects but complicates site assembly. The outer frames of the casting cells are also complex, and the bottom platforms can be suspended only after launching the main girders over the abutment span, which requires air work and further increases assembly duration.

7.11.4 Structure–equipment interaction

Loads and deflections are major concerns with such heavy machines. The main girders are flexible on such long spans, but the segments are cast individually and supported by the hammer after application of top slab prestressing. The casting cells are stiff and the segments may be wider and much longer than those achievable with form travellers. The structure–MSS interaction at the application of prestressing is also minimised, as the residual support action exerted by the casting cells is limited to the two end segments of the hammer.

A 1.5-span suspension-girder MSS is supported on the leading pier table and the rear pier during hammer erection to minimise the structure–MSS interaction. No loads are applied to the hammer during construction, and no materials are stored on the hammer, as they are delivered through the MSS. When the MSS stabilises the hammer during construction, it also relieves most of load imbalance of staged construction in the leading pier and its foundations. Launching requires temporary loading of the front deck cantilever (a full two-span MSS would be too heavy for such long spans), but the launch loads do not cause time-dependent effects.

These machines are very heavy and the loads applied to the piers and foundations may affect their design and could cause differential settlement, especially when the piers are short and the weight of deck and MSS is a major component of the foundation load. As the deck is erected directionally from one abutment toward the other, progressive settlement of foundations may generate locked-in stresses throughout the bridge.

Prior to midspan closure, the mass of the hammer and the tributary mass of the MSS lengthen the vibration periods of tall piers and make them prone to wind-induced vibrations. Although this problem is common to all types of balanced cantilever construction, the MSS provides much more tributary mass than a pair of travellers, but the MSS can also be used to stabilise the torsional and longitudinal modes of the hammer.

REFERENCES

AASHTO (American Association of State Highway and Transportation Officials) (2003) *Guide Specifications for Design and Construction of Segmental Concrete Bridges*. AASHTO, Washington, DC, USA.

ASBI (American Segmental Bridge Institute) (2008) *Construction Practices Handbook for Concrete Segmental and Cable-supported Bridges*. ASBI, Buda, TX, USA.

Gimsing NJ and Georgakis CT (2012) *Cable Supported Bridges*, 3rd edn. Wiley, New York, NY, USA.

Harridge S (2011) Launching gantries for building precast segmental balanced cantilever bridges. *Structural Engineering International* 21(4): 406–412.

Hewson NR (2003) *Prestressed Concrete Bridges: Design and Construction*. Thomas Telford, London.

Homsi EH (2012) Management of specialized erection equipment: selection and organization. *Structural Engineering International* 22(3): 349–358.

IABSE (International Association for Bridge and Structural Engineering) (1997) Composite Construction: Composite and Innovative, International Conference, Innsbruck.

Matyassy L and Palossy M (2006) Koroshegy Viaduct. *Structural Engineering International* 16(1): 36–38.

Meyer M (2011) Under-slung and overhead gantries for span by span erection of precast segmental bridge decks. *Structural Engineering International* 21(4): 399–405.

Rosignoli M (2002) *Bridge Launching*. Thomas Telford, London.

Rosignoli M (2007) Robustness and stability of launching gantries and movable shuttering systems – lessons learned. *Structural Engineering International* 17(2): 133–140.

Rosignoli M (2010) Self-launching erection machines for precast concrete bridges. *PCI Journal* **2010(Winter)**: 36–57.

Rosignoli M (2011) Bridge erection machines. In *Encyclopedia of Life Support Systems*, Chapter 6.37.40, 1–56. United Nations Educational, Scientific and Cultural Organization (UNESCO), Paris, France.

VSL (2008) *Over 50 Years of Excellence*. VSL International Ltd, Köniz, Switzerland.

Bridge Construction Equipment
ISBN 978-0-7277-5808-8

ICE Publishing: All rights reserved
http://dx.doi.org/10.1680/bce.58088.269

Chapter 8
Special equipment for precast spans

8.1. Technology of full-span precasting

Several major bridges for high-speed railway (HSR) projects have been built in recent years. Prestressed-concrete bridges are primary components of new HSR lines: the 113 km Beijing–Tianjin route includes 100 km (88%) of bridges, the 1318 km Beijing–Shanghai route includes 1140 km (86%) of bridges, the 904 km Haerbin–Dalian route includes 663 km (73%) of bridges and the 995 km Wuhan–Guangzhou route includes 402 km (41%) of bridges. Most recent LRT projects are also based on prestressed concrete elevated guideways. More than 35 000 spans for HSR and LRT bridges have been built in China in recent years (Liu, 2012). Large investments in HSR infrastructure have also been made in Europe, Japan, Korea and Taiwan. Bridges of such dimensions are built using highly mechanised methods to lower the construction cost and to accelerate project completion.

Many HSR bridges and some LRT bridges have been built with full-length precast spans transported into place and erected span-by-span. Box girders are well suited for dual-track bridges, while single-track U-spans offer the additional advantages of noise reduction, train containment in the case of derailment, optimum integration with the urban environment, and easier handling due to the lighter weight (Vion and Joing, 2011). Long-span box girders comprising trussed webs, railroad tracks on the bottom slab, and a roadway platform on the top slab have been erected using floating cranes when the project dimensions permitted such investments.

Full-span precasting offers rapid construction and repetitive high-quality casting processes in factory-like conditions. The spans are erected all year round in almost any weather conditions. The maximum span length for ground delivery depends on the capacity of the erection equipment, even if in the large-scale projects, where this construction method is used, it is common to build custom equipment for the length and weight of the spans to be handled.

The lead time of full-span precasting and the investment needed to set up large precasting facilities and to provide special transportation and means of placement limit the cost-effectiveness of this construction method to long bridges comprising hundreds of equal-length spans. A low gradient of the bridge and suitable access routes are also important factors for span delivery on the deck. These conditions are frequently met in long HSR bridges. Ground transportation is rarely used for spans longer than 35–40 m due to the cost of the erection machines and the prohibitive loads that they apply to the deck.

8.1.1 Full-span precasting of LRT decks

HSR bridges are typically built in rural environments, while LRT bridges are built in urban environments. Dual-track precast segmental box girders have been used extensively for LRT bridges; single-track U-spans are being used more and more frequently because of the high quality and fast erection rate ensured by full-span precasting. Dual-track U-spans are erected using match-cast precast segments and epoxy, as they are too wide for ground transportation.

Single-track U-spans comprise two edge girders and a bottom slab. The edge girders include a web and a top flange that is used as a maintenance walkway and for passenger evacuation from the train. The combined lateral stiffness of web and top flange is often enough to keep the train on the bridge in the case of derailment. The need to have access to the evacuation walkways from the train limits the depth of the webs. The elevated guideway is often uninterrupted throughout aerial stations, and the need to have access to the passenger platforms implies a similar depth restriction in the webs.

Single- and dual-track U-spans reach 30–35 m, while stiffer dual-track box girders reach spans of 40–45 m. Longer spans offer several advantages in urban environments due to the fewer columns and foundations required. Train-induced resonant vibrations and limitations on the vertical peak acceleration (which is related to the comfort of passengers and the durability of rolling stock) preclude the use of simple spans longer than 40–45 m in most cases. Two-span continuous decks may reach longer spans but are complex and expensive to erect, and are therefore used only for special crossings.

Compared to span-by-span erection of precast segmental dual-track box girders, full-span precasting of two adjacent single-track U-spans offers several advantages in LRT applications (Vion and Joing, 2011).

1   Simple forming and casting operations: two 35 m U-spans are cast in one operation, while a precast segmental box girder requires match-casting of 11–13 segments. Casting is simpler and faster, and geometry control is much simpler.
2   Rapid and inexpensive delivery and erection: single-track precast spans may be transported on public roads overnight, and two ground cranes typically erect four U-spans (so, two full dual-track spans) per night, while a gantry erects a precast segmental box girder in 2–3 days. Gantry erection is also a linear process, while the precast spans may be erected according to pier availability. On the other hand, twin U-spans require more expensive pier caps, the precasting facility must be designed for such production, and erection with ground crane requires accessible areas and short piers throughout the bridge.
3   Optimal integration with the urban environment: a reduction in the visual impact, built-in sound barrier (the inner faces of the webs act as a sound barrier), built-in cable support and system functions (cables and systems are attached to the inner faces of the webs), and integration of walkways for maintenance and passenger evacuation.
4   Built-in structural elements capable of maintaining the train on the bridge in the case of derailment. Transverse spacing of bearings controls uplift under derailment loads without the need for tie-downs.
5   Possibility of lowering vertical profile and aerial stations by 1.5–2.0 m: a lower centre of mass reduces the seismic demand on the columns and foundations, and escalators can sometimes be avoided.

A single-track U-span is 4.5–5.5 m wide, may weigh from 150 ton for 25 m spans to 200–220 ton for 35 m spans, and can often be transported on the ground using trucks and rear steering trolleys (Figure 8.1). The weight of an 11–12 m-wide dual-track U-span ranges from 400 to 550 ton and

Figure 8.1. Truck delivery of an LRT single-track U-span (Photo: Systra)

the span is too wide for ground delivery. Dual-track spans can be lifted onto the deck using portal cranes and delivered on the deck using tyre trolleys or portal carriers; in most cases, however, the spans are made using match-cast precast segments glued with epoxy and erected with an overhead gantry.

Different proportions of pre- and post-tensioning are possible with full-span precasting. Pre-tensioning simplifies forming, improves durability, and reduces the cost of labour and prestressing. Pre-tensioning, however, requires investment in end anchor bulkheads and full-length reaction beams designed for the load of many strands (Figure 8.2). Self-reacting casting cells are rarely used in full-span precasting for reasons of cost.

The anchor bulkheads for pre-tensioned strands are placed at the ends of casting lines comprising multiple casting beds (Figure 8.3). Pre-tensioning of casting lines requires a matrix organisation of the precasting facility so that all the spans in a casting line have the same age at transfer of prestressing. Pre-tensioning designed for span lifting and transportation to storage may be combined with integrative post-tensioning applied prior to final span delivery, in order to reduce the cost of strand anchor bulkheads and reaction beams and to increase the efficiency of prestressing.

The bottom slab can be post-tensioned transversely. Replaceable transverse mono-strand anchored at the top flanges increases the flexural capacity of the bottom slab, the shear capacity of the webs, and the train confinement capability. Rail plinths are cast in two phases to relax the geometry tolerances: the base is cast in the precasting facility during span curing, and a top adjustment layer is cast after span placement. The slab is devoid of crossfall: rainfall flows longitudinally along the plinths and is drained at the piers.

The precasting facility is located near the bridge site, and there are different possibilities for the fabrication of the steel cages. When the cages are assembled on the casting beds, each casting line includes 2–6 beds where the spans are cast at once (Vion and Joing, 2011). Span erection using ground cranes typically requires 6-day cycles, and delivery of four spans every night

271

Figure 8.2. End anchor block for pre-tensioning of a four-bed casting line (Photo: Systra)

therefore requires six casting lines with four beds each. The cycle time depends on the materials and fabrication procedures used, and can be shorter than 6 days, and the number of casting lines required will be reduced accordingly. The fabrication process must be consistent with the delivery means and the rate of erection of the ground cranes.

The precasting facility is set up as a rectangular matrix of casting beds. The number of casting lines corresponds to the cycle time for a span, and the number of casting beds in each line corresponds to the planned number of spans for just-in-time (JIT) daily delivery according to the productivity of the erection lines. Storage is necessary only when the spans are delivered on the deck, and is expensive due to storage platforms and double handling. Crane erection requires only 2–3 day curing, as the spans are not loaded after lifting and the curing time can be completed once on the piers.

With a 6-day cycle time, a casting line contains spans ready for overnight delivery, a second line is being cleaned and prepared for cage assembly, a third line is occupied by the ironworkers for cage fabrication and pre-tensioning, a fourth line is being cast, and the two remaining lines are occupied by curing spans. The curing days may be used to cast the plinths for the rail fasteners. A full-length set of web shutters is moved from line to line by means of portal cranes or gantry cranes. The casting beds are cambered, the web shutters are rectilinear, the end bulkheads are rotated to the local plan radius, and the curved alignment of the rail fasteners is achieved by shifting the plinths laterally on the bottom slab. A few storage platforms for span stacking provide some flexibility in case of delivery delays or defective spans (Figure 8.4).

Figure 8.3. Four-bed casting line (Photo: Systra)

In optimal conditions, the spans are lifted from the casting bed and loaded onto the truck for JIT delivery. Large-scale storage of precast spans is necessary when the spans are lifted on the deck and delivered on tyre trolleys, because the spans are loaded immediately after placement for delivery of the next spans (Figure 8.5). Portal cranes on rails or rubber tyres are typically used for handling the spans in the precasting facility.

8.1.2   Full-span precasting of HSR decks
The precasting facilities for HSR spans are organised in a different way. The units are too heavy for ground transportation and crane lifting, and are delivered on the deck and positioned using dedicated machines. The transporters load the spans immediately after their placement for delivery of the subsequent spans, and the curing time is therefore longer. The productivity of precasting facility and erection lines is calibrated, and the number of storage platforms is chosen based on the curing time required for delivery, which depends on the type of transporter.

The portal carriers spread the load to two spans, the spans may have 7–8 days of curing, and each casting line for daily span production therefore requires 7–8 storage platforms (Rosignoli, 2012). Tyre trolleys do not spread the load, the spans often need full 28-day curing to resist the load of the trolley, and each casting line requires 28 storage platforms. Three casting lines require

Figure 8.4. Span stacking on support frames (Photo: Systra)

84 storage platforms, a few platforms for emergencies and atypical spans lead to 100 platforms or more, and the dimensions of the storage facility grow so much that stacking becomes unavoidable.

The availability of heavy lifters in the precasting facility suggests cage prefabrication in full-span rebar jigs (Figure 8.6). The use of rebar jigs provides additional flexibility to the organisation of

Figure 8.5. Large-scale storage of single-track U-spans (Photo: Systra)

Figure 8.6. Rebar jig for full-span cage prefabrication of an HSR single-track U-span

the casting lines, as the number of jigs and casting cells can be calibrated to the different productivities of the two working areas (Rosignoli, 2011).

The spans are often designed for post-tensioning to facilitate cage handling and to shorten the curing time within the casting cell. The end bulkheads are applied to the rebar jig according to span geometry, the tendon anchorages are applied to the bulkheads, the outer grid of the cage is fabricated, watertight plastic ducts are fabricated and tested, the rest of the cage is assembled, and all the embedded items are fixed to the cage. The strand is inserted into the ducts prior to and during span casting to lighten the cage during transfer into the casting cell and to extract activities from the critical path (Rosignoli, 2011). A stiffening truss is applied to the cage prior to lifting in order to avoid distortion (Figure 8.7).

Combinations of pre- and post-tensioning are also possible (VSL, 2008). Rectilinear pre-tensioned strands are used in the bottom slab, and anchorages and parabolic ducts are embedded in the webs for application of integrative post-tensioning during span storage. Hydraulic stressing blocks are used at one end of the casting line for initial pre-tensioning and transfer of prestressing. The reaction beams and rear anchor blocks resist the load of the strands and stiffen the foundations of the casting cells (Figure 8.8). The forming systems are designed using conventional criteria.

The number of rebar jigs in a casting line is chosen in relation to the productivity of casting cells and rebar jigs (Rosignoli, 2011, 2012). Typically, a casting line produces a span per day, and the cycle time is 2–3 days depending on the span dimensions and the type of cross-section and prestressing. If a casting cell produces a span every 3 days, three casting cells are needed in the casting line. If a rebar jig produces a cage every 4 days, four jigs are needed in the casting line. The use of rebar jigs requires additional investment, but different cycle times for casting cells

Figure 8.7. Cage delivery with stiffening truss

and rebar jigs provide great flexibility in fine-tuning span production after the learning curves have been completed. If necessary, additional jigs can be installed at minor cost.

Partial post-tensioning is applied at 12–18 hour curing to make the span self-supporting for transfer to the storage platform; transfer of pre-tensioning often requires a longer curing time. Parallel casting lines are separated by transportation routes for delivery of the steel cages and removal of the precast spans. Three casting cells have been used in the casting lines for single-track U-spans in combination with an inner shutter that rolls on rails to serve the entire casting line (Figure 8.9). The casting cells for U-spans are complete enclosures and only the top of the webs is hand finished. After cage delivery, the inner shutter is moved over the casting cell to close the mould. The runways of the inner shutter are extended at one end of the casting line to create a maintenance area.

The inner forms for one-phase casting of dual-track box girders are more complex to operate and to extract from the span. The form includes multiple hydraulic systems to drive the support truss and to collapse the form modules for stripping and to clear the pier diaphragm during extraction. When the spans are not pre-tensioned, the inner form is stored between two casting cells and serves them alternately; each casting line therefore includes two or four casting cells. Portal cranes move along the casting line for longitudinal delivery of the steel cages and span removal. Casting and handling equipment for dual-track box girders is more expensive than that for single-track U-spans due to the weight and complexity of the precast units, but the number of units halves and this shortens the project duration or reduces the number or machines needed.

Figure 8.8. Casting cell for combined pre- and post-tensioning (Photo: VSL)

Figure 8.9. Inner shutter for U-spans with concrete distribution arms

Figure 8.10. Lateral transfer of a 900 ton dual-track span with a 64-wheel portal crane (Photo: BWM)

Concrete is pumped from the batching plant to the casting cells through fixed pipelines. Each line includes a mixer, an agitator, a pump, hydraulic switches at the nodes of the network, and a tower-mounted distribution arm. Two feeding lines are used for each casting cell to accelerate filling (Rosignoli, 2011). One batching plant may be enough for two casting lines for U-spans, while two plants are used for multiple casting lines and dual-track spans. Casting the span in one pour shortens the cycle time and enhances quality. Long spans delivered on barges and positioned using floating cranes are cast in multiple phases using long-line techniques.

Portal cranes on rails or steering wheels are used to move the spans around the precasting facility. Wheeled cranes are more complex to design, operate and maintain but offer unrivalled flexibility of operations. Portal cranes with orthogonal tractors and paired steering wheels facilitate transverse access to casting cells and storage platforms, and portal tractors also allow longitudinal access (Figure 8.10). These cranes are very stable when handling stacked spans; computerised wheel rotation allows movements in every direction, but wide transportation routes reduce the volumetric efficiency of the storage area and sometimes require concrete runways for load distribution. Stacking rarely exceeds two levels. Portal carriers with pivoted tractors may be used within the precasting facility and for final delivery and placement of the spans in combination with an underbridge (Figure 8.11).

The permanent bearings are positioned over the support blocks of the storage platform before lowering the span onto them. Good ground conditions are required at the precasting facility to

Figure 8.11. Span transfer to the storage platform with a portal carrier with pivoted tractors

support the spans. Pile foundations may be necessary for casting cells and storage platforms to prevent the spans from twisting or warping. Two or three days of storage are necessary to complete post-tensioning and finishing; the rest of the curing time is dictated by strength requirements at span delivery.

The HSR spans are transported on the deck using portal carriers or tyre trolleys. The portal carriers position the spans using special underbridges, while the tyre trolleys feed heavy span launchers for placement. For faster rate of erection multiple trolleys are often used to feed the launcher when crossover areas are available along the delivery route.

8.2. Tyre trolleys and span launchers

Heavy span launchers are used for placing HSR spans in combination with tyre trolleys that transport the spans along delivery routes and then along the completed deck to the rear end of the launcher. The tyre trolleys have 600–900 ton capacity and the operating speed varies significantly from machine to machine.

The tyre trolleys are monolithic machines comprising a full-length main beam and a large number of force beams on either side. Each force beam carries a vertical pivot for 360° rotation of two paired wheels. Long force beams ensure lateral stability during steering and transportation, and align the wheels with the webs of the box girder to minimise transverse bending in the completed deck. Steering computers control wheel alignment, all wheels are motorised, the

Figure 8.12. Automatic drive of a 16-axle trolley for 900 ton spans (Photo: BWM)

Figure 8.13. Front support block of a tyre trolley for 400 ton single-track box girders (Photo: BWM)

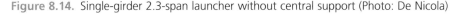

Figure 8.14. Single-girder 2.3-span launcher without central support (Photo: De Nicola)

wheel load is equalised hydraulically or electronically, and position sensors keep the wheels aligned with the deck webs during the automatic drive along the bridge (Figure 8.12).

An articulated extraction saddle runs along the main beam for the entire length of the trolley. The saddle supports the rear end of the span under the webs; the span is laid on elastomeric pads to prevent surface damage. The saddle is equipped with PTFE skids or cast-iron wheels on equalising beams for low-friction movement along the main beam of the trolley. The articulation allows free three-dimensional rotations so that the saddle works like a spherical hinge during span transportation and extraction. A fixed front block supports the span on elastomeric pads during transportation and provides the torsional restraint (Figure 8.13).

Several configurations of span launchers have been tested, not always with fully satisfactory results. Single-girder 2.3-span launchers without central support are tall and heavily loaded, as the erection span of the gantry is twice as long as the typical span of the bridge; these machines are expensive and complex to operate and have not found widespread use (Figure 8.14). Twin-girder 2.3-span launchers with a central support frame have been used to erect twin single-track box girders; their use has been limited by the infrequent recourse to this type of cross-section for the deck (Figure 8.15).

A twin-girder telescopic span launcher comprises a rear main frame for span placement and a short 1.2-span front underbridge for launching. The underbridge is supported at the leading

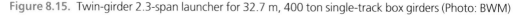

Figure 8.15. Twin-girder 2.3-span launcher for 32.7 m, 400 ton single-track box girders (Photo: BWM)

pier of the span to be erected and at the next pier. The front end of the main frame is supported on the underbridge during launching and on the front half of the leading pier cap during span placement. A trapezoidal C-frame supports the rear end of the main frame on the completed deck and allows the precast span to pass through.

The main frame comprises two overhead box girders connected by a front crossbeam at the root of the front nose. The main girders cantilever out behind the rear support frame to store two winch trolleys for span lifting from the tyre trolley. A rear platform connects the main girders at the end of the rear cantilever (Figure 8.16). C-frames designed to allow the winch trolleys to pass through connect the main girders at midspan, at the rear support frame and along the rear cantilever to further stiffen the frame. The main C-frame closes the rear support frame to enhance the torsional stiffness of the assembly. Box sections are used in all main structural members, and braced I-girders are used only for the front nose.

The front nose carries an auxiliary underslung winch trolley for repositioning of the underbridge at the end of launching of the main frame. A two-span self-launching underbridge may be used to avoid the need for a front nose in the main frame (Figure 8.17). When the main frame is equipped with foldable rear support legs, the main frame and underbridge can be repositioned with the tyre trolley used for span delivery, with minor dismantling (Figure 8.18). After reaching the approach embankment at the leading end of the bridge, the underbridge is launched over the trolley, the main frame is moved over the underbridge, and the rear support frame is folded (Figure 8.19).

Figure 8.16. Pivoted span launcher for 32.7 m, 900 ton dual-track spans (Photo: BWM)

Figure 8.17. Short launcher with a full two-span underbridge for 32.7 m, 900 ton dual-track spans (Photo: BWM)

Figure 8.18. Span launcher for 32.7 m, 900 ton dual-track spans disassembled on a tyre trolley (Photo: BWM)

Two winch trolleys span between the main girders to handle the precast span. The winch trolleys are long in order to spread the load over long sections of the main girders, and the rear cantilever of the main frame is therefore very long to be able store two trolleys for span lifting. The winch trolleys pick up the span as close as possible to the rear support frame to reduce negative bending in the main frame and to control overturning.

Figure 8.19. Span launcher and full two-span underbridge repositioned with a tyre trolley (Photo: BWM)

The main girders are located far apart to enhance stability and to reduce transverse bending in the rear support frame. The crane bridges of the winch trolleys are long and heavy, and translation bogies, hoist crab, reeved hoisting ropes, sheave blocks and lifting beam are also heavy. Hoist winches and PPU are therefore applied to the rear platform of the main frame to lighten the winch trolleys and to reduce the clearance demand at the C-frames bridging between the main girders. Hydraulic motors or capstans drive the trolleys along the launcher and oppose the line pull in the hoisting ropes.

Enclosed fan-cooled industrial generators energise double axial displacement pumps that supply open-loop, high-flow-rate hydraulic systems, allowing most functions to be operated simultaneously. Heat exchangers remove heat from the bypass oil to provide cooler operation and stabilise viscosity. Return-line filters remove contaminants before allowing return oil to flow back into the tank. The PPU and the fluid tanks are located on the rear platform to reduce the load applied to the underbridge during launching and to increase the support reaction at the rear frame for enhanced lateral stability. Hydraulic launch motors are applied to the bogies of the rear support frame for enhanced traction and to provide simple connection to the main hydraulic systems.

The underbridge has a box section and closely spaced diaphragms for enhanced stiffness to torsion and distortion. Short 1.2-span and full two-span underbridges can be interchanged on the same machine; in both cases the front support leg of the main frame takes support on the front half of the leading pier cap at the end of launching. Long-stroke double-acting support cylinders are used in the front leg for geometry adjustment and to disengage the underbridge.

Short 1.2-span underbridges are not self-launching and must be repositioned using the main frame. The front winch trolley picks up the rear end of the underbridge, an auxiliary winch trolley suspended from the front nose lifts the underbridge at midspan, and the two hoists move the underbridge to the next span to clear the area under the main frame for placement of a new span. The underbridge has two support legs: the rear leg is braced to the underbridge to resist the launch loads of the main frame, and a front pendular leg is moved along the underbridge to adjust the support geometry to shorter spans. Long-stroke double-acting support cylinders are used in both legs to provide vertical geometry adjustment.

Full two-span underbridges are not prone to overturning during launching and are repositioned using capstans. Two or three pier towers equipped with articulated bogies support the underbridge during operations and launching. The pier towers may be suspended from the underbridge and driven with motorised wheels to be repositioned without the use of ground cranes during launching of the underbridge.

8.2.1 Loading, kinematics, typical features

Loading a span launcher is conceptually similar to loading a beam launcher when the precast beam is delivered on the deck. The tyre trolley is driven under the rear cantilever of the launcher to bring the front lifting point of the span as close as possible to the rear support frame of the launcher. The operator cab is often placed at the rear end of the tyre trolley for faster drive back to the precasting facility and so as not to interfere with span extraction.

The front winch trolley picks up the front end of the span and moves forward, and the articulated extraction saddle of the tyre trolley supports the rear end of the span and rolls along the trolley.

When the saddle reaches the front end of the trolley, the rear winch trolley picks up the rear end of the span to release the transporter.

The span is moved out and lowered to the deck level. Four load cells and stainless-steel shims at the multidirectional bearing are used to adjust the support reactions. Placing the span takes 2–3 hours and repositioning the launcher takes other 2–3 hours, so 2–3 spans can be placed every day when there are crossover embankments along the delivery route and the precasting facility is designed for such production.

Spreader beams are anchored to the span using large-diameter threaded bars that cross the span and are anchored to the bottom slab. The lifting beams are permanently connected to the winch trolleys, and two lifting points are used at each beam. The lifting points are interconnected at the rear winch trolley to create a torsional hinge that replicates the articulation provided by the extraction saddle of the tyre trolley. Independent lifting points at the front winch trolley provide the torsional restraint during extraction and placement of the span.

Although large-scale HSR projects are designed for a typical span, some shorter spans are often unavoidable. The distance between the winch trolleys is not restrained, and the front support legs of underbridge and main frame slide longitudinally for geometry adjustment to shorter spans. Walkways are applied to the main frame, rear platform, winch trolleys, C-frames and underbridge to facilitate access, maintenance and inspection.

The rear support frame is wider than the precast span to allow the span to pass through. A closed trapezoidal frame resists transverse bending and provides most of the torsional stiffness to the launcher. The weight of main frame, PPU, fluid tanks, hydraulic systems, hoist winches, winch trolleys and precast span generates high bending and shear stresses in the rear support frame. Support cylinders and launch bogies at the base of the frame are located over the deck webs for direct load transfer.

Lifting a deep HSR span from a tyre trolley requires ample vertical clearance under the main girders. The main girders are also tall, and the height of these machines may pose problems of lateral stability during span erection and launching. Multiple stressed bars are used to anchor the main frame and underbridge to the deck and pier caps to provide stability and to resist lateral loads from wind and operations. The tie-downs are typically oversized, and their number increases on small pier caps due to the short lever arm.

8.2.2   Support, launch and lock systems

The main frame of most telescopic launchers is supported on a rear trapezoidal frame and a lighter mobile front leg. The rear frame rolls on the new span during launching and provides most of the torsional stiffness. The front leg is supported on the underbridge during launching and on the leading pier cap during span placement. The runway for the main winch trolleys goes from the rear end platform of the main frame to a front crossbeam beyond the front leg.

A bottom crossbeam closes the rear frame to enhance stability and load distribution. Loading is demanding, the load conditions are complex, and the frame and its connections with the main girders are typically overdesigned. The frame does not rotate in the horizontal plane due to the large plan radii of HSR bridges.

Figure 8.20. Launcher configuration at span extraction from the tyre trolley (Photo: De Nicola)

During span placement, the front leg of the main frame and the rear leg of the underbridge are both supported on the front half of the leading pier cap to lower the span into position without conflicts with the support systems of the launcher (Figure 8.20). The front support leg of the main frame rotates to the local radius of plan curvature and slides longitudinally along the main frame to adjust the support geometry to different span lengths. The same set of longitudinal double-acting cylinders drives shifting and rotation. Long-stroke double-acting support cylinders with mechanical locknuts are housed within the rear support frame and the front leg to set the launcher as horizontal as possible for span placement and to lower the launch bogies on the deck and underbridge for launching.

Hydraulic motors at the rear support frame push the main frame over the underbridge in the first phase of launching (Figure 8.21). Vertical cylinders at the rear launch bogies adjust the support geometry. Linear displacement sensors and pressure valves provide feedback to the launch computer for control of operations. After supporting the front leg on the new pier cap, the front winch trolley and the auxiliary winch trolley reposition the 1.2-span underbridge. Full two-span underbridges are self-launching.

8.2.3 Performance and productivity

Placing a precast span takes 2–3 hours. Repositioning the launcher takes another 2–3 hours, and three spans per day have been erected with one launcher. The long-term productivity of the erection line, however, depends on multiple factors such as productivity of the precasting facility, the number and speed of the tyre trolleys, the length of the bridge and the delivery route, and the availability of crossover embankments that allow the loaded trolley to move forward prior to the passage of the empty trolley. The launcher is the last link in the delivery chain and its productivity rarely governs the erection rate.

The spans are erected as soon as they can resist the load applied by the trolley for placement of the subsequent span. The load allowed on the span governs the curing time and the number of storage platforms. Tall portal cranes are required in the precasting facility to double the storage volume

Figure 8.21. The front support leg of the main frame moves along the underbridge during the first phase of launching (Photo: De Nicola)

by stacking and to move the spans over curing spans for the loading of tyre trolleys (Figure 8.22). The erection lines are designed for JIT delivery, and the storage platforms are designed for the curing time, and not to provide a buffer to the production of the precasting facility.

The speed of the trolleys often governs the erection rate of long bridges without crossover embankments. A tyre trolley is faster than a portal carrier, as the machine is lighter, the precast span is supported instead of suspended and does not oscillate, and the trolley has a

Figure 8.22. Loading of a tyre trolley for 43.5 m, 900 ton dual-track spans (Photo: BWM)

short interaction time with the span launcher. After span delivery, the trolley is driven back to the casting yard during span placement, and the cycle time is shorter. Speed is a basic issue because, without crossover embankments, the availability of multiple trolleys cannot shorten the time window between having access to the bridge and leaving the bridge for access of the next trolley.

Telescopic span launchers fed by tyre trolleys are preferred for very long bridges as the launcher is not used for span transportation and the trolleys are stable and fast. Multiple shorter bridges are better handled with portal carriers, because the carriers pick up the underbridge from the landing embankment and transport it to the next bridge. Some portal carriers have been designed to move through tunnels and also to deliver the spans in tunnels (Liu, 2012). No dismantling is necessary to reposition the erection line to another bridge, and immediate restart of span delivery minimises disruption of the precasting facility.

Portal carriers rarely require heavy lifters in the precasting facility and reduce the number of storage platforms due to the shorter curing time. The first-generation carriers had large steering radii; the tractors of the new-generation machines are pivoted to the main girder or equipped with $\pm 90°$ steering wheels controlled by steering computers that facilitate driving and do not require tractor rotation for the lateral movements over the span for lifting.

Last but not least, span placement is not the last task in bridge construction. HSR bridges require finishing work after span placement. Span delivery with tyre trolleys prevents the presence of other machines on the deck, which may delay the finishing work or make it slower and more expensive.

8.2.4   Structure–equipment interaction

Structure–equipment interaction is relatively simple with such sophisticated machines. The precast span is supported at the ends during transportation and suspended at the ends during placement, which avoids negative bending whatever the flexibility of the tyre trolley, bridge and span launcher may be. A torsional hinge at the rear end of the span avoids twisting during transportation and placement, and the dynamic load amplification is typically low with such slow machines on regular support surfaces.

The load on the axle lines of the trolley is equalised hydraulically or electronically. The weight of span and trolley is not spread along the deck, and, although a tyre trolley is lighter than a portal carrier, localised loading typically requires a longer curing time of the spans. Position sensors control the wheel location to load the deck over the webs and to minimise transverse bending and torsion. Some trolleys are fully automated and run along the deck without a driver.

The front pier diaphragm of every span is detailed for the loads applied by the rear support frame of the launcher and includes tie-down systems for anchoring. The loads applied by the launcher and tyre trolley rarely govern the design of piers and foundations of bridges designed for HSR loads.

The rear support frame of the launcher rolls on the new span during repositioning. Launching during strong wind is avoided due to the absence of tie-down bogies. The main frame of the launcher is a simple span with a rear cantilever, and the rear support reaction is predicted accurately. The positive moment in the span is typically smaller than the moment generated by the tyre trolleys, but the localised load may result in higher shear. The rear support reaction may be reduced by moving the winch trolleys forward prior to launching, although this operation increases the load applied to the underbridge.

8.3. Portal carriers with underbridge

A portal carrier for precast spans comprises two wheeled tractors connected by a box girder that supports two hoists. The carrier lifts the span by means of brackets overhanging from the main girder or winch trolleys sitting on the top flange. The tractors have motorised steering wheels controlled by steering computers. In some machines the tractors rotate by 90° for lateral movements of the carrier; in other machines the wheels slew by ±90° and the main girder is rigidly framed to the tractors. Both types of carrier are used in combination with a full two-span underbridge for span erection (Rosignoli, 2011, 2012).

The PPU, hydraulic systems, fluid tanks and hoist winches are applied to the rear end of the carrier to lighten the front tractor during operations on the underbridge (Figure 8.23). Picking up the span from the casting cell involves a complex sequence of operations. The carrier is moved alongside the span and the tractors are lifted and rotated by 90° on hydraulic props. The tractors are rotated individually to ensure lateral stability; tractor rotation is not necessary when the carrier is equipped with ±90° steering wheels. The carrier is moved laterally over the span, lifts the span from the casting cell, is moved back to the transportation route, and is realigned with an inverse sequence of operations. The same operations are repeated to release the span on the storage platform and to pick it up for final delivery.

Automatic drive systems govern the movement of the carrier over the webs of box girders or within single-track U-spans (Rosignoli, 2011). The carrier spreads its self-weight and payload over two spans, and the load on the axle lines is equalised hydraulically or electronically to

Figure 8.23. A 58 m, 321 ton carrier with pivoted tractors for placement of 31.5 m, 740 ton single-track HSR spans

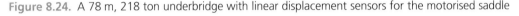

Figure 8.24. A 78 m, 218 ton underbridge with linear displacement sensors for the motorised saddle

compensate for span deflections. At the front end of the deck, the front tractor of the carrier reaches the rear end of a full two-span underbridge.

The underbridge is a stiff box girder supported on adjustable crossbeams at the leading pier of the span to be erected and at the next pier (Rosignoli, 2012). Long-stroke cylinders or framed legs support the rear end of the underbridge on the rear pier. The crossbeams include support cylinders for geometry adjustment and carry articulated support bogies for launching of the underbridge. Double-acting cylinders shift the saddles laterally along the crossbeams to align the rear end of the underbridge with the front tractor of the carrier (Figure 8.24). Crossbeams and rear supports are suspended from the underbridge for self-repositioning without the use of ground cranes during launching.

A motorised saddle rolls on the underbridge and houses vertical support cylinders for the front tractor of the carrier. The underbridge has a rear cantilever and the saddle is moved back over the leading span to receive the tractor (Figure 8.25). The saddle carries the PPU and the hydraulic systems for independent operations; full-length linear displacement sensors monitor the position of the saddle along the underbridge.

The front tractor of the carrier is driven over the saddle, and the vertical cylinders of the saddle are extracted to transfer the load to the saddle and to lift the wheels of the tractor from the deck. When the tractor is long and the rear cantilever of the underbridge is, therefore, also long, the rear support cylinders of the underbridge are replaced with a more stable braced frame. The

Figure 8.25. Support saddle on the rear cantilever of the underbridge

hydraulic motors of support saddle and rear tractor are synchronised to move the carrier over the underbridge until the front tractor is beyond the leading pier (Figure 8.26).

After reaching the span-lowering position, the support bogies of the underbridge are unlocked, the rear support cylinders are retracted, and the motors of the saddle are inverted to push the underbridge forward until the area under the carrier is cleared. The parking brakes of the rear tractor provide the longitudinal restraint during repositioning of the underbridge (Figure 8.27). A full two-span underbridge is necessary to reposition the front support crossbeams during this operation. The PPU of the motorised saddle supplies the hydraulic systems of the support cross-beams for self-repositioning and alignment adjustment.

The bottom counterplates of the bridge bearings are installed during pier erection. Three plates are set at the design elevation and the fourth plate is set 4–5 mm low. The span is lowered onto four load cells placed on the counterplates. After adjusting the support reactions with stainless-steel shims at the fourth plate, the span is lifted, the load cells are removed, and the span is lowered into position (Figure 8.28). Finally, the motorised saddle moves the underbridge back-ward to release the front tractor of the carrier onto the new span for a new placement cycle (Rosignoli, 2012).

Operating the underbridge requires a tall and wide clearance between the wheel groups of the tractors. This geometry requirement turns out to be one of the main advantages of portal carriers over the tyre trolleys. Also the trolleys apply the load over the deck webs, but the taller clearance

Figure 8.26. Portal carrier positioned for span lowering

Figure 8.27. Repositioning of the underbridge

Figure 8.28. Final lowering of the span

of the carriers permits the presence of finishing equipment and materials on the deck, provided that the machines are parked at the deck centreline and on the side wings on the arrival of the carrier.

8.3.1 Loading, kinematics, typical features

Four lifting beams permanently connected to the hoists of the carrier are applied to the span with large-diameter through bolts anchored at the bottom surface. The lifting beams at one end of the span are interconnected to create a torsional hinge, while the lifting beams at the other end are independent to provide the torsional restraint. The lifting beams are articulated, and typically use two blocks of sheaves each.

The carrier is equipped with two hoists: one hoist is fixed and the other may be shifted longitudinally along the main girder to cope with shorter spans. Long-stroke double-acting cylinders are used to shift the mobile hoist on PTFE skids. Hydraulic hoist winches are applied to the rear overhang of the carrier to maximise the efficiency of the PPU and hydraulic systems, and to reduce the load applied to the underbridge. The carriers move the spans within the precasting facility and on long delivery routes, the structural and mechanical components are fully loaded for long periods of time, and high quality is a must in these machines.

The portal carriers are taller than the span launchers. The underbridge and motorised saddle are taller than a tyre trolley, and the span of the main girder is longer due to the length of the tractors.

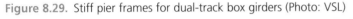

Figure 8.29. Stiff pier frames for dual-track box girders (Photo: VSL)

The precast span is kept close to the main girder during transportation to avoid oscillations, and the combined centre of gravity of carrier and span is high on the ground. The wheels of the tractors are located over the deck webs and, combined with a narrow support base, the height of the centre of gravity may pose problems of lateral stability during span transportation and placement. Stability is more critical with single-track U-spans, as the two lines of wheels are closely spaced to run within the U-section. Wheel flexibility further lowers the first lateral frequency, and the wheels are often filled with water to provide additional stiffness.

Most HSR decks have symmetrical crossfall, and geometry control ensures flat delivery routes for the carriers. The transportation routes along precasting facilities, access embankments and crossover areas are accurately levelled and kept well maintained to minimise crossfall of the support base.

The support systems of the underbridge are designed for maximum stiffness. The pier frames for dual-track box girders are stiffer than those for single-track U-spans because of the greater depth and the heavier weight of the span (Figure 8.29). Anchoring the rear overhang of the underbridge to the deck provides additional stability when necessary.

The lateral stability and limited span length are the weak points of these machines. The length of the precast spans erected with launchers and tyre trolleys is limited only by the load capacity of the deck. When using portal carriers, longer spans require taller underbridges, and a higher centre of gravity soon leads to potentially unstable delivery conditions. Tall pier frames for the underbridge are also less stable, and longer launch bogies complicate supporting the rear end of the underbridge on the front half of the leading pier cap during span lowering.

8.3.2 Support, launch and lock systems

Both tractors are motorised. The rear tractor is electronically synchronised with the motorised saddle of the underbridge to minimise the longitudinal load applied to the underbridge during movements of the carrier. Pressure valves and full-length linear displacement sensors along the underbridge provide feedback to the launch computer for PLC control of actuators.

Because of its primary role in ensuring torsional stiffness, the underbridge has a box section and closely spaced diaphragms to control distortion (see Figure 8.24). The underbridge is supported on inverted cast-iron roll assemblies on equalising beams. The central pier frame is the most loaded support and is anchored to the front half of the leading pier cap so as not to interfere with span lowering. During span lowering, the load of the front tractor is mostly transferred through the central pier frame. Pier brackets positioned with ground cranes may be necessary to support the pier frames when the piers are slender.

A full two-span underbridge equipped with a front support frame is necessary for self-repositioning of the pier frames without the use of ground cranes. Hydraulic motors or capstans drive the pier frames along the underbridge, and vertical support cylinders and lateral shift cylinders provide geometry adjustment. The pier frames are anchored to the pier caps with stressed bars to prevent uplift and ensure stability. Operations are powered by the PPU of the motorised saddle.

The saddle shuttles back and forth along the underbridge, with four articulated assemblies of cast-iron rolls and balancing bogies under the PPU. Four support cylinders for the front tractor of the carrier are located over the central articulations of the support bogies for direct load transfer. The PPU supplies closed-loop high-flow-rate hydraulic systems, and several wheels are motorised to avoid slippage under uphill traction.

8.3.3 Performance and productivity

Most of the service life of a portal carrier is spent on span transportation. Span lifting takes 1–2 hours, placement takes 3–4 hours, and transportation may take an entire day when the delivery route is long. The unloaded speed of the carrier may be increased for faster return to maintain the daily cycle time on long routes. Faster machines are expensive, and may be unsafe due to load oscillations.

When the precasting facility has two casting lines, each carrier serves a casting line. Two carriers and two underbridges may be used to erect two bridges simultaneously. Two carriers may also work with a common underbridge, to double the erection rate of one bridge, when crossover embankments exist along the delivery route. The daily operations are filling the oil tanks, moving a fresh span from the casting cell to the storage platform, and then picking up a span from a storage platform and leaving for the delivery route.

Portal carriers are the first-choice solution for the construction of multiple HSR bridges separated by embankments or tunnels. The carrier picks up the underbridge from the landing embankment and moves it to the next abutment, without the need for dismantling or the use of ground cranes (Figure 8.30). Immediate restart of span placement minimises the disruption of precasting facilities designed for JIT delivery.

The tractors may have single or paired wheels. Single wheels require large-diameter tyres, and wheel spacing increases the axle load. In the transverse direction, however, single wheels maximise

Figure 8.30. Transportation of the underbridge from bridge to bridge

the clearance between the wheels for easier storage of machinery and materials at the centre of the deck during span delivery. Both types of wheels may be designed for 360° independent steering, or for a small steering angle combined with 90° rotation of the tractor for lateral access to casting cells and storage platforms.

In most cases the portal carriers do not require heavy lifters in the precasting facility; the number of storage platforms is also smaller due to a more favourable load distribution on the completed deck. Carrier operations require minimal crossfall and smooth gradients for reasons of equilibrium and power. The large steering radius of carriers devoid of 360° steering wheels requires a radial distribution of the storage platforms in place of the typical rectangular matrix distribution of precasting facilities served by tyre trolleys.

When access to the bridge abutment is impossible or the precasting facility is at a lower elevation, twin portal cranes on rails may be used to lift the spans on the deck (Figure 8.31). In this case tyre trolleys are the preferred delivery means, as the portal carriers require double handling while tyre trolleys can be loaded directly. The portal cranes lift four spans into position to establish a working platform for assembly of the span launcher and tyre trolley. The tyre trolley shuttles back and forth along the bridge to feed the launcher at the leading end of the elevated erection line (Figure 8.32). Multiple tyre trolleys may be used if there are crossover embankments along the delivery route.

8.3.4   Structure–equipment interaction

A portal carrier suspends the span at the ends during transportation and placement. This avoids negative bending, whatever the flexibility of carrier, deck and underbridge may be. A torsional hinge at one end of the span avoids twisting, and dynamic load amplification is typically low with such slow machines on regular support surfaces.

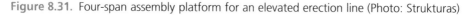

Figure 8.31. Four-span assembly platform for an elevated erection line (Photo: Strukturas)

The load applied by the carrier rarely governs the design of the piers and foundations of bridges designed for HSR loads, and the structure–equipment interaction is mainly related to the load applied to the deck. The tractors are far apart and, although a portal carrier is heavier than a tyre trolley, the weight of the span and carrier is spread over two spans of the bridge. The length of tractors and axle lines are designed to meet the deck capacity at the given curing time, and electronic equalisation of the axle line loads avoids load peaks due to span flexibility.

Sensors control the lateral position of the wheels to align the load with the deck webs and to minimise transverse bending. Some carriers are fully automated and run along the deck without a driver. The portal carriers are prone to wind-induced oscillations and therefore span delivery is interrupted in high-wind conditions.

Structure–equipment interaction may also derive from adjacent structures. In mountain areas with steep valleys, the bridge abutments are often incorporated in the tunnel portals. Special carriers may move through tunnels but there may be no room to load the front tractor on the underbridge. In such circumstances, an underslung platform may be applied to the bottom flange of the underbridge to carry the wheels of the tractor during access to the motorised saddle (Figure 8.33). The platform does not interfere with the typical operations of the underbridge.

8.4. SPMTs

Tyre trolleys and portal carriers are used to deliver hundreds of spans in large-scale HSR projects. When the bridge includes only one span or a few spans to be delivered on a flat and regular

Figure 8.32. Loading of an on-deck tyre trolley (Photo: Strukturas)

surface, the spans may be cast close to bridge and moved into position using SPMTs. SPMTs are often used for replacement of existing spans, as the transporter is used to remove the old span and to place the new span.

SPMTs are robust computer-controlled platform vehicles that are able to lift, transport and place heavy and large loads. They are used for many ultra-heavy transportation operations in the industrial field; in the bridge industry SPMTs are used for removal and replacement of existing spans or for the installation of new spans (FHWA, 2007). The spans typically have modest skew and may weigh from a few hundred tons to more than 10 000 ton.

SPMTs do not prevent span twisting like both tyre trolleys and portal carriers do; they are, therefore, used in applications where the potential span twist induced by loss of or unequal or uneven support is acceptable. By limiting the deflections of the structure, a span may often be moved without the need for complex structural analyses (UDT, 2008).

In skewed spans the SPMTs are arranged to be parallel to the end bents. Three-dimensional structural models are necessary in the most complex cases to investigate the stress distribution during

Figure 8.33. Tractor loading platform for 32.7 m, 900 ton dual-track spans (Photo: BWM)

lifting, transportation and placement of the span. Refined structural analysis provides a higher level of understanding, and a lower load factor for self-weight is sometimes allowed for twist analysis.

SPMT construction facilitates the prefabrication and demolition of bridge spans off-site under controlled conditions; rapid span removal and installation are additional advantages. Impacts on traffic may be reduced to a few hours, compared to the months required for conventional on-site construction or demolition. Other benefits include improved safety, minimised environmental impact and enhanced quality (FHWA, 2007).

SPMT operations consist in moving the span from one location to another. Operations require commitment to time saving through coordination, concurrent preparation, construction and prefabrication on- and off-site, traffic planning and control operations. They also require a flat and accessible area under the span, and therefore SPMT construction is mostly used for replacement of existing spans over highways.

The basic module of an SPMT system is a top load-carrying platform designed for direct loading and supported on multiple axle lines with independently steered pairs of wheels. Each pair of wheels is connected to the frame of the module by a hydraulic jacking system, enabling the whole platform to be raised or lowered under load and driven in any direction under computerised control conditions. The load on the axle lines is equalised by three- or four-point hydraulic suspension systems, and small ground settlements are compensated for by the hydraulic system (UDT, 2008).

Figure 8.34. Longitudinal coupling of two six-axle modules (Photo: Fagioli)

SPMT units may be coupled longitudinally, side-by-side laterally, orthogonally or freely (only a data link) to form a larger, integrated platform having the same attributes of lift, travel and manoeuvrability (Figure 8.34). The maximum load per axle line is 30–40 ton for 2.4 m wide units. The required number of axle lines is estimated starting from the payload and weight of the PPU, and is often based on 80% of the load capacity of the axle line to allow for weight of super-works and out-of-service axles.

Several types of SPMTs are available from various manufacturers. SPMT units have two, three, four or six axle lines. Each axle line consists of four or eight wheels arranged in pairs: units with four wheels per axle line are 2.3–2.4 m wide, and units with eight wheels per axle line are 3.0 m wide. Each pair of wheels can pivot 360° about its support point. The SPMT units can be coupled mechanically at the platform level. Vehicles of different generations are often combinable.

The units are connected by way of a data link to a central computer (the controller). The driver operates the controller while walking with the units. The controller has four basic commands: steer, lift, drive and brake. Computerised steering systems allow pre-programed or manually controlled movements in any direction, as well as carousel steering. All functions are synchronised: one unit is selected as the master unit, and the other units receive control commands for steering, drive, braking and lifting from this master. Loaded SPMTs travel at 4–5 km/h: some axles are motorised, others are braked, and most are free.

Figure 8.35. SPMT delivery of the Rialto Bridge in Venice (Photo: Fagioli)

A four-axle SPMT unit is 5–6 m long without the PPU, and a six-axle unit is 8–9 m long. The PPU adds another 2.5–4.5 m to the length of the unit, depending on the number of axle lines. The PPU includes a diesel engine that provides power to the hydraulic systems that control axle height, traction, steering and braking, as well as to the electrical systems. The PPU is connected to the front unit of the assembly by hydraulic cylinders for adjusting the approach angle when moving on ramps.

Lift capacity and lift range vary by manufacturer, configuration and ground conditions. SPMT units are transported to the bridge site on flatbed trailers or shipped in containers. For highway transport, the six-axle units often require overweight permits, and 3.0 m wide units require permits for both width and weight.

Load magnitude, tyre pressure and ground surface variations along the delivery route affect the travel height of the loaded SPMT platform. The preferred travel height is 1.1–1.5 m, but the platform may be as low as 0.9–1.3 m during travel. The vertical stroke of the platform is typically 0.6 m, and an available stroke of 0.4–0.5 m is assumed for operational purposes; the reserve stroke is useful where there is minor ground settlement.

An SPMT platform can be adjusted vertically up to 0.6 m to keep the load horizontal without distortion on uneven or sloping ground surfaces. SPMTs can travel on uneven terrain with surface variations up to 0.4 m and on gradients of up to 8–9%, depending on the ground surface friction. Additional equipment for vertical lifting can be mounted on the SPMT platform as needed.

A set of SPMT units under one lifting point can be connected to a set of SPMT units under a second lifting point. Data-link interconnections are also used for large spans that require four support points. Elevation monitoring of an expanded system may become critical in these applications.

The wheels exert a maximum ground pressure of 70–100 kPa. Heavy spans can create excessive ground bearing pressures, and it is not always possible to make use of the maximum load capacity. Additional SPMT units may be required to reduce the bearing pressure, or steel plates may be placed along the delivery route to spread the load. Using the maximum load per axle line is also not recommended when the structure–SPMT assembly has a high centre of gravity.

To remove a bridge span, the SPMT units are assembled into an integrated platform, and blocking is placed on top of the platform. Typically, blocking supports each of the beams in the span. The SPMT platform is driven under the span, and the span is lifted using the SPMT hydraulic systems. The loaded platform travels to an off-site staging area and the span is lowered onto temporary blocking for demolition. A reversed sequence of operations is used to move a new span into place; SPMT platforms can also be mounted on barges for marine operations (Figure 8.35).

The use of SPMTs poses specific problems of structural integrity. The support lines of the span during lifting are about 3 m inward from the final bearing lines due to access requirements for the SPMT and super-works. During lifting, the cantilever ends of the span deflect downward and the midspan deflects upward. Analysis of deflections before and after lifting is necessary to determine the minimum jacking stroke required by the SPMT system to clear the bearings.

Increasing the cantilever overhangs and closing the distance between the lift points decreases stability and increases deck vibrations. A dynamic allowance of 5–15% is applied to the self-weight of the span to account for uneven support surfaces in addition to the imposed twist (UDT, 2008).

The results of structural analysis are controlled by surveying the profile of the span before and after lifting. Real-time monitoring of deflections is recommended during lifting, transportation and placement to ensure structural integrity. Stainless-steel levelling bolts, rivets or inserts are applied to the deck over the edge beams at the permanent bearing alignments, the SPMT supports and midspan. Strain gages or other means of monitoring may also be used.

Uneven support may cause structural warping in the span. Twist occurs when one corner of the span deflects from the plane defined by the other three corners. Twist can occur at any time during transportation due to uneven support or settlement, and can also occur when the span is transferred from the SPMT platform to the permanent bearings.

Ideally, a span should be lifted, transported and positioned with no twist to maintain a stress-free state. Twist calculations are performed by computing the difference in crossfall at the ends of the span. Twist is corrected by shimming the bearings at one end of the span to recreate the as-cast differential in crossfall.

REFERENCES

FHWA (Federal Highways Administration) (2007) *Manual on Use of Self-propelled Modular Transporters to Move Bridges.* Publication No. FHWA-HIF-07–022. FHWA, Washington, DC, USA.

Liu Y (2012) Erecting prefabricated beam bridges in a mountain area and the technology of launching machines. *Structural Engineering International* **22(3)**: 401–407.

Rosignoli M (2011) Industrialized construction of large scale HSR projects: the Modena bridges in Italy. *Structural Engineering International* **21(4)**: 392–398.

Rosignoli M (2012) Modena viaducts for Milan–Naples high-speed railway in Italy. *PCI Journal* **50(Fall)**: 50–61.

UDT (Utah Department of Transportation) (2008) *Manual for the Moving of Utah Bridges Using Self Propelled Modular Transporters (SPMTs).* UDT, Salt Lake City, UT, USA.

Vion P and Joing J (2011) Fabrication and erection of U-through section bridges. *Structural Engineering International* **21(4)**: 426–432.

VSL (2008) *Over 50 Years of Excellence.* VSL International Ltd, Köniz, Switzerland.

Bridge Construction Equipment
ISBN 978-0-7277-5808-8

ICE Publishing: All rights reserved
http://dx.doi.org/10.1680/bce.58088.305

Chapter 9
Design loads and load combinations

9.1.   Introduction

Standardised design criteria for bridge construction equipment would improve the safety and economy of bridge construction. The availability of internationally recognised design standards should therefore be the concern of owners, contractors, unions, insurance carriers and the general public.

Researchers and educators neglect bridge construction equipment because of the complexity of the topic and the number of technological aspects. Many bridge designers are unaware of the intricacies of these machines, and are not interested in the subject. Uniform design criteria could limit the reuse of existing machines, and those (few) contractors that handle the erection equipment like the 'competitive edge' of the business could hardly hide the risks they take during construction. As a result, minimal guidance is available on this field of the bridge engineering practice (Andre *et al.*, 2012).

Bridge construction equipment is subjected to multiple load conditions and multiple support configurations during operations and repositioning. Some machines are anchored during operations and only the cranes they carry are free to move; other machines move during full-load operations. Mechanical or hydraulic systems force major components of the machine to move relative to other components during repositioning.

Permanent bridge structures are designed with varying loads applied to a fixed geometry. Bridge construction equipment is designed for similar loads, but the additional complication of movement requires consideration of multiple ultimate limit states (ULSs) in relation to stability of equilibrium, and a few extreme-event limit states (EELSs). Redundancy of systems and controls should be considered, along with structural redundancy, when determining the factored load combinations. Serviceability limit states (SLSs) are typically not related to structural safety, while fatigue and fracture limit states (FFLSs) are in an intermediate position.

Lower safety factors have been used in the past for the design of bridge construction equipment in relation to its temporary nature. The long history of failures and the low cost of equipment compared with the costs associated with its failure suggest a risk-based approach to safety of design (Sexsmith and Reid, 2003). In the light of the complexity of the new-generation machines, the interaction of mechanical and hydraulic systems with electronic control systems, the high stress levels used in the design, the long-term application of full design loads, the great number of load reversals during launching and operations, and the presence of multiple risk scenarios during most operations, the safety factors for construction machinery should actually be higher than those required by the design standards for permanent bridge structures.

On the other hand, however, construction loads are more predictable and controllable than the live loads applied to a bridge over decades of use, and bridge construction equipment is used only for short periods of time during its service life. In the absence of specific design standards, therefore, a reasonable approach is to design, detail, fabricate and operate bridge construction equipment according to design standards for permanent steel structures subjected to similar loads.

Using the same suite of standards for the design of the bridge and the erection equipment ensures calibration of load and resistance factors, the same serviceability criteria, coherent external actions, uniform levels of safety and robustness, and more predictable structure–equipment interaction. This approach also facilitates the reuse of the machines in future bridge projects. All major bridge design standards (AASHTO, BS, CSA, DIN, Eurocode, etc.) include provisions for both steel and prestressed concrete structures.

Modern bridge design standards are based on load and resistance factor design (LRFD), a reliability-based design methodology in which the effects of factored loads are not permitted to exceed the factored resistance of members. Each component and connection must satisfy the following condition for each limit state (AASHTO, 2012a):

$$\Sigma \eta_i \gamma_i Q_i \le \phi R_n$$

The design standards define the load factors γ_i to be applied to the load effects Q_i in relation to the limit state and the load combination examined. The load factors reflect the variability of loads and the effects of assumptions and approximations made during the analysis. The load factor for permanent loads is $\gamma_i \ge 1.0$ if the load effect increases the demand, and $\gamma_i \le 1.0$ if it diminishes it. Stability of equilibrium is assessed like any other ULS condition.

Many factors are considered in determining an acceptable level of safety. These considerations include potential loss of life and other consequences of failure, type and warning of failure, and the existence of alternative load paths. Some standards provide load multipliers η_i that increase or diminish the design demand in relation to features of the structure that may affect its behaviour at failure. The stakeholders can thus manage the target reliability level of the structure.

The load multipliers prescribed by BS EN 1990: 2005 (BSI, 2005c) are related to the consequences of failure and the levels of quality control and inspections implemented. The load multipliers prescribed by AASHTO (2012b) are related to ductility, redundancy and operational classification of the bridge, and are used only for ULS checks. For loads for which a maximum value of γ_i is appropriate, the total load multiplier (AASHTO, 2012b) is:

$$\eta_i = \eta_D \, \eta_R \, \eta_I \ge 0.95$$

For loads for which a minimum value of γ_i is appropriate, the load multiplier is:

$$\eta_i = \frac{1}{\eta_D \, \eta_R \, \eta_I} \le 1.0$$

The load multiplier η_I is related to the operational importance of the bridge and does not affect the design of erection equipment. The structural components should be designed and detailed to ensure the development of significant and visible plastic deformations before failure. The load multiplier related to ductility is $\eta_D \geq 1.05$ for non-ductile members and connections, $\eta_D = 1.0$ for design and detailing complying with AASHTO, and $\eta_D \geq 0.95$ for members and connections with enhanced ductility. Ductility of connections may be enhanced with strength excess to force plasticity at locations designed to provide a ductile, energy-absorbing response.

Multiple load paths and continuous structural systems enhance redundancy and the robustness of erection equipment. Redundancy is the capability of the structural system to carry load after the failure of one or more members. Member redundancy classification is based on the member contribution to the overall safety of the machine. The AASHTO (2012b) load multiplier related to redundancy is $\eta_R \geq 1.05$ for non-redundant members, $\eta_R = 1.0$ for conventional levels of redundancy, and $\eta_R \geq 0.95$ for exceptional levels of redundancy. Different load multipliers are used for the design of the individual members; for example, a pendular leg is designed with $\eta_R \geq 1.05$, while its redundant in-plane bracing is designed with $\eta_R = 1.0$.

FEM 1.001 (FEM, 1987) requires the use of two additional load multipliers for the design of heavy lifters. The load multiplier for the mechanical components varies in the range $1.0 \leq \eta_M \leq 1.3$ in relation to the mechanical classification of the lifter, and the load multiplier for the structural components varies in the range $1.0 \leq \eta_S \leq 1.2$ in relation to the structural classification.

The design standards also define the resistance factors $\phi \leq 1$ to be applied to the nominal resistance R_n of members and connections to determine the design capacity for the type of stress and the expected response at the limit state under consideration. The resistance factors reflect the variability of the strength of materials, the difference between as-built and as-designed dimensions, and the effects of the assumptions made in determining the member resistance.

The structural checks are expressed in terms of demand-to-capacity ratio (D/C) to assess whether, due to the variability of loads and resistances, a weaker than average structure subjected to higher than average loads may withstand failure:

$$\frac{D}{C} = \frac{\Sigma \eta_i \gamma_i Q_i}{\phi R_n} \leq 1.0$$

Because of the absence of comprehensive design standards for bridge construction equipment, calibration of load and resistance factors from different industry standards is sometimes necessary to ensure that the load effects are handled with consistent levels of reliability (CIRIA, 1977).

Load Q and resistance R are taken as random variables with a Gaussian distribution. Unless R exceeds Q by a large amount, there will be some probability that R may be less than Q, which identifies structural failure as $R/Q < 1$, or $\ln(R/Q) < 0$. The mean value of the function $\ln(R/Q)$ is defined as a multiple β of the standard deviation of the function. The multiplier β is called the reliability index of the design standard (Ratay, 2009).

Monte Carlo simulations are used to relate R and Q in primary components of the machine, and to calibrate load and resistance factors to the reliability index of the design standard. If the design standard provides resistance factors for all the types of members and connections of the machine, only the load factors for the types of loads not covered by the standard are calibrated. Calibration is performed at a local level under the assumption that the resistance of the individual members ensures adequate overall resistance of the machine (Sexsmith, 1988).

The reliability index indicates the consistency of safety of components and systems designed with different methods, and may be used to establish new methods that will have consistent margins of safety. The reliability index achieved when designing as per AASHTO (2012b) is 3.5. BS EN 1990: 2005 (BSI, 2005c) is based on a reliability index of 5.2 for a 1-year reference period and a high consequence class. The reliability index suggested by ASCE/SEI 7-10 (ASCE, 2010) is 4.8 for structures whose failure is sudden and results in widespread progression of damage and to which 5–500 persons are exposed.

Engineers typically resort to design standards to make decisions. Design standards are developed to address areas where significant past experience exists and to prevent failures experienced in the past. In the absence of a recognised design standard for bridge construction equipment, frequently encountered pathologies may be avoided or mitigated by means of accurate design.

1 The temporary nature of bridge construction equipment should not be an excuse for reduced quality standards and safety levels. Respect of design standards and compliance of fabrication to design should be as rigorous as for a permanent structure.
2 Structural members, mechanical components and control systems that are critical for the stability of equilibrium should be redundant.
3 Design should be as clear as possible. Simple structural schemes show the flow of forces and the purpose of every element, are easier to assemble and inspect, and are more intuitive to operate.
4 Design should allow for fabrication and assembly tolerances.
5 Detailing must materialise the design assumptions. The safety and performance of bridge construction equipment are often a matter of careful detailing.
6 Non-stiffened steel shapes have low capacity to localised loads. Web stiffeners should be used at every node and at fixed application points of localised loads.
7 All welds should be executed by certified welders and inspected by experienced technicians. Critical welds and all field welds with structural relevance should be examined using radiography or ultrasound imaging. Penetrating liquids do not detect internal defects, but can be used for contrast or to check fatigue cracking.
8 Every machine should be equipped with systems that prevent collapse in case of human error. Mechanical end-of-stroke blocks should be used as much as possible to limit load application to areas of the machine that can safely sustain those loads. Automatic blocks on control systems and actuators are not recommended, as they may prevent or complicate emergency operations: overloading should be allowed and signalled by alarms.

9.2.   Design loads common to most bridge construction equipment

Self-weight, effects of geometry imperfections, loads on walkways and working platforms, and meteorological loads are common to most types of bridge construction equipment. Other loads are specific to specific types of equipment.

The permanent loads include self-weight and superimposed dead load. These two loads are distinguished in the design of permanent structures, as the superimposed dead load has higher statistical scatter and the load factors are therefore more demanding. In a bridge erection machine, the superimposed dead load (PPU), fluid tanks, hydraulic systems, etc.) is applied permanently to the machine and is necessary for proper functioning. In most cases, therefore, all permanent loads may be treated using the less demanding load factors for self-weight.

Geometry imperfections may have a substantial impact on the strength and stability of some types of equipment. Geometry imperfections may be expressed in terms of equivalent loads or may be incorporated in the structural models by modifying the coordinates of nodes and the alignment of members.

Meteorological loads include snow, temperature and wind. Snow and ice are rarely considered in temperate climates; in cold weather they may increase the surface exposed to wind and diminish the first lateral frequency with additional mass. Meteorological loads are often determined without reductions in relation to the short return period related to project duration (Boggs and Peterka, 1992). For bridge construction machines it is generally possible to identify precise meteorological limit conditions for their use, and the load combinations distinguish normal operational conditions and out-of-service conditions.

Settlement of foundations is rarely a major concern, as most bridge erection machines are loaded as simple spans.

9.2.1 Self-weight

Bridge design standards specify different load factors for self-weight and superimposed dead load. AASHTO (2012b) requires the use of $\gamma = 1.25$ for the self-weight and $\gamma = 1.50$ for the superimposed dead load when their effects increase the demand, and of $\gamma = 0.90$ and $\gamma = 0.65$, respectively, when their effects diminish the demand. Different load factors reflect the statistical scatter of the two types of load and the different probability of full application or removal.

Distinguishing between self-weight and superimposed dead load is rarely necessary for a bridge erection machine, because most non-structural loads are applied permanently to the machine and are necessary for functioning. The weight and mass of the PPU, fluid tanks, hydraulic systems, motors and electric systems, vibrators and pneumatic systems, walkways, working platforms, stairs and ladders, and insulation and covering are therefore added to the self-weight, and the load thus determined is processed as a characteristic load. Materials stored on the machine could conceptually be processed as superimposed dead loads, but their light weight rarely justifies a specific load case.

The weight and mass resulting from the cross-sectional area of structural members are increased by 30–40% to allow for rolling tolerances, welds, stiffeners, end plates, lap plates, bolts, pins and the like. Heavier localised loads and masses such as the PPU, hydraulic systems and mechanical components are modelled as concentrated loads and masses. The accuracy of numerical modelling may be enhanced during final design by adjusting the member weight as per the plans. The sub-assemblies are weighted with the crane load sensors during site assembly for an additional check.

Self-weight is analysed in operational conditions and during repositioning. Launching is analysed in a sequence of structural schemes where the support points are shifted longitudinally along the machine and their geometry and connectivity are adjusted according to launch operations (Rosignoli, 2002). The launch stresses thus determined are enveloped to determine the design-governing demand for members and connections.

Deflections calculated at multiple control points for the main phases of launching are included in the operation manual, together with their allowed range, to provide the equipment supervisor with multiple control parameters. Support reactions and realignment loads are expressed in terms of pressure in the hydraulic systems to allow for immediate comparison. Redundant checks provide alternative solutions if such are required, and redundant information if all checks are possible.

9.2.2 Geometry imperfections

Design standards provide instructions to allow for the effects of geometry imperfections in steel structures. The design provisions are calibrated with the fabrication and assembly tolerances allowed in the standard, and may therefore vary from standard to standard. BS EN 1993-1-1: 2005 (BSI, 2005a) recognises three levels of tolerance: normal tolerances are used for structures loaded statically, special tolerances are used for structures different from conventional buildings and where fatigue phenomena prevail, and more restrictive tolerances are used for crane runways.

The effects of geometry imperfections are considered in the global analysis, in the analysis of the bracing systems, and in the design of members. The tolerances in verticality are considered in the analysis with an equivalent angular deviation from the verticality ϕ given by (BSI, 2005a)

$$\phi = 0.005 k_c k_s$$

with

$$k_c = \sqrt{0.5 + 1/n_c} \leq 1.0$$

and

$$k_s = \sqrt{0.2 + 1/n_s} \leq 1.0$$

where n_c is the number of columns and n_s is number of bracing panels in the tower or temporary pier. The equivalent angular deviation is applied in the longitudinal direction and in the transverse direction, and the effects are not superimposed. When the temporary pier has four columns, the angular deviation is also applied with opposite sign to two planes of columns to induce torsional effects.

The bracing systems that ensure lateral stability of compression members are also designed allowing for geometry imperfections. According to BS EN 1993-1-1: 2005 (BSI, 2005a), the alignment tolerances may be analysed using a lateral midspan deflection e_0 given by:

$$e_0 = 0.002 k_r L$$

where L is the span length of the bracing system and

$$k_r = \sqrt{0.2 + 1/n_r} \leq 1.0$$

where n_r is the number of braced members.

The geometry imperfections may be modelled using equivalent systems of lateral forces or by modifying the node coordinates in the structural models. Some software packages for structural analysis export the model definition into an Excel-compatible file to make changes to the model within an Excel environment prior to reimporting. Other packages allow the correction of node coordinates through text files. Other packages force the operator to displace the nodes individually within the graphical user interface, and are therefore more time consuming.

9.2.3   Loads on walkways and working platforms

Walkways and working platforms are often designed for a distributed load of 1.5 kN/m². A maximum value of 20 kN is sometimes stipulated for every walkway and working platform, and maximum values of 60 kN during concrete pouring and 20 kN during launching may be stipulated for the entire machine. Upper bounds for personnel loads are justified by the controlled access to bridge construction equipment and generate more realistic design demands, especially when stability of equilibrium is concerned. The loads allowed on walkways and working platforms should be indicated on the plans and by signs on the machine.

Storage of materials and equipment should preferably be avoided; in some cases, however, reconfiguring the working locations after every repositioning of the machine is impractical. When materials and equipment are stored on the machine during launching, the storage areas should be identified on the plans and by signs on the machine, and the total loads allowed on the storage areas are considered in the structural analysis without any upper bounds.

According to BS EN 1991-3: 2006 (BSI, 2006a), access walkways, stairs and working platforms are designed for a characteristic vertical load spread over a square surface with a side length of 0.3 m. A characteristic load of 3 kN is used where materials can be stored, and a load of 1.5 kN is used if the walkways, stairs and working platforms are provided for access only. The handrails are designed for a single horizontal load of 0.3 kN.

9.2.4   Thermal loads

Rise and fall in temperature are calculated starting from the temperature assumed at the end of repositioning, when the machine is anchored to the bridge. Most design standards provide design temperature ranges for steel and PC structures, and for cold and moderate climates. The design temperature range represents the maximum difference in temperature between summer and winter, and the daily variations are much smaller.

Rise and fall in temperature may damage rigid members that prevent thermal movements of long portions of steel structures. The most vulnerable elements of gantries and MSSs are the pier brackets and the tower–crossbeam assemblies. Breakaway friction of PTFE skids causes P-delta effects in tall towers and horizontal bending in the crossbeams. The equivalent friction coefficient of launch bogies is lower, but the main girders are mostly lifted from the launch saddles for span construction, and the guided steel–PTFE sliders used at the top of the support cylinders have a higher friction coefficient.

The longitudinal thermal loads on cylinder seals and stiff pier brackets may be demanding. When a girder made with steel with $f_y = 355 \, \text{N/mm}^2$ is subjected to a 10°C rise in temperature, full restraint of thermal expansion would cause a compressive stress of 7% of the yield point. Full restraint is improbable and the actual locked-in stresses will be smaller (Rosignoli, 2010). Structural analysis with accurate representation of the internal releases of degrees of freedom is necessary to check the stresses and displacements.

In underslung gantries and MSSs propped from foundations, thermal stresses may arise in the support systems due to the different thermal inertia of the steel props and the permanent reinforced-concrete piers. The flexural stiffness of the continuous deck of span-by-span macro-segmental bridges prevents thermal expansion of the temporary piers, and additional compressive stresses in the columns diminish the buckling factor (Rosignoli, 2011). When a self-launching gantry is used to stabilise the hammers during balanced cantilever erection of precast segments, thermal gradients in the trusses may overload strong-backs at the midspan closure pours.

The thermal displacements of a continuous deck interact with the support systems of an MSS for span-by-span casting. The deck is typically fixed at the starter abutment and supported on sliding bearings at the piers. With the progress of construction, the leading end of the deck is subjected to larger and larger thermal displacements. The rear end of the casting cell is anchored to the deck, and, if the front support of the MSS were allowed to slide, the outer form and form soffit should also slide at the pier. In most cases, therefore, the pier support is prevented from sliding: forming is simpler, but MSS and support systems must be designed to oppose the thermal movements in the deck.

These issues are amplified when an MSS is used to cast the closure span between two bridge portions anchored to the opposite abutments (Pacheco *et al.*, 2011). In reality, however, the thermal inertia of a box girder mitigates the effects of daily temperature changes, and friction at the sliding bearings further diminishes the thermal displacements of the leading end of the deck.

9.2.5 Wind loads

Bridge construction equipment is typically designed to be reused, and the meteorological loads should therefore be determined without reductions in the return period in relation to project duration (Hill, 2004; Rosowsky, 1995). It is generally possible to identify precise limit wind conditions for the use of the machines, and the load combinations distinguish between normal operational conditions and out-of-service wind conditions.

Operations are allowed or suspended based on the wind speed measured by anemometers attached to the machine. The operation manual indicates the maximum wind speed allowed during operations and the type of speed to be measured (average speed over a certain period of time, gust speed, or both) (CNC, 2007). The gust speed is typically considered for the structural analysis of normal operations. A lower wind speed is specified for launching in order to enhance the safety of operations that involve less stable structural configurations; open MSSs are also more flexible during launching and more prone to wind-induced vibrations.

Depending on the orientation of the surface to the wind direction, the wind load is applied through drag forces and lift forces. Turbulence and vortices caused by the edges of the girders increase wind pressure locally, and the wind speed at the edges is higher than the average wind speed because air accelerates as it takes a longer path to move around an obstacle.

Because of the complexity of the response of structures to dynamic wind loading, several design standards specify a kinetic wind pressure that is only related to the wind speed, the height of the structure from the ground, and environmental features. CNR 10027 (CNR, 1985b) specifies a lower-bound wind pressure of $0.2\,kN/m^2$ for MSSs in operational conditions and $0.8\,kN/m^2$ for out-of-service conditions. CNR 10021 (CNR, 1985a) specifies a lower-bound wind pressure of $0.25\,kN/m^2$ for heavy lifters in operational conditions, while the out-of-service pressure is related to the height of the machine from the ground. For piers shorter than 20 m the wind pressure in out-of-service conditions is $0.8\,kN/m^2$. For piers in the 20–100 m range it is $1.1\,kN/m^2$, and for taller piers it is $1.3\,kN/m^2$. These kinetic wind pressures correspond to wind velocities of 130, 151 and 166 km/h, and are therefore conservative in most practical cases.

Some design standards for permanent structures specify higher wind pressures. DIN 1072 (DIN, 1985) specifies 1.75 kN/m^2 for piers shorter than 20 m in the absence of other loads, 2.10 kN/m^2 for piers in the 20–50 m range, and 2.50 kN/m^2 for piers in the 50–100 m range. These wind loads, however, may be reduced to 70% for analysis of the erection loads, and to 20% when an operation takes less than one day and the wind speed is less than 20 m/s.

Several design standards specify shape factors for direct wind exposure and sheltered locations; in the box girders the shape factor also depends on the depth-to-breadth ratio. Load reduction on leeward trusses is often neglected in the design of bridge construction equipment, as wind loading at small angles from the orthogonal can eliminate the potential for the windward truss to shield the leeward truss. DIN 1055 Part 4 (DIN, 1986) and BS EN 1991-1-4: 2005 (BSI, 2005b) provide shape factors for multiple types of steel shapes and for triangular and rectangular trusses.

The wind load for normal operations is applied to all possible working configurations of the machine, while the out-of-service wind load is applied to specific configurations with the machine anchored to the deck as specified in the operation manual. Typical wind speeds for the interruption of operations are 10–12 m/s as an average value over 10 minutes, and 16–18 m/s as an instantaneous gust value.

Wind loading is treated as a live load and is assumed to act only on the portions of the machine that produce the maximum demand in the member considered. Design standards specify limits for the slenderness ratio for primary and secondary members of a truss. When the minimum-size members of lateral bracing are selected to satisfy the slenderness ratio, the calculated lateral wind demand is typically below capacity.

In most types of bridge construction equipment, lateral wind acts above the supports and causes torsion and overturning couple. These forces are transferred to the bridge through the stiffer members, irrespective of the intent of design, and support reactions and uplift loads will balance accordingly. MSSs are prone to wind loading due to the large wind drag area and the frequent presence of advertising banners. Portal carriers for precast spans are also affected due to the high combined centre of gravity, and the low lateral frequency resulting from tyre flexibility makes them prone to wind-induced oscillations.

Bridge construction equipment and permanent piers may respond dynamically to wind loading. Dynamic amplification occurs when frequencies in the wind speed spectrum excite natural frequencies of the structure–equipment system. Aeroelastic phenomena such as galloping, vortex-shedding excitation and flutter are also possible in such flexible structures. BS EN 1991-1-4: 2005 (BSI, 2005b) provides comprehensive information on vortex shedding and aeroelastic instability. DIN 1055 Part 4 (DIN, 1986) defines a cantilever pier of width b orthogonal to the wind direction and height h as susceptible to wind-induced vibrations when the resonance frequency is

$$f_r \le \sqrt{\frac{0.10}{\delta}} \left[\frac{44\sqrt{(h/b+1)/20}}{h} + 0.05 \right]$$

The equivalent damping δ may be taken as 2% for steel structures with slip-critical field splices and 4% for reinforced-concrete columns. Low values of damping accentuate the perception of vibrations, and may cause concern to construction personnel.

Lateral wind loading on the erection equipment should always be considered in the design of tall piers. The taller the piers, the bigger the advertising banner the contractor will want to install on the machine. Piers and erection equipment are often designed with wind loads applied to most of the solid area between the chords, which minimises one of the advantages of trusses over braced I-girders.

Trusses of medium length are unlikely to sustain severe damage from wind-induced vibrations. The wind loads specified by the design standards are typically conservative, designers often disregard the load capacity of components that are not part of the intended lateral-load-resisting system, launching is avoided in the presence of strong wind, and the dynamic response to out-of-service wind rarely superimposes on the full service load (Rosignoli, 2007). The number of load cycles related to wind-induced vibrations, however, may be very large and may cause fatigue distress.

Dynamic analysis may be necessary when the launch nose is very long. This is often the case in new-generation single-truss overhead MSSs and twin-truss underslung MSSs, as a long launch nose allows earlier access to the next pier for preparatory activities. These machines are used on 70–75 m spans, and light and long cantilevers with low first lateral frequency may be exposed to large wind-induced vibrations.

Analysis of wind-induced resonance is performed by applying different distributions of lateral wind pressure to the truss with sinusoidal load time functions. The loading frequency is adjusted until lateral resonance in the truss is reached. If the resonant frequency is realistic in relation to the frequency range of the wind spectrum, lateral bracing is checked using the peak dynamic loads at resonance.

9.2.6 Exceptional loads

The exceptional loads are out-of-service wind, load testing, collision with buffers or fixed obstacles, and the design-level earthquake. Impacts against fixed obstacles are rare with some types of machine and less rare with others. Collision with buffers is often disregarded for translation speeds lower than 0.5–0.7 m/s, provided that the runway is equipped with end-of-stroke switches. Buffers must always be applied to crane runways as most erection equipment follows the bridge gradient and the runways are therefore inclined planes. When not in use, winch trolleys and portal cranes are locked by means of parking brakes. In case of brake failure, the buffers must halt the crane without the intervention of end-of-stroke switches, which cannot operate in the absence of electric traction.

According to BS EN 1991-3: 2006 (BSI, 2006a), the forces applied by a collision with the buffers are calculated from the kinetic energy of the crane moving at 70–100% of the nominal speed. The longitudinal buffer force F_B is determined as

$$F_B = \varphi v \sqrt{m_C S_B}$$

where v is 70% of the peak longitudinal velocity, m_C is the mass of the crane and the hoist load (kg), and S_B is the spring constant of the buffer (N/m). The buffer characteristic ξ_B is determined in relation to the response force to displacement as

$$\xi_B = \frac{1}{Fu} \int_0^u F \, du$$

The dynamic allowance is $\varphi = 1.25$ for $0 \le \xi_B \le 0.5$ and $\varphi = 1.25 + 0.7(\xi_B - 0.5)$ for $0.5 \le \xi_B \le 1$. The dynamic allowance specified by FEM 1.001 (FEM, 1987) is $\varphi = 1.25$ for linear-spring buffers and $\varphi = 1.60$ for constant-force hydraulic buffers. Dynamic analysis can provide more accurate information about the effects of collision.

Movements of the hoist crab along the crane bridge of the winch trolley or of the gantry along the support crossbeams may also cause transverse buffer loads. If the payload is free to swing, the buffer forces may be taken as 10% of the hoist load and crab weight. In other cases the buffer forces can be determined as for the longitudinal movement of the winch trolley.

Bridge construction equipment is typically designed for a linear-elastic response to the design-level earthquake. The design demand is determined from the seismic design criteria of the project with a reduced return period corresponding to the construction duration. Setting p. as the probability of exceeding the seismic design event during bridge construction (BS EN 1998-2: 2005 (BSI, 2005d) recommends that this probability does not exceed 5%) and t_c as the duration of construction of the project (in years), BS EN 1998-2: 2005 (BSI, 2005d) defines the return period of the seismic design event during construction as

$$t_{rc} = \frac{t_c}{p}$$

Set a_g as the peak ground acceleration corresponding to a 475-year return period, the peak ground acceleration of the seismic design event during construction is defined as

$$a_{gc} = a_g \left(\frac{t_{rc}}{475} \right)^k$$

The constant $k = 0.30$ to 0.45, depending on the reliability of the seismological data. Seismic loads are often applied to unloaded gantries and MSSs; analysis of the portal carriers for precast spans should include the weight and mass of the span because these machines are fully loaded for most of their service life.

9.2.7   Human error

Most bridge failures occur during construction and are related to the erection equipment. Advances in technology, the pressure of time and cost-cutting driven by competition are contributing factors, but the most direct cause of failure is human error (Reason, 1990). Carelessness, incompetence, ignorance, poor management and communication, disregarding standards and specifications, and the deliberate choice of accepting greater risks are the most common forms of non-adherence to good practice. Even not anticipating the confluence of events or circumstances is a human error and, as such, it could be prevented.

The effects of incorrect operations should be considered as part of the design process. Identifying the effects of human error requires envisaging the shortcuts that can be invented on the field to solve problems with the construction equipment and, most of all, to recover delays in the time schedule. The structural analysis of the effects of human error requires examining the operation manual and identifying the possible voluntary violations of the prescribed operations.

- In many twin-girder gantries and MSSs the trusses are not connected to each other and can be repositioned independently. A frequent error is advancing one truss excessively relative to the other when something delays one launch line. Connecting the trusses with end braces improves stability and prevents staggered launching, but is typically impossible with underslung machines. When staggered launching is allowed, the transverse couple applied to the support crossbeams may be calculated by increasing the launch differential allowed by the operation manual by 50% or to the maximum extent physically possible. Anchorages and tie-downs will be sized accordingly.

■ Most machines carry mobile cranes. When not used, the cranes are locked to the runways. Brakes may fail or be forgotten, the runways are often inclined planes, and the strength of members and stability of equilibrium should be checked by placing loaded cranes in the most demanding locations allowed by the geometry of the machine or the end-of-stroke buffers.

■ Redundant anchorages increase robustness but complicate operations. When the tie-downs are designed without margin for human error, load sensors can be applied to the anchor bars for remote control of bar stressing prior to critical operations.

■ If the equipment supervisor is not used to working with spirit levels, consider some degree of oblique hoisting being due to misalignment of the load and hoist.

Hydraulic systems may be overloaded because of human error but also intentionally, in response to emergency situations. Overloading should not be prevented by control systems that must be 'hammered to death' to allow emergency operations. Overloading should be allowed and signalled by lights and sound systems, and the hydraulic systems should be designed to tolerate overloading.

The PLC, microcontrollers, programmable subsystems and servo-systems, process analysis and supervision systems, and terminal operator controls/commands may all be affected by human error related to the control software or the integration of programmable subsystems from different producers. These forms of human error are typically avoided by good quality assurance/quality control procedures.

9.3.   Design loads of MSSs

Beam launchers, self-launching gantries, lifting frames, span launchers and portal carriers with underbridge are designed using standards for cranes and heavy lifters. The nature of loading is less dynamic in an MSS, the structural components of the machine are designed using standards for permanent steel structures, and the forming systems are designed using specific industry standards after calibration of load and resistance factors to a common reliability index.

The design loads of an MSS are separated into fixed loads and varying loads. Fixed loads include dead and superimposed dead loads, which are usually combined and referred to as 'self-weight'. Varying loads include the weight and pressure of fresh concrete and reinforcement, load redistribution at the application of prestress, actions applied by carried cranes, actions of snow and wind, loading of personnel on walkways and working platforms, and thermal loading.

9.3.1   Casting cell

The casting cell is designed for the weight of forms and reinforcement, the filling procedure, and the load redistribution at the application of prestress. Geometry tolerances (extra weight) in the segment or the span may be allowed for by using a higher value of the characteristic density of concrete.

The form panels are designed for the pressure of fluid concrete and the weight of workers and reinforcement. Bottom crossbeams, inner and outer rib frames, articulations and geometry adjustment systems are designed for the stress distribution generated by the support cones of the inner tunnel form, the anti-floating tie-downs, and the horizontal ties between the inner and outer rib frames in the webs. The lateral pressure of fluid concrete may be calculated using the hydrostatic distribution in the upper portion and constant value in the lower portion. Form design for full hydrostatic pressure is specified for fast raising rates or when retarding admixtures or self-compacting concrete are used; full hydrostatic loads are also used for the design of tie-downs that prevent floating.

Lateral load imbalance between the webs of a box girder can displace the inner tunnel form or twist the MSS, and the difference in elevation of fluid concrete in the webs is therefore kept controlled during filling. Form designers typically specify the maximum difference in elevation of fluid concrete between the webs.

When the geometry tolerances are not stringent or strictly enforced and the webs and slabs are thin, a characteristic concrete pressure may be selected for the design of local members with a small loaded area such as form panels, rib frames and bottom crossbeams. The characteristic pressure rarely exceeds the nominal pressure by more than 15%. Characteristic pressure is also used for the slab areas that collect concrete during pumping. Self-compacting concrete typically does not require any load increase due to its fluidity.

The nominal density of concrete and the load distribution per deck geometry are typically used for the design of the carrying structure of an MSS and for camber analysis. Form stiffness is rarely considered in the camber analysis, although it may make a significant contribution and may lead to excess of camber and an overestimated structure–MSS interaction at the application of prestress. In order to include the effects of form stiffness in the casting curve, the outer form should be applied to the MSS and configured for casting prior to load testing.

Span-by-span casting requires analysis of the structure–MSS interaction during application of prestressing and release of the span, and structural checks on the deck and the MSS. As prestress is applied, the deck tends to lift up from the casting cell. As the span weight is relieved, however, the MSS tends to recover the deflection and realign upwards. As an MSS is more flexible than a PC span, the casting cell recovers only a portion of the initial deflection. After application of prestressing, therefore, the casting cell exerts a residual support action on the deck. The more flexible the MSS, the larger the deflection of the casting cell during filling, and the higher the residual support action. The residual support action can cause an excess of prestress and cracking in the young concrete if not controlled by staged tensioning of tendons.

The structure–MSS interaction is particularly complex when the first-phase U-span is prestressed prior to casting the top slab. The deflections of the casting cell are often surveyed in the first spans of the bridge to fine-tune the first-phase post-tensioning. The structure–MSS interaction is assessed at the SLS in the deck (cracking in the young concrete) and the MSS (overloading); ULS checks are also necessary to ensure adequate structural safety. The use of the same suite of design standards for the bridge, MSS and forming systems facilitates interaction checks; when different industry standards are used, the load and resistance factor may be calibrated to a common reliability index.

9.3.2 Load combinations

The loads applied to an MSS are normally grouped into two load conditions.

- Load condition I is the normal operational condition. It combines in the least favourable way self-weight, service load (the pressure of fluid concrete and the weight of the span or the segment before and after application of prestress), design loads on walkways and working platforms, dynamic loads applied by carried cranes, wind loads and thermal loads. Load condition I is also used to check the launch stresses.
- Load condition II is the out-of-service condition, which combines self-weight, partial design loads on walkways and working platforms, snow if more demanding, and out-of-service wind. Wind loading for normal operations is resisted in any possible work and launch configuration of the machine, while out-of-service wind is resisted in anchored configurations with a closed casting cell.

SLSs, ULSs, EELSs and FFLSs are all assigned the same importance. The SLSs are related to the loss of functionality of the machine and to events such as misalignment of the casting cell, excessive form deflections, and rotations of launch bogies and hydraulic hinges. The ULSs are related to critical conditions of stability and strength, such as stability of equilibrium (overturning, uncontrolled sliding), out-of-plane buckling, local buckling of compression members, rupture of connections, and yielding of structural members. Some standards consider FFLSs as additional types of ULSs. EELSs are rarely distinguished from ULSs in the design of bridge construction equipment.

For the SLS checks, CNR 10027 (CNR, 1985b) specifies a load factor $\gamma = 1.0$ for self-weight and primary varying load (typically concrete weight and pressure) and a load factor $\gamma = 0.8$ for the accompanying independent varying loads. When the effects of self-weight are opposite in sign to the effects of the primary varying load, a lower-bound value is used for self-weight. The remaining independent varying loads are considered only if their effects amplify the effects of the primary varying load.

For the ULS checks, CNR 10027 (CNR, 1985b) specifies $\gamma = 1.38$ for self-weight and primary varying load, and $\gamma = 0.97$ for the accompanying independent varying loads. Other industry recommendations specify $\gamma = 1.35$ for self-weight, $\gamma = 1.5$ for the primary varying load, and lower factors for the accompanying varying loads. Where the effects of self-weight are opposite in sign to the effects of the primary varying load, a load factor $\gamma = 1.0$ is applied to self-weight. Upper- and lower-bound load factors are applied to overturning and stabilising couples to assess stability of equilibrium.

SLS and ULS checks are also performed for the launch operations. These checks are the same as those for the heavy lifters, and are discussed in Section 9.4.3. Wind loading during launching is applied to the open casting cell: full wind pressure is applied to the windward sectors of the casting cell, and wind suction is applied to the leeward sectors. Vertical wind loading is also considered in the analysis.

9.4. Design loads of heavy lifters

The load combinations used for the design of heavy lifters are more complex than those used for machines for in-place casting due to the dynamic nature of loading. The following sections mainly refer to FEM 1.001 (FEM, 1987) because of the historical role of this design standard for cranes and heavy lifters. Provisions harmonised with the suite of Eurocodes may be found in more recent design standards for cranes, BS EN 13001-1: 2004 (BSI, 2004a) and BS EN 13001-2 (BSI, 2004b), and for crane runways, BS EN 1991-3: 2006 (BSI, 2006).

Heavy lifters are subjected to inertial forces due to the application or removal of the load, positive or negative accelerations during vertical and horizontal movements of the load, load oscillations, collision with buffers, collision of lifting attachments with obstacles, and other such causes. The dynamic load components induced by vibrations due to inertial and damping forces are accounted for by applying a dynamic allowance φ to the static value of the force or load in movement. Setting F as the characteristic static value of the force or load, the characteristic value of the dynamic action F_φ is expressed as

$$F_\varphi = \varphi F$$

BS EN 1991-3: 2006 (BSI, 2006) identifies accidental actions and related dynamic factors on crane-carrying structures; as this standard does not deal with the hoists themselves, classification is more approximate than the FEM 1.001 (FEM, 1987) approach. The accidental actions are classified as follows.

1 Vibration of the crane structure due to lifting the hoist load. The dynamic factor φ_1 is applied to the self-weight of the crane.
2 Dynamic effects of transferring the hoist load to the crane. The dynamic factor φ_2 is applied to the hoist load.
3 Dynamic effects of sudden release of the hoist load. This load condition occurs in the case of accidental release of a precast beam or segment. The dynamic factor φ_3 is also applied to the hoist load.
4 Dynamic effects due to carried cranes travelling on runways. The dynamic factor φ_4 is applied to the self-weight of the carried crane and the hoist load. When the specified tolerances for runway tracks are respected, it may be taken as $\varphi_4 = 1$.
5 Dynamic effects caused by drive forces. The dynamic factor φ_5 is applied to the drive forces.
6 Dynamic effects of a test load moved by the drive systems of the crane in the way the crane is used. The dynamic factor φ_6 is applied to the test load.
7 Dynamic elastic effects of impacts on buffers. The dynamic factor φ_7 is applied to the buffer loads.

Winch trolleys that are required to operate together are treated as a single crane action. The simultaneous presence of different crane load components is taken into account by considering groups of loads as defining one characteristic crane action for the combination with non-crane loads. The load groups for the characteristic crane actions are defined in Table 2.2 of BS EN 1991-3: 2006 (BSI, 2006).

9.4.1 Classification

Heavy lifters are grouped into classes in relation to the tasks they will perform during their service life. The class of a lifter is determined based on the number of load cycles and the loading level. The number of load cycles is the expected number of operations that the machine will perform during its service life. A load cycle starts when the load is lifted and ends when after load placement, the machine is ready to lift a new load.

The FEM 1.001 (FEM, 1987) classification is based on eight classes, A_1 to A_8, and governs the load multiplier to be used for the design of the structural components of the machine. The load multiplier varies from $\eta_S = 1.00$ for A_1 class lifters to $\eta_S = 1.20$ for A_8 class lifters, and represents the probability of exceeding the calculated stresses due to approximations in the analysis and unforeseen events. The class of the lifter is determined on the basis of 10 classes of utilisation, U_0 to U_9, and four classes of load spectrum, Q_1 to Q_4. The number of load cycles is less than 16 000 for the U_0 classification and progressively increases to more than 4 million for the U_9 classification. The heavy lifters used in bridge construction are typically in the U_0 class.

The loading level is defined by a spectrum that relates the entity of loading with the number of load cycles at that load level. Let P_i be a generic load level, n_i be the number of times the load P_i is lifted during the service life of the lifter, n_{tot} be the total number of load cycles, and P_D be the design load of the lifter. FEM 1.001 (FEM, 1987) defines the spectral factor with the following expression:

$$K_P = \sum_i \frac{n_i}{n_{tot}} \left(\frac{P_i}{P_D} \right)^3$$

Most load cycles of the lifters for bridge construction take place at or near the load capacity, and the spectral factor is therefore in the range $0.50 < K_P < 1.00$ that characterises the most demanding Q_4 classification. As a result of this classification, U_0 and Q_4, bridge erection machines are typically designed as A_2 class lifters, and the load multiplier for the design of the structural components is $\eta_S = 1.02$.

A similar classification applies for the design of the mechanical components, which are classified into eight groups, M_1 to M_8, on the basis of 10 classes of utilisation, T_0 to T_9, and four classes of load spectrum, L_1 to L_4. The classification governs the load multiplier to be used for the design of the mechanical components, which varies from $\eta_M = 1.00$ for an M_1 class unit to $\eta_M = 1.30$ for an M_8 class unit. Bridge erection equipment is usually designed as an M_2 or M_3 class lifter (although the load level of the individual components may vary significantly), and the load multipliers are therefore $\eta_M = 1.04$ and $\eta_M = 1.08$, respectively.

9.4.2   Design loads

The main difference between a heavy lifter and an MSS is that, during normal operations, the lifter receives most of the design load at once, in contrast with the progressive loading of in-place casting. The dynamic effects are therefore more pronounced in heavy lifters.

According to FEM 1.001 (FEM, 1987), the structural components of a lifter are designed for static loads distributed in the least favourable way, inertial effects due to vertical and horizontal movements of the load or the machine, and meteorological loads. The design loads are divided into four groups: forces that act regularly during normal operations, forces that arise occasionally in the machine in service, exceptional forces in service and out-of-service conditions, and special load cases during assembly and decommissioning.

The regular forces include self-weight, hoist load and the inertial effects generated by movements or oscillations of the load. The hoist load includes the payload and lifted accessories such as hoisting rope, sheave block, hook, lifting beam, spreader beam and hangers. Heavy loads, such as macro-segments and precast spans, are often restrained with auxiliary diagonal ropes to avoid oscillations.

When the load is lifted at four points, it is possible that only two lifting points on a diagonal resist the entire load. Overloading of the lifting systems may be avoided in different ways.

■ For light loads of small dimensions handled with one central hoist, a pull cylinder may be applied to one of the lifting slings. The cylinder is controlled by a hydraulic system that balances the load on the slings.
■ For heavy loads of medium dimensions handled with one hoist, long-stroke double-acting cylinders with parachute valves may be applied to the four slings. Computerised hydraulic systems facilitate control of the load position, ensure high-precision load manoeuvring and real-time pull monitoring for each sling, and reduce load oscillations.
■ For heavy and long loads handled with two or four hoists, such as macro-segments and precast spans, a torsional hinge is used at one end of the load and two independent lifting points are used at the other end.

The weight of all types of precast elements depends on the geometry tolerances of the forming systems. Newly cast segments can also be 3–5% heavier than when the concrete is fully dry.

9.4.2.1 Vertical inertial forces

Vertical inertial forces derive from the sudden application or removal of the load and from the positive or negative accelerations during the vertical movement of the load. The dynamic load components induced by vibration due to inertial and damping forces are accounted for by applying a vertical dynamic allowance φ_V to the static value of the load in movement.

Impacts related to the movement of cranes along the runways are another cause of vibrations. The dynamic load components due to irregular field splices in the rails may be demanding for the structural and mechanical elements of a gantry. The field splices in the rails are therefore welded at the end of site assembly, and the welds are ground away at decommissioning of the machine.

CNR 10021 (CNR, 1985a) specifies the vertical dynamic allowance φ_V to be applied to the static hoist load in relation to the lifting speed of the load:

$$\varphi_V = 1 + 0.6 v_V$$

The steady lifting speed v_V is expressed in meters per second. The vertical dynamic allowance is subjected to the following limitations:

$$1.15 \leq \varphi_V \leq 1.60$$

Bridge construction equipment is typically designed with $\varphi_V = 1.15$ because of the slow lifting speed. BS EN 1991-3: 2006 (BSI, 2006) specifies the vertical dynamic allowance for crane-carrying structures as a combination of two factors. A first factor, $\varphi_1 = 0.9$ or $\varphi_1 = 1.1$, is related to the vibration of the gantry due to lifting the hoist load, and is applied to the self-weight of the gantry. The two values reflect the lower and upper values of the vibrational pulses. A second factor

$$\varphi_2 = \varphi_{2.min} + \beta_2 v_V$$

is related to the dynamic effects of transferring the hoist load to the crane, and is applied to the hoist load. The factor $\varphi_{2.min}$ is 1.05 for HC1 hoists, 1.10 for HC2 hoists, 1.15 for HC3 hoists and 1.20 for HC4 hoists. The factor β_2 is 0.17 for HC1 hoists, 0.34 for HC2 hoists, 0.51 for HC3 hoists and 0.68 for HC4 hoists. Cranes are assigned to hoisting classes HC1 to HC4 according to Annex B of BS EN 1991-3: 2006 (BSI, 2006).

A different dynamic allowance is used when the load is released suddenly for accidental reasons. Let P be the hoist load and ΔP be the load portion that is released. CNR 10021 (CNR, 1985a) specifies the vertical dynamic allowance to be applied to P as

$$\varphi_V = 1 - 2\frac{\Delta P}{P}$$

If the entire load is lost suddenly (e.g. breakage of the hook or the lifting beam) the dynamic effects are analysed by applying the hoist load as directed upward. If the crane is light, sliding clamps or counter-rolls may be necessary to prevent uplift of the crane or its components. BS EN 1991-3: 2006 (BSI, 2006) specifies the same approach to determine the dynamic factor φ_3 to be applied to the hoist load in case of accidental release.

The vertical dynamic allowance for the design of cable cranes is determined using large-displacement dynamic analysis because of the influence of the sag in the catenaries of the carrying and hoisting ropes and the longitudinal distribution of suspenders on the dynamic response of the system. Load testing of cable cranes often leads to unexpected results because of the marked non-linearity of the response of such flexible systems. The lifting platforms of suspension bridges are often designed as conventional cranes because of the lower flexibility of the suspension cables.

9.4.2.2 Horizontal inertial forces

Longitudinal inertial forces derive from accelerations or decelerations of carried cranes along the runways. According to FEM 1.001 (FEM, 1987), they are expressed as a longitudinal dynamic allowance φ_L to be applied to the load in movement, which includes the hoist load and the self-weight of the crane.

The heavy lifters are physical systems that comprise suspended masses (hoist load), driven masses (trolleys), auxiliary masses (PPU, hydraulic systems, winches, bogies), flexible support systems (trusses) and flexible connections between some of these systems (capstans, hauling ropes, hoisting ropes). When such a system is subjected to a new load condition, its response depends on the characteristics of the new load condition and on the previous equilibrium state. If the new load condition is steady, the system tends progressively toward a new equilibrium state. In most practical cases, the system responds with oscillations about the new equilibrium state. Because of these oscillations, the stresses may amply exceed the static stresses generated by the new load condition.

Such situations also arise during the acceleration or deceleration (braking) of the longitudinal movement of carried cranes. Inertial forces arise in the elements subjected to acceleration, and, when the full translation velocity is reached, acceleration becomes null and the inertial forces disappear.

Let v_L be the steady longitudinal velocity of the crane at the end of the acceleration period or before braking, M_D be the mass of the hoist load, and F_L be a longitudinal force applied to the crane that generates the same effects on the movement as the accelerating or decelerating couple applied by motors or brakes. Accelerating and decelerating couples resist the translational inertial forces of the crane and hoist load and the rotational inertial forces of the motors, wheels and intermediate mechanisms. The total inertia of the elements in movement (hoist load excluded) can be represented as an equivalent mass M_E applied to the crane, expressed as

$$M_E = M_L + \Sigma_i \frac{I_i \omega_i^2}{v_L^2}$$

where M_L is the mass of the elements that are subjected to the same longitudinal movement as the crane, I_i is the polar moment of inertia of the rotating parts about the axis of rotation, and ω_i is the angular velocity of the rotating parts corresponding to the velocity v_L of the crane. In a system comprising masses rigidly connected to each other, the mean acceleration or deceleration is

$$a_{mean} = \frac{F_L}{M_E + M_D}$$

If the acceleration is constant, the mean duration of the accelerated motion is

$$T_{mean} = \frac{v_L}{a_{mean}}$$

Under the influence of the longitudinal acceleration, the hoisting rope loses the vertical position and reaches a maximum inclination from the vertical equal to

$$\alpha_{max} = \text{arctg} \frac{a_{mean}}{g}$$

As a result, the rope applies a longitudinal inertial force to the crane equal to

$$F_L = M_D a_{mean}$$

In reality the system is not rigid, the acceleration is not constant, the load follows an oscillatory motion, and the peak longitudinal force applied to the crane is higher than the mean value. The dynamic amplification due to the elasticity of the system is represented by the longitudinal dynamic allowance φ_L, and the design inertial force is

$$F_D = \varphi_L F_L$$

FEM 1.001 (FEM, 1987) provides graphs for φ_L based on the relative mass

$$\mu = \frac{M_D}{M_E}$$

and the relative period

$$\beta = \frac{T_{mean}}{T_L}$$

Setting L as the load suspension length in the translation configuration, the longitudinal load oscillation period with a fixed crane is

$$T_L = 2\pi \sqrt{\frac{L}{g}}$$

The graphs for φ_L are based on the following expression for the longitudinal dynamic allowance:

$$\varphi_L = \sqrt{\left(1 - \cos \phi\right)^2 + \left(1 + \mu\right) \sin \phi^2}$$

The variable ϕ is obtained by trial and error from the following equation:

$$\frac{\phi + \mu \sin \phi}{2\mu\beta\sqrt{1 + \mu}} = 1$$

If the mass M_D of the hoist load is smaller than the equivalent mass M_E of the crane, $\mu < 1$ and $\varphi_L \leq 2$. The value $\varphi_L = 2$ is reached beyond a critical value for the relative period that is a function of μ. For $\mu > 1$, beyond another critical value for the relative period $\varphi_L > 2$, and the maximum value for the dynamic allowance

$$\varphi_{L.max} = \sqrt{2 + \mu + \frac{1}{\mu}}$$

is reached during the load oscillations that follow the immobilisation of the crane at the end of braking. The critical value for the relative period is reached more rapidly for short suspension lengths, and the highest values for φ_L are therefore reached when the load is transferred close to the crane.

Most winch trolleys have longitudinal speed $v_L \leq 1.0$ m/s and mean acceleration in the range $a_{mean} = 0.1$ to 0.2 m/s^2, and the dynamic allowance during acceleration is therefore low. The portal cranes are faster and can generate larger inertial effects. Much higher accelerations are reached in the cable cranes as the haulage motors are on the ground and can be heavier and more powerful. In all cases, however, the longitudinal dynamic loads are typically governed by braking. Multiple disk brakes can generate very demanding decelerations, especially during the emergency response to power blackouts.

According to BS EN 1991-3: 2006 (BSI, 2006), the longitudinal dynamic allowance is expressed by the dynamic factor φ_5. Unless the dynamic factor is defined by the crane supplier, $1.0 \leq \varphi_5 \leq 1.5$ is used for systems where forces change smoothly, $1.5 \leq \varphi_5 \leq 2.0$ is used for cases where sudden changes can occur, and $\varphi_5 = 3.0$ is used for drives with considerable backlash. The dynamic factor is applied to the drive force of the crane.

Transverse inertial forces should also be considered in relation to lateral movements of the hoist crab along the crane bridge and of the gantry along the crossbeams for traversing. These movements are typically so slow that vibrations are minimal, but sudden movements may result from imbrication of capstans. Longitudinal and transverse inertial forces do not include the effects of oblique hoisting due to misalignment of load and hoist, because oblique hoisting is generally forbidden.

9.4.3 Load combinations

FEM 1.001 (FEM, 1987) combines the loads applied to a heavy lifter into four load conditions.

- Load condition I is the normal operational condition with regular forces. It combines in the least favourable way self-weight, design loads on walkways and working platforms, hoist load (treated as a live load) multiplied by the vertical dynamic allowance φ_V, and longitudinal inertial effects on hoist load and crane weight (both treated as live loads) multiplied by the dynamic allowance φ_L. When the gantry carries two cranes that operate independently, they are treated as random live loads. The effects of these loads are multiplied by the structural load multiplier η_S of the machine and the load factors and the additional load multipliers prescribed by the structural design standard for the limit state under consideration.
- Load condition II is the operational condition in the presence of occasional actions. It combines in the least favourable way the actions of load condition I with wind in service,

snow and ice, and thermal differences. The load multiplier η_S is not applied to the effects of occasional actions. In the presence of strong wind the longitudinal dynamic allowance φ_L may be different from the value used for load condition I, because the action of wind can affect the starting and braking times of carried cranes. Load condition II is also used to check the launch stresses; launching is typically avoided in the presence of strong wind.

■ Load condition III represents the action of exceptional loads. It considers the least favourable of the following combinations: (1) self-weight and out-of-service wind; (2) self-weight, hoist load and collision with buffers; (3) self-weight, hoist load and design-level earthquake; and (4) self-weight, wind in service, and operations of site assembly or decommissioning. Although conceptually similar to site assembly and decommissioning, self launching is assessed using load condition II, as these operations are performed more frequently.

■ Load testing requires specific checks only when the static test load is greater than $1.4P_D$, the dynamic test load is greater than $1.2P_D$, or the static test load is in the range $1.25P_D$ to $1.4P_D$ or the dynamic test load is in the range $1.1P_D$ to $1.2P_D$, and the load test stresses are opposite in sign to those from self-weight. The load combination for load testing includes self-weight, test load and dynamic allowance in the case of dynamic tests.

BS EN 1991-3: 2006 (BSI, 2006) requires considering the simultaneous action of accidental load components by means of groups of loads. Each group of loads defines one crane action for the combination with non-crane loads. Crane- and non-crane loads are then combined using standard rules for steel structures. The following horizontal loads from travelling cranes are taken into account: (1) longitudinal forces caused by acceleration or deceleration of the crane along the runway; (2) transverse forces caused by acceleration or deceleration of the hoist crab along the crane bridge; (3) transverse forces caused by skewing of the crane along the runway; (4) longitudinal collision with buffers related to crane movements; and (5) transverse collision with buffers related to crab movements. Only one of these horizontal forces is included in the same group of simultaneous crane load components.

SLSs, ULSs, EELSs and FFLSs are all assigned the same importance. The SLSs are related to the loss of functionality of the lifter and to events such as rotation of hydraulic hinges, displacement of components or excessive deflections or vibrations. The ULS are related to critical conditions of stability and strength such as stability of equilibrium, out-of-plane buckling, local buckling of compression members, rupture of connections and yielding of structural members. Some standards consider FFLSs and EELSs are rarely distinguished from ULSs in the design of heavy lifters.

The design standards for heavy lifters specify load factors for the different limit states; for ULS checks, for example, CNR 10021 (CNR, 1985a) specifies $\gamma_{LC1} = 1.50$ for load condition I, $\gamma_{LC2} = 1.34$ for load condition II and $\gamma_{LC3} = 1.20$ for load condition III. The design standards for steel structures specify capacity reduction factors for the different types of stress and limit state; AASHTO (2012a), for example, specifies for ULS checks $\phi = 1.00$ for flexure, shear and bearing on pins and milled surfaces, $\phi = 0.95$ for tension when yielding occurs in the gross section, $\phi = 0.90$ for axial compression, $\phi = 0.85$ for complete penetration welds, and $\phi = 0.80$ for tension when fracture occurs at the net section and for bolts in tension or bearing on material, block shear, web crippling, partial penetration welds and fillet welds. Combining resistance factors from design standards for general steel structures with load factors from industry standards for cranes and heavy lifters may require calibration to the stipulated reliability index.

CNR 10021 (CNR, 1985a) and FEM 1.001 (FEM, 1987) do not distinguish between local and global (out-of-plane) buckling, and both conditions are checked like any other ULS condition. Out-of-plane buckling, however, is a more risky event than local buckling of a web panel or a secondary compression member. No post-critical domain exists in most cases, and the critical load is influenced by geometry imperfections that are difficult to detect. It is therefore recommended to check out-of-plane buckling using a much higher load factor, of the order of $\gamma_{LC1,2,3} = 2.2$ to 2.5.

CNR 10011 (CNR, 1988) specifies several load combinations for checking stability of equilibrium. The overturning moment is calculated by enveloping the effects of five load combinations.

1   Static combination: self-weight and 1.6 times the design hoist load in the most demanding configuration.
2   Dynamic combination with loaded machine: self-weight, 1.35 times the design hoist load and 1.10 times the weight of the winch trolley inclusive of longitudinal dynamic allowance.
3   Dynamic combination with unloaded machine: self-weight, -0.10 times the design hoist load, 1.10 times the weight of the winch trolley inclusive of longitudinal dynamic allowance, and 1.10 times wind in service.
4   Out-of-service wind: self-weight and 1.20 times the out-of-service wind.
5   Rupture of the hoisting rope: self-weight, -0.30 times the design hoist load, 1.10 times the weight of the winch trolley inclusive of longitudinal dynamic allowance, and 1.10 times the wind in service.

Stability of equilibrium in out-of-service wind conditions often requires anchoring the winch trolleys at precise locations. The trolleys are typically moved over the pier crossbeams and anchored to the runways with mechanical clamps or to the crossbeams with diagonal crossed ropes.

9.5.   Launch and restraint systems

In lightweight machines such as form travellers and forming carriages, the launch stresses are low and the operations are simple and intuitive. In heavier machines such as self-launching gantries and MSSs, the launch stresses may be so demanding as to govern the design of primary components of the machine. The launch stresses and the loads applied to the bridge depend on the launch procedure, and repositioning heavyweight machines typically involves complex sequences of operations.

Most gantries and MSSs for span-by-span construction are supported on hydraulic cylinders during span erection. The cylinders lift the main girders to the elevation for span erection and lower them back on the launch bogies for span release, and the launch saddles are designed for the launch loads only. In other cases the launch saddles support the main girders also during span erection: the support cylinders are inserted between crossbeams and pier towers or between W-frames and through girders, and the saddles are designed for the entire load of span and erection equipment.

The main girders are flexible, and large flexural rotations at the supports dictate the use of rocking saddles. Additional demand for fitting of supports results from a varying deck gradient along the bridge. Dispersal of the support reactions within non-stiffened web panels of I- and box girders requires long patch loads devoid of stress concentration. Trusses also require long patch loads to control local bending and shear in the bottom chords. The geometry of the launch saddles is therefore very complex and, as the saddles are designed to also shift laterally and to rotate in the horizontal plane, the height of the assembly is often significant.

Most launch saddles for trusses and plate girders include longitudinal and transverse articulated bogies. The bogies are assemblies of cast-iron rolls on equalising beams. The number of rolls depends on the load, and the diameter and is usually 2, 4 or 8, with progressively higher costs. According to CNR 10011 (CNR, 1988) the linear contact pressure transferred by a roll with radius r_R and transverse contact breadth b_R under a vertical load P is

$$\sigma_{LC} = \sqrt{\frac{0.18EP}{r_R b_R}}$$

CNR 10011 (CNR, 1988) limits the linear contact pressure to $\sigma_{LC} \leq 4.0\sigma_{all}$ for SLS and $\eta_i \gamma_i \sigma_{LC} \leq 4.0f_{td}$ for ULS. The mean linear contact pressure for a pair of rolls is slightly higher to allow for imperfect load distribution:

$$\sigma_{LC} = \sqrt{\frac{0.20EP}{2r_R b_R}}$$

In the case or four or more rolls it is

$$\sigma_{LC} = \sqrt{\frac{0.24EP}{nr_R b_R}}$$

where n is the number of rolls. The rolls are mounted on the equalising beam with bearing brasses or roller bearings for lower friction. The equivalent coefficient of friction with roller bearings is in the range 1–3% in ideal conditions, but 5% is typically used for design purposes.

Rectangular rails welded to the flanges of trusses and crossbeams to engage the flanges of the rolls are used in most cases to provide guidance and transfer of lateral loads. Less expensive PTFE skids often replace the bottom bogies in new-generation machines, especially when the gantry is traversed only for launching. In this case, the lateral guides act against the edges of the flange. Integrated launch saddles include support cylinders and paired long-stroke double-acting launch cylinders (see Figure 3.19).

Cable bearings are rarely used in bridge construction equipment. In these devices a tensioned ring cable supports the rolls and distributes the load. The bearing is thinner than an articulated bogie, and load distribution is more uniform. The cable bearings are placed onto orientation ball-plate bearings and shifting bearings that, respectively, rotate the bogie in the horizontal plane and shift it laterally for traversing. The assembly may be placed onto interconnected hydraulic cylinders to balance the support reactions with a torsional hinge.

Sliding saddles are based on PTFE skids (Figure 9.1). The truss slides on a lubricated PTFE surface without interposition of polished stainless-steel sheets, friction can exceed 10%, and the poor torsional constant of paired crossbeams provides rotational flexibility to the support. Load distribution may be enhanced by means of multi-layered elastomeric blocks supporting a dimpled PTFE plate. The blocks are housed within an articulated rocking arm; in the most

Figure 9.1. Sliding saddles with PTFE skids

delicate cases, the elastomeric blocks may be supported on spherical springs for enhanced load distribution.

At the current state of practice, launch bogies and PTFE skids are complementary. Bogies are suitable for medium loads and fast launching, and sliding saddles are suitable for slow launching of heavy loads when higher friction is tolerable. Transverse PTFE skids provide an excellent and relatively inexpensive support to longitudinal launch bogies for traversing and simultaneous erection of adjacent decks, and the high frictional loads are balanced within the crossbeam.

The launch and restraint systems have a substantial influence on the stability and safety of bridge construction equipment (Rosignoli, 2007). Several methods are available to drive launching. In the simplest machines, a winch applied to the rear end of the truss pulls a haulage rope anchored to the front end of the deck, and a chain system prevents backward rolling of the truss in case of failure of the drive winch. This solution can be used only for uphill launching.

Restraint redundancy during launching and operations has pros and cons. Safety is a major advantage, as double restraint diminishes the risks associated with incorrect operations and mechanical failure. Locked-in forces are a possible disadvantage, but they are considered anyway due to friction at the launch supports. Double restraint is also more labour intensive.

When only one longitudinal restraint is used, great care must be exerted to ensure that the machine is always properly restrained.

In many beam launchers and light gantries, a haulage winch applied to a winch trolley pushes the truss forward by acting on a capstan anchored to the ends of the truss. The winch trolley is brought over the leading pier and anchored to the crossbeam with crossed ropes that provide the longitudinal restraint during launching. Lever counterweights at the ends of the truss keep the capstan in tension to avoid slippage of the hauling rope at the drive winch. At the end of launching, bidirectional automatic clamps restrain the trusses to the crossbeams, the crossed ropes are removed, and the capstan is used to move the winch trolley along the gantry for normal operations.

The main advantage of the capstans over the tow winches is the possibility of moving the trusses in both directions, which facilitates control of downhill launching. Downhill launching is often unavoidable with the beam launchers because, even if the beams are erected uphill, the unloaded launcher must be moved downhill to pick up a new beam. The capstans provide no redundancy and breakage of any component of the drive/brake system would leave the gantry unrestrained on a low-friction inclined plane. Automatic brakes may be incorporated in the bogies of the haulage winch trolley for braking of the truss in case of failure of the capstan.

In heavier machines, long-stroke double-acting cylinders housed within the launch saddles push or pull the truss forward by acting within perforated rails fixed to the truss. These launch systems are very efficient and safe, but are also slower and more expensive than capstans. Paired launch cylinders are used for redundancy, and to be able to lock the truss with one cylinder during repositioning of the other. The cylinders may be parallel or may work on the opposite sides of the launch saddle, so that the rear cylinder pulls the truss forward and the front cylinder pushes it. Hydraulic extractors can be used to insert and remove the two pins from the perforated rail for labourless operations; end-of-stroke switches applied to the extractors confirm pin insertion to the PLC for consent to removal of the other pin. The design stroke of the launch cylinders is oversized for the cylinders to provide the longitudinal restraint also after lifting the trusses from the launch bogies for span erection.

In other machines, one or two hydraulic launchers apply the thrust force by friction. In order for the girder to move forward, the ratio of the thrust force to the vertical load must be smaller than the friction coefficient of the contact surface between the launcher and the girder flange (Rosignoli, 2000). The vertical load is the support reaction of the girder at the launcher, which varies during launching (Rosignoli, 2002). Two synchronised launchers are often necessary to overcome the reduction in the rear support reaction with the progress of launching. Vibrations may further diminish the net support reaction, and the launchers may be equipped with hydraulic clamps that increase friction capacity.

The friction launchers offer high intrinsic safety (the worst consequence of hydraulic failure is launch stoppage) but are more expensive than paired launch cylinders. The launch systems can be designed for the launch loads and can be overloaded without excessive worries in case of need. Computerised controls permit synchronisation and labourless operations, and increase the launch speed.

The tip of the cantilever truss is deflected downward on landing at the new pier. In many beam launchers and light gantries the deflection is recovered by inclining the bottom chords of the front panel of the truss to force progressive realignment with the progress of launching. The alignment force is small initially and the peak negative moment in the truss increases further after landing. The launch saddles are anchored to the landing crossbeam to control over-turning, and their articulations accommodate large rotations to cope with the alignment wedge. The same solution is used at the rear end of the gantry to progressively release the support reaction during forward launching and to recover the downward deflection during backward launching.

In heavier machines, realignment is achieved with long-stroke double-acting cylinders that rotate lifting arms pinned to the nose tip. When the lifting arm reaches the landing pier, it is lowered down and forced until recovery of deflection. The launch saddles at the landing crossbeam are subjected to minor rotations, mechanical articulations may sometimes be avoided, and the peak negative moment in the main girder is better controlled. Hybrid solutions are also possible where the bottom chords are rounded and the alignment cylinders recover only a portion of the deflection.

Many overhead gantries and MSSs are designed to reposition the support systems without ground cranes. After landing at the new pier, an auxiliary leg at the front end of the gantry supports the front cantilever during placement of the hammerhead segment, and the winch trolley then repositions the front crossbeam onto the segment. A similar solution is used to position the front friction launcher.

Some underslung machines are designed to reposition the pier brackets. Hydraulic motors drive the brackets along the trusses and the geometry adjustment systems of the brackets are used to insert the shear-keys into recesses in the pier wall. One bracket carries the clamping bars for immediate coupling after application to the pier.

9.6.   Winch trolleys

Winch trolleys are travelling cranes devoid of counterweights. The crane bridge spans between the top flanges of the twin-girder overhead machines and between the bottom flanges of the single-girder machines. The winch trolleys of single-girder beam launchers and portal carriers sit on top of the main girder and lift the load with side hoists on either side.

The main mechanical components of a winch trolley are the hoist systems and the translation systems. The hoist systems include strand jacks, winches, pull cylinders and hoist bars with through pins. The translation systems include articulated bogies driven by braked motorised wheels or capstans.

The winch trolleys of most gantries carry a PPU that supplies the hoist and translation systems, the tilt and rotation systems of the lifting beam, and the electric control systems. Light winch trolleys of beam launchers are often supplied by a PPU carried by a third trolley to diminish the travelling load on the trusses. The winch trolleys of MSSs and lifting frames are supplied by the main PPU of the machine by way of bus-bars or retractable cables. The hoist winches of portal carriers and span launchers are mounted on the rear platform of the machine and are fed directly by the PPU, and the hoist trolleys are devoid of winches.

The winch trolleys are among the most expensive components of a self-launching gantry (Hewson, 2003). Most components and control systems are manufactured by third parties, and proper selection is necessary to optimise weight, cost and cycle time of the assembly. The hoist is often the most expensive part of the assembly, and the main parameters that drive its choice are load capacity and lifting speed.

Early winch trolleys used strand jacks to lift the load. Strand jacks are relatively inexpensive, although the strand must be replaced frequently due to the damage inflicted by the anchor wedges. Lowering the load is less safe than lifting it, as the bottom safety grips have to be held open. The jack is gimbal-mounted on the hoist crab for uniform load distribution between the strands and to avoid bending should the load sway during lifting. The gripping mechanisms are accessible at any time.

The hoist capacity ranges from 15 to 750 ton based upon lift cables of 1–50 strands 18 mm diameter, seven-wire prestressing strand of guaranteed breaking load of 38 ton is used for lifting. The stroke of the jacks varies between 250 and 500 mm. The load is held mechanically when movement is stopped at any point of the stroke.

The cycle time of a strand jack is easy to measure; hoisting, however, is often slower due to the initial setup of the manoeuvres and the need for maintenance, re-lubrication of gripping mechanisms, and coordination of personnel. Strand jacking is slow compared with other hoisting methods, but the lifting height is practically unlimited if strand reeling gear is provided on the winch trolley. The lifting speed depends on the power of the PPU and is in the range 2–10 m/h, with a typical estimate of 5 m/h. The lifting speed of paired jacks can be synchronised irrespective of load. Nowadays strand jacking is used only for long lifts of heavy segments or entire spans.

When hoisting of segments is on the critical path, no lifting speed is fast enough, and investing in PPUs and high-quality mechanical components is often the best choice. This is particularly true when launching gantries are used on tall piers in large marine operations with segment delivery on barges. When the segments are handled using lifting frames, the repositioning speed of the frames from hammer to hammer is often more important than the cycle time of segment handling. Self-contained lifting frames that can be moved to the next hammer in a single lift are particularly cost-effective. If precast segments or beams are lifted from the sea or ground level, the most common hoisting method uses sheave blocks to reeve rope that lifts a large load with a small line pull. An even number of rope lines is typically used to anchor the dead end of the hoisting rope to the winch trolley for easier inspection.

A hydraulic winch supplied by a powerful PPU can lift a typical 40 ton segment for dual-track LRT bridges with four lines of rope at 300 m/h, which is 60 times faster than strand jacking. Lifting speeds of 3 m/min with heavy segments and 6 m/min with light segments are easily achieved, with the transition between light and heavy segments being at lifting-power-to-load ratios of about 5. If one assumes a 20 m average lift, lifting a heavy segment may take 7 minutes and a cycle time of 30 minutes is easily attained with gantries for span-by-span erection, which typically avoids the crane use becoming the bottleneck.

Heavy loads require many sheaves and falls of rope and a large winch drum to store the rope. The hook block needs significant weight to overcome friction when the hoisting rope passes through the sheaves when lowering with no load. This increases the design hoist load unless the designer can be sure that the lifting beam will never be disconnected from the hook block.

According to FEM 1.001 (FEM, 1987), the diameter of hoisting rope, winch drum and sheaves is chosen based on the mechanical classification of the lifter. The design line pull in the rope is calculated taking into account several aspects.

■ The net lifting capacity of the crane is higher than the design payload to account for geometry tolerances and the heavier density of young concrete.
■ The weight of hoisting rope, lower sheave block, hook block, lifting beam, spreader beam, hangers and stressing platform applied to the segment is added to determine the gross hoist load.
■ Vertical inertial forces are applied to the gross hoist load by means of the vertical dynamic allowance φ_V to determine the design hoist load.
■ The reeving ratio of the rope and the efficiency of the reeving systems also influence the design line pull.

Starting from the design line pull in the rope, the load factor applied to the rope breaking strength guaranteed by the manufacturer is $\gamma = 3.35$ for M_2 units and $\gamma = 3.55$ for M_3 units. The power of the motor is transferred to the winch drum through a torque, and the available line pull diminishes with the increasing number of rope layers on the drum.

The service life of the rope depends on the diameter of winch drum and sheaves, which governs the bending stresses in the rope, and on the radial pressure exerted by the rope. The minimum diameter $d_{D.min}$ of winch drum and sheaves measured along the rope centreline is

$$d_{D.min} = k d_{HR}$$

where d_{HR} is the notional diameter of the rope. According to FEM 1.001 (FEM, 1987), $k_{M2} = 12.5$ and $k_{M3} = 14.0$ for the drum of M_2 and M_3 lifters, and $k_{M2} = 14.0$ and $k_{M3} = 16.0$ for the sheaves. The factor k is greater for the sheaves than for the drum because during every passage the sheave folds the rope twice and the drum only once. Heavy hoist loads and the need for keeping the rope as small as possible require large reeving ratios and several sheaves. Increasing the design line pull reduces reeving and increases the lifting speed but requires a more powerful winch and a larger winch drum and sheaves. Larger sheaves increase the minimum closer distance between the upper and lower blocks of sheaves, and the geometry of the winch trolley must often be modified from a flat frame to a portal.

The fixing devices at the ends of the hoisting rope are designed for 2.5 times the design line pull of the rope. The anchorages at the drum may be designed by allowing for friction, provided that at least two complete turns of rope remain on the drum at maximum lowering of the load. The friction coefficient between the drum and the rope may be taken as $\mu = 0.1$. The dead end of the rope may be anchored to a pull cylinder for minimal speed movements of the load.

The winches are designed for compact dimensions and easy assembly on the hoist crab. The hydraulic system supports the load, and a multi-disk holding brake provides additional safety. New-generation winches use pressure-controlled, variable-flow hydraulic motors and planetary gearboxes in an oil bath. Sensors adjust oil flow to maximise winch speed depending on the load, and the rope speed may amply exceed 100 m/min. Free-fall systems with clutches and multi-disk brakes are rarely used nowadays.

An alternative way of lifting precast segments or beams over short heights is by the use of pull cylinders. This solution is inexpensive, simple and reliable, but works only when the segments or beams are delivered on the deck. The practical limit of a single-stage cylinder is 5–6 m; multi-stage cylinders may be longer, but they are more expensive. The lifting beam is gimbal-mounted to an end eye in the plunger. The cylinder is gimbal-mounted to the hoist crab to avoid bending, which could damage the plunger and the cylinder's seals.

Long rectangular bars with multiple holes may be used instead of pull cylinders. The lifting beam is gimbal-mounted to an end eye at the bottom of the bar. A gimbal-mounted saddle is applied to the hoist crab. The hoist bar passes through the saddle and is locked with a through pin. Two double-acting cylinders supported on the saddle move an upper saddle vertically along the hoist bar. The upper saddle houses another through pin. After reaching the end-of-stroke of the vertical cylinders, the lower pin is inserted into the hoist bar and the upper saddle is repositioned at a lower hole for lifting and at an upper hole for lowering. Hydraulic extractors can be used to facilitate pin handling.

All types of hoist are assembled on a crab that rolls or slides transversely along the crane bridge. Shifting the lifting point within the winch trolley is faster and simpler than traversing the gantry along the crossbeams, avoids diagonal lifting, and facilitates accurate positioning of the load. Capstans, motorised braked wheels, or long-stroke double-acting cylinders drive the hoist crab along the crane bridge, with progressively lower speeds.

When pull cylinders or hoist bars are used to handle precast segments, 90° rotation of the segment is achieved by rotating the hoist within the crab. Double-acting lateral cylinders or motors operating pinions that engage with toothed gears are the typical solutions for hoist rotation. When winch and reeved rope are used, a mechanical slewing ring is applied to the lifting beam to rotate the segment.

The winch trolley typically carries a PPU that supplies the hydraulic systems and provides electric power to auxiliary and control systems. An enclosed fan-cooled industrial generator drives double axial displacement pumps that supply open-loop high-flow-rate hydraulic systems, allowing all functions to be operated simultaneously. Automatic working pressure cut-offs are often integrated in the pumps to minimise pressure peaks. Heat exchangers remove heat from the bypass oil to provide cooler operation and stabilise viscosity. Return-line filters remove contaminants before allowing the return oil flow back into the tank. LCD displays provide self-testing, diagnostic and readout capability. Pressure transducers are preferred to analogue gauges, as they are more accurate and durable against mechanical and hydraulic shocks.

The frame of the winch trolley includes two crossbeams and two longitudinal beams over the runways; the hoist crab spans between the crossbeams, and all members typically have a box section. Three-hinge portals are heavier and more expensive than rectangular flat frames, but facilitate reuse of the crane in different applications, provide more lifting height, and permit lifting beyond the main girders of the gantry. Both types of frame may carry one or two independent hoist crabs. Assemblies of cast-iron wheels on equalising beams are used for the translation bogies to spread the load applied to the runway. The diameter of the wheels is determined based on the design load, the quality of material, the type of rail, the rotation velocity, and the class of the mechanism.

Figure 9.2. Design load conditions for the translation bogies

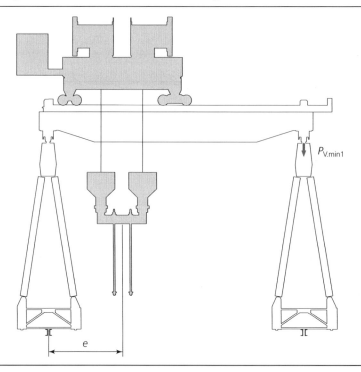

According to FEM 1.001 (FEM, 1987), the design load for the translation bogies is determined using two load conditions. The loaded crane condition is used to obtain the maximum load $P_{V.max1}$ on the runway beam, and the unloaded crane condition is used to obtain the minimum load $P_{V.min1}$ (Figure 9.2). The design load is then calculated by averaging the minimum and maximum load:

$$P_{V.mean} = \frac{P_{V.min1} + 2P_{V.max1}}{3}$$

The vertical loads for checking load conditions I and II do not include dynamic allowance. Setting n as the number of wheels in the bogie, b_R as the effective breadth of the rail (breadth of the central flat portion between rounded corners) and d_W as the diameter of the wheel (both in millimetres), FEM 1.001 requires the following check for load conditions I and II:

$$\frac{P_{V.mean}}{n} \leq k_1 k_2 b_R d_W p_{W.all}$$

The first coefficient depends on the rotation rate of the wheel and varies in the range $0.66 \leq k_1 \leq 1.17$ for speeds ranging from 200 to 5 rpm. The second coefficient depends on the class of the mechanism, and for a typical gantry (M_1 to M_3 class lifter) it is $k_2 = 1.12$. The notional contact pressure depends on the strength of the cast-iron used for the wheel, and increases from $p_{W.all} = 5.0\ \mathrm{N/mm^2}$ for $f_t > 500\ \mathrm{N/mm^2}$ to $p_{W.all} = 7.2\ \mathrm{N/mm^2}$ for $f_t > 800\ \mathrm{N/mm^2}$. Load condition III is checked using $k_1 = 1.20$ and $k_2 = 1.15$.

BS EN 1991-3: 2006 (BSI, 2006) specifies that the top web–flange weld of the crane runway beam must be checked for transverse eccentricity of the vertical wheel load equal to

$$e \geq 0.25b$$

where b is the total width of the rail, including the edge corners. The minimum vertical load on the runway is also used to determine the drive force P_D of the trolley at the contact surface between the rail and the driven wheels under the assumption that wheel spin is prevented:

$$P_D = \mu n_D \frac{P_{V.min1}}{n}$$

The friction coefficient is $\mu = 0.2$ for cast-iron wheels on a steel rail and $\mu = 0.5$ for rubber wheels, n is the number of wheels on each runway, and n_D is the number of single wheel drives on each runway. The drive force thus determined is increased using the longitudinal dynamic allowance: φ_L according to FEM 1.001 (FEM, 1987) and φ_5 according to BS EN 1991-3: 2006 (BSI, 2006).

The PPU, hydraulic systems and fluid tanks are applied to one side of the winch trolley so as not to load the crane bridge, and the hoist crab shifts laterally along the crane bridge. The centre of gravity of the winch-trolley is therefore offset from the runway midspan, and equal drive forces P_D at the runways generate a moment about the centre of gravity in the horizontal plane. Transverse forces at the wheels equilibrate this moment (Figure 9.3), and the lateral loads applied to the runways are

$$H_{T.1} = \varphi_5 \left(1 - \frac{P_{V.max1}}{P_{V.max1} + P_{V.max2}} \right) \frac{2P_D e}{a}$$

and

$$H_{T.2} = \varphi_5 \frac{P_{V.max1}}{P_{V.max1} + P_{V.max2}} \frac{2P_D e}{a}$$

BS EN 1991-3: 2006 (BSI, 2006) also specifies the transverse guide forces caused by skewing of the winch trolley. These forces are proportional to a 'non-positive factor' that depends on the skewing angle. The skewing angle is chosen from arbitrary locations of the instantaneous centre of

Figure 9.3. Lateral loads applied to the runways

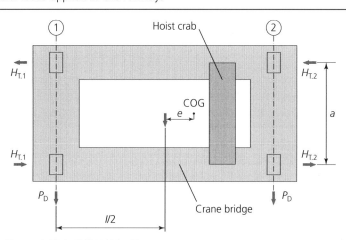

rotation, taking into account the lateral clearance between the guidance means (guide rolls or wheel flanges) and the rails, as well as wear and dimensional variation of wheels and rails. The equations disregard lateral flexibility of the runways and geometry tolerances, which may be pronounced in several types of bridge construction equipment. The approach of BS EN 1991-3: 2006 (BSI, 2006) is therefore very conservative in practical cases.

Capstans and motorised wheels are designed for rolling friction, the maximum gradient of the runway, and the total vertical load (winch trolley and design hoist load). The parking brakes should be redundant and oversized. The emergency brakes are typically based on springs loaded by the hydraulic pressure, and the springs are automatically released in case of pressure collapse. The elastic constant of the springs should be calibrated to the levels of deceleration incorporated in the longitudinal dynamic allowance used in the design.

9.7.   Service life and reconditioning

The service life of well-maintained bridge construction equipment may be 20–40 years. Erecting a precast segmental bridge may take less than one year, and in-place casting with MSSs and form travellers may take a couple of years. If a machine is designed with modularity and handled properly, the investment can be depreciated over several projects.

Bridge construction equipment is mostly designed for a specific application. Reusing the equipment in future projects often requires new assembly configurations, where second-hand members are combined with new members. The used members should be clearly identified, and are sometimes checked with lower resistance factors to allow for the wear and tear accumulated in previous uses.

Different types of machine have different reuse flexibility, and ultimately a different service life. Beam launchers, launching gantries and lifting frames perform simple tasks and can often be reused multiple times without major reconditioning. Form travellers, forming carriages and MSSs require geometry adjustment of the casting cell and often also of the framing system to be reused in a new bridge. Custom machines such as heavy gantries for macro-segmental construction, portal carriers with underbridge, tyre trolleys and span launchers are so specialised and expensive that redesigning the bridge to reuse an entire precasting line is often less expensive than purchasing new equipment. Value engineering is not infrequent in design-build pursuits, where the owner can get a lower cost for the project and lower risk profiles from reusing time-tested construction methods, and the contractor can get better amortisation conditions for the initial investment.

The number of second-hand machines available on the market ensures prompt availability in many cases, although some reconditioning is often necessary. Changes in the design standards and original design according to standards of other countries may increase the cost of reconditioning. Reconditioning is simpler when the original design identifies the critical points for proper functioning of the machine.

The feasibility analysis for reconditioning examines the bridges built in the past and the use limitations of the machine. Analysis of previous bridges includes the number and length of spans, gradient and crossfall, plan and vertical radii, design loads, structure–equipment interaction, and assembly geometry of the machine for the individual applications. Analysis of use limitations includes the age of the machine, the number of reuses, the design standards used for the original

Figure 9.4. Proper support conditions for welded square truss modules

design and the subsequent modifications, traceability of materials, weld certifications, availability and completeness of previous operation manuals, limitations of the launch systems, the possibility of longitudinal and transverse widening, and the possibility of reusing winch trolleys and support systems.

The next step is a thorough inspection of the machine to evaluate the maintenance conditions. Inspecting an operational machine is preferable, but in most cases the machine will be in a storage yard. The controls to be performed during the inspection depend on the condition of the machine, the role of the individual components, the number of reuses, the stress levels reached during previous uses, and previous anomalous work conditions. The order and cleanliness of the storage yard, members and components kept far from the ground and water, the correct support of stacked members, the presence of identification tags, oiled bolts in wooden boxes, and sheltered mechanical components and hydraulic systems are additional indicators of proper handling (Figure 9.4).

A reconditioning estimate is prepared, and the contractor ultimately decides whether to recondition the machine or purchase a new one. Reconditioning is based on some or all of the following steps.

- A new assembly configuration is designed for the bridge to be erected. The new configuration may include only second-hand members or may be a combination of used and new members.
- The machine is modelled and analysed to determine the stress envelopes in members and connections.

Figure 9.5. Loss of section under the washers, paint applied over rust

- The original design of members, connections and mechanical components is checked for the new work conditions. If the required modifications affect stiffness distribution, the structural models are updated and the structural checks are iterated to convergence.
- The new assembly geometry of the machine, loads applied to the bridge during operations and launching, and corresponding deflections are submitted to the bridge designer for approval. This is a fundamental step, as the use of a reconditioned machine almost always implies some compromise. A modular machine may have the right length but may be too heavy, or too flexible, or both. Pier brackets may be too invasive, pier towers may be incompatible with the geometry of pier caps and pier diaphragms, and so on.
- New inspections are performed in the storage yard. Structural members, connections and field splices are inspected to detect corrosion, loss of cross-section, fatigue cracks, and buckling and distortion due to improper handling during assembly and decommissioning. Figures 9.5 and 9.6 show an example of the findings. Non-destructive controls may be performed on critical welds, and steel specimens may be extracted for laboratory tests aimed at checking the reliability of the quality control documents.
- Slip-critical splices and frontal connections with stressed bars often require re-sandblasting and painting as per RCSC specifications. New bolts and well-maintained old spare bolts are typically used in new assemblies. Pins and threaded bars are often recycled after visual inspection and checking of the quality control marks.
- The mechanical components are inspected and all movements are tested. If properly maintained, launch bogies and articulated saddles often just require greasing. Electric motors may also last a long time if properly stored. Hoisting and hauling ropes of winches and capstans are typically replaced. Air and electric vibrators are tested individually.
- The hydraulic systems are inspected, emptied and cleaned to remove deposits. Hydraulic cylinders often require new seals; the growing cost of labour sometimes suggests

Figure 9.6. Distortion in web panel

replacement in borderline conditions. Hoses and oil filters are replaced, rigid pipes and manifolds are replaced if damaged during decommissioning, and the circuits are filled with new oil and tested.

- Electric circuits and electronic control systems are inspected and tested. Electric circuits and boxes are easily damaged during decommissioning and storage. The electronic control systems have rapid obsolescence and should be replaced as soon as spare parts are no longer available.

Winch trolleys and portal cranes are often reused as they are. Reconditioning may require different hoist systems or a new PPU. Mechanical and hydraulic components rarely require modifications. Portal cranes and winch trolleys may be easily adapted to a different spacing of the runways by adjusting the length of the crane bridge.

REFERENCES

AASHTO (American Association of State Highway and Transportation Officials) (2012a) *LRFD Bridge Design Specifications*. AASHTO, Washington, DC, USA.

AASHTO (2012b) *LRFD Bridge Construction Specifications*. AASHTO Washington, DC, USA.

Andre J, Beale R and Baptista A (2012) Bridge construction equipment: an overview of the existing design guidance. *Structural Engineering International* 22: 365–379.

ASCE (American Society of Civil Engineers) (2010) ASCE/SEI 7–10: Minimum design loads for bridges and other structures. ASCE Press, Reston, VA, USA.

Boggs D and Peterka J (1992) Wind speeds for design of temporary structures. *Proceedings of the 10th ASCE Structures Congress, San Antonio*. American Society of Civil Engineers (ASCE), Reston, VA, USA.

BSI (2004a) BS EN 13001-1: 2004. Cranes – General design – Part 1: General principles and requirements. BSI, London.

BSI (2004b) BS EN 13001-2. Cranes – General design – Part 2: Load effects. BSI, London.

BSI (2005a) BS EN 1993-1-1: 2005. Eurocode 3. Design of steel structures – Part 1.1: General rules and rules for buildings. BSI, London.

BSI (2005b) BS EN 991-1-4: 2005. Eurocode 1. Actions on structures – Part 1–4: General actions – Wind actions. BSI, London.

BSI (2005c) BS EN 1990: 2002 + A1: 2005. Eurocode – Basis of structural design. BSI, London.

BSI (2005d) BS EN 1998-2: 2005. Eurocode 8. Design of structures for earthquake resistance. Part 2: Bridges. BSI, London.

BSI (2006) BS EN 1991-3: 2006. Eurocode 1. Actions on structures – Part 3: Actions induced by cranes and machinery. BSI, London.

CIRIA (Construction Industry Research and Information Association) (1977) *Rationalization of Safety and Serviceability Factors in Structural Codes*. CIRIA, London, UK.

CNC (2007) *Manual de Cimbras Autolanzables*. CNC, Madrid, Spain.

CNR (Consiglio Nazionale delle Ricerche) (1985a) CNR 10021. Strutture di acciaio per apparecchi di sollevamento. istruzioni per il calcolo, l'esecuzione, il collaudo e la manutenzione. CNR, Rome, Italy.

CNR (1985b) CNR 10027. Strutture di acciaio per opere provvisionali. Istruzioni per il calcolo, l'esecuzione, il collaudo e la manutenzione. CNR, Rome, Italy.

CNR (1988) CNR 10011. Strutture di acciaio. Istruzioni per il calcolo, l'esecuzione, il collaudo e la manutenzione. CNR, Rome, Italy.

DIN (Deutsches Institut für Normung) (1985) DIN 1072. Road and foot bridges. Design loads. DIN, Berlin, Germany.

DIN (1986) DIN 1055 Part 4. Design loads for buildings. Imposed loads. Wind loads on structures unsusceptible to vibration. DIN, Berlin, Germany.

FEM (Fédération Européenne de la Manutention/European Federation of Materials Handling) (1987) FEM 1.001. Règles pour le calcul des appareils de levage. FEM, Brussels, Belgium.

Hewson NR (2003) *Prestressed Concrete Bridges: Design and Construction*. Thomas Telford, London.

Hill H (2004) Rational and irrational design loads for 'temporary' structures. *Practice Periodical on Structural Design and Construction* 9: 125–129.

Pacheco P, Coelho H, Borges P and Guerra A (2011) Technical challenges of large movable scaffolding systems. *Structural Engineering International* 21(4): 450–455.

Ratay R (2009) *Forensic Structural Engineering Handbook*, 2nd edn. McGraw Hill Professional, New York, NY, USA.

Reason J (1990) *Human Error*. Cambridge University Press, Cambridge, UK.

Rosignoli M (2000) Thrust and guide devices for launched bridges. *ASCE Journal of Bridge Engineering* 5(1): 75–83.

Rosignoli M (2002) *Bridge Launching*. Thomas Telford, London.

Rosignoli M (2007) Robustness and stability of launching gantries and movable shuttering systems – lessons learned. *Structural Engineering International* 17(2): 133–140.

Rosignoli M (2010) Self-launching erection machines for precast concrete bridges. *PCI Journal* 2010(Winter): 36–57.

Rosignoli M (2011) Bridge erection machines. In *Encyclopedia of Life Support Systems*, Chapter 6.37.40, 1–56. United Nations Educational, Scientific and Cultural Organization (UNESCO), Paris, France.

Rosowsky D (1995) Estimation of design loads for reduced reference periods. *Structural Safety* **17**: 17–32.

Sexsmith R (1988) Reliability during temporary erection phases. *Engineering Structures* **20**: 999–1003.

Sexsmith R and Reid S (2003) Safety factors for bridge falsework by risk management. *Structural Safety* **25**: 227–243.

Bridge Construction Equipment
ISBN 978-0-7277-5808-8

ICE Publishing: All rights reserved
http://dx.doi.org/10.1680/bce.58088.343

Chapter 10
Analysis and design

10.1. Numerical modelling

The optimal level of detail for the numerical model of a machine depends on numerous factors. The more refined the model, the more accurate the results of analysis, and the deeper the understanding of the behaviour of the machine. On the other hand, however, the complexity and quantity of load cases and support conditions to examine suggest the use of simple models to investigate all the possible combinations rapidly.

Simple beam models facilitate the research of the design-governing conditions, provide the stress magnitude to be expected from more refined analyses and the boundary conditions to assign to local models, and are often accurate enough for underslung box girders supported on stiff pier brackets. Three-dimensional (3D) truss models are necessary with machines comprising trusses on flexible crossbeams or when only some trusses sustain the load. Figure 10.1 shows the 5300 frame-element model of a twin-truss overhead MSS with square trusses, and Figure 10.2 shows the contribution of the lateral modes of tall pier towers to the first lateral modes of the machine. Modelling the supports is also necessary when the machine is propped from foundations.

In a telescopic overhead gantry the front legs of the main frame are closely spaced to take support on the pier cap, while the rear C-frame is wider to feed segments from the deck. In some single-girder overhead MSSs, and in most span launchers, the rear C-frame is wider to insert full-width prefabricated cages or precast spans. Most support legs are equipped with multiple hydraulic systems for geometry adjustment. In the presence of so many technological constraints, the structural nodes are so complex that local 3D models with shell or solid finite elements are often necessary to investigate stress distribution. Beam models are used to analyse span erection, structure–equipment interaction at the application of prestressing and launching, and to identify the boundary conditions to apply to local 3D solid models.

10.2. Ideal and actual trusses

An ideal truss should meet three conditions (Rosignoli, 2007).

1   The members of diagonals, verticals and chords are perfectly hinged at the truss nodes. This condition is never respected in bridge erection machines. Long continuous chord segments are used for fast site assembly, and the field splices between the segments are located at the centre of the panel in order to use the same end node design for the diagonals throughout the truss. The nodes are mostly continuous and, when the field splices of diagonals and verticals are pinned, the pins are located far from the geometric nodes of the panel.

Figure 10.1. Self-weight deflection in the 5300 frame-element model of an overhead MSS

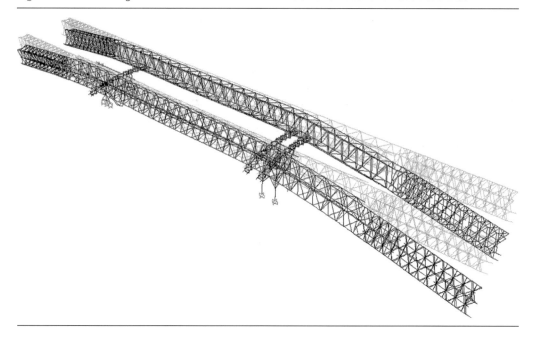

Figure 10.2. Tall pier towers add flexibility to the first lateral modes

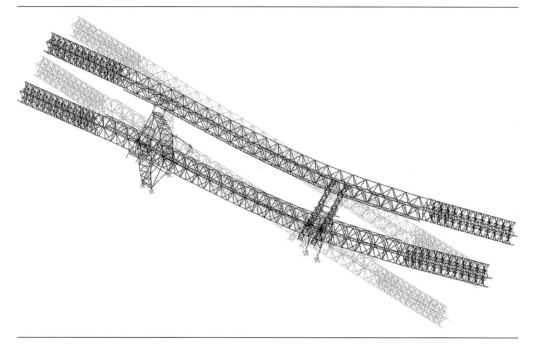

2 The loads are applied at the truss nodes. Also this condition is never respected. Technological requirements dictate the position of the support points of precast segments or casting cells, the loads applied by carried cranes travel along the top chord, the support reactions travel along the bottom chords during launching, and the trusses are often supported with out-of-node eccentricity also during operations.

3 The axes of gravity of all members converge to full-load the geometric panel node. This condition could actually be respected, although in most cases the convergence points of diagonals and verticals into the chords are staggered in order to minimise welding.

Depicting stresses generated by out-of-node eccentricity, geometry imperfections and flexibility of supports requires accurate modelling (Rosignoli, 2010). The numerical models should describe the entire machine (trusses, winch trolleys if capable to exert a lateral restraint action on the compression chords, launch saddles and tower–crossbeam assemblies), and out-of-node eccentricity should be modelled at permanent connections, field splices and support points.

The flanges of top and bottom chords are kept flush along their full length for smooth travelling of winch trolleys during operations and of the truss on the launch bogies during repositioning. Vertical steps may arise in the axis of gravity of the chord at section changes, which is another cause of local load eccentricity.

Special joints are modelled at points of discontinuity, such as structural boundaries, internal releases of degrees of freedom, changes in cross-section, field splices, support points, points of application of localised loads, and points where the deflections are to be determined. Auxiliary joints are moved along the bottom chords to provide travelling internal restraints for the analysis of launch stresses (Rosignoli, 2011).

Additional loads and masses are applied to represent non-structural attachments, such as hydraulic systems, mechanical components and PPU. Scale factors are used to modify the cross-sectional properties to allow for distributed masses or additional stiffness. End offsets are used to account for the finite size of diagonals, verticals and chords at the connections.

The reliability of internal releases of degrees of freedom should be reviewed critically. For example, the overhead trusses are often supported on tower–crossbeam assemblies. One tower restrains the trusses longitudinally, and the launch bogies at the other tower allow thermal displacements. Before overcoming the breakaway friction of the launch bogies, the thermal deformations of the trusses generate horizontal bending in the crossbeams and P-delta effects in the towers. If the towers are flexible, the longitudinal thermal load at the bogies may be insufficient to overcome the breakaway friction. Before releasing translational degrees of freedom in the model, it is therefore necessary to check that the longitudinal stiffness of the tower–crossbeam assemblies exceeds the breakaway resistance. Breakaway frictional loads are then applied to the crossbeams to investigate the P-delta effects.

Similar considerations apply for the release of rotational degrees of freedom. When launch bogies or friction launchers are used, the truss is supported on articulated beams that may be modelled as eccentric cylindrical hinges. When the pier towers have four distant legs and use two sets of crossbeams, the crossbeams are supported on interconnected hydraulic cylinders to generate a flexural hinge during launching. Similar solutions are also used in the W-frames on through girders. The cylinders are equipped with mechanical locknuts that are released prior to lowering the MSS for span release, and are tightened at the end of repositioning. Different restraint conditions are therefore modelled for the supports during launching and operations.

During launching, the equalising beams of the launch bogies are modelled to equalise the support reactions applied to the truss, in spite of different deflections of the support points. During span casting, the support cylinders are modelled as vertical frame-elements and the rotational degrees of freedom at the top are released to model the tilt saddles. Analysis with tightened locknuts will show that inner crossbeams and W-frames are more heavily loaded than the outer ones due to the flexural rotations in the trusses. When one line of cylinders supports PTFE saddles for thermal sliding, additional bending results from breakaway friction at the sliders (Rosignoli, 2007).

10.3. Truss instability

The stability of a freestanding truss depends on the degree of fixity of supports, a support condition prompting the truss to twist as it deflects laterally, the lateral restraint exerted by inclined diagonals and lateral bracing on the compression chords, the location of the applied loads in relation to the centre of shear of the cross-section, and the level of imperfections in the initial geometry of the truss.

In most trusses for bridge construction equipment, these factors coexist and coalesce (Rosignoli, 2010).

1 The degree of fixity at the supports is low, as the truss is supported on flexible crossbeams, with out-of-node eccentricity, and often without diaphragms that prevent distortion.
2 The truss bends laterally and twists because of crossbeam deflections, and the depth of the launch saddles and truss amplifies the lateral displacements of the top chords.
3 Many triangular trusses are tall and narrow, and the lateral restraint that such inclined diagonals exert on the top chord is poor. Narrow rectangular trusses are geometrically more efficient, but the great number of field splices, the flexibility of lateral bracing, and the high depth-to-width ratio of the cross-section lead to similar levels of stability. Wide square trusses of new-generation MSSs are definitely more stable.
4 Overhead travelling cranes load the truss above the centre of shear of the cross-section. Torsion due to eccentric vertical and lateral loads relative to the centre of shear is a typical problem in crane-supporting structures.
5 The geometry imperfections may be significant due to the large number of field splices and distortion accumulated in second-hand machines.

Most design standards recognise two forms of instability. The first form is related to the sway of the entire system (out-of-plane buckling) and is typically catastrophic in nature. The second form is due to deformations of members between the end nodes (local buckling) or to cross-section distortion (buckling of flanges, webs and stiffeners). Both forms of instability are related to member slenderness and are characterised by a rapid change in geometry and member distortion. Inelastic deformation typically follows elastic instability.

Buckling is the only failure mode of a steel structure that depends primarily on the geometry and only secondarily on the properties of steel; buckling is also the only failure mode that occurs under compressive loads (Ratay, 2009). Structural analysis programmes detect linear buckling through P-delta analysis, and are able to depict the instability in every member of the model. Investigating many buckling modes with detailed structural models, however, involves a long computation time, even with new-generation workstations, and rarely warrants the effort. In many cases only out-of-plane buckling is analysed numerically, and local instability is checked using equations specified by the design standards in relation to the cross-section geometry and the effective length and end restraint of members. Load factors are sometimes specified to allow for eccentricity, distortion and fabrication-induced residual stresses.

10.4.   P-delta effect

When the loads applied to a structure are small and generate small deflections, the load–deflection relationship is linear, the equilibrium equations can be written for the undeformed geometry of the structure, the equilibrium conditions are independent of the loads, and the results of different analyses can be superposed (CSI, 2012).

If the loads on the structure and/or the resulting deflections are large, the load–deflection relationship may become nonlinear, and equilibrium equations written for the original and the deformed geometry may differ significantly. Bridge construction equipment is typically fabricated from high-grade steel and is heavily loaded during operations and launching, and because of the large deflections of many of these machines, the nonlinearity of the load–deflection relationship is often non-negligible.

The nonlinearity generated by the axial force upon transverse bending and shear behaviour is called the 'P-delta effect'. A compression force tends to make a structural member more flexible in transverse bending and shear, and a tension force tends to stiffen it against transverse deformations. The P-delta effect can be illustrated with the following example (CSI, 2012).

Considerer a cantilever beam of length L subjected to a tensile axial load P and a transverse tip load F. If equilibrium is examined in the undeformed geometry, the moment at the root of the cantilever is $M = FL$ and it decreases linearly to zero at the loaded end. If equilibrium is examined in the deformed geometry, there is an additional moment caused by the axial force P acting on the transverse tip deflection δ generated by the load F. The moment no longer varies linearly along the length, the variation depends on the deflected shape, and the moment at the root is $M = FL - P\delta$.

If the beam is in tension, the moment throughout the member reduces, the transverse deflection δ also reduces, and the beam is therefore stiffer against the transverse load F. If the beam is in compression, the transverse deflection increases and the beam is more flexible.

The P-delta effect applies locally to individual members and globally to the entire truss. In a member loaded at the ends, the stiffness matrix for axial force and bending depends only on the mechanical properties of the member:

$$\begin{bmatrix} F_i \\ M_i \\ F_j \\ M_j \end{bmatrix} = \frac{EI}{L^3} \begin{bmatrix} 12 & 6L & -12 & 6L \\ 6L & 4L^2 & -6L & -2L^2 \\ -12 & -6L & 12 & -6L \\ -6L & -2L^2 & -6L & 4L^2 \end{bmatrix} \begin{bmatrix} \eta_i \\ \phi_i \\ \eta_j \\ \phi_j \end{bmatrix}$$

If the member is subjected to a constant tension T, additional forces ΔF_i and ΔF_j develop at the end nodes for the member to be in equilibrium under the lateral displacements η_i and η_j of its ends.

Additional bending moments ΔM_i and ΔM_j also develop in response to the end rotations ϕ_i and ϕ_j. If the deformed shape is assumed to be a cubic function, the force–displacement relationship is

$$\begin{bmatrix} \Delta F_i \\ \Delta M_i \\ \Delta F_j \\ \Delta M_j \end{bmatrix} = \frac{T}{30L} \begin{bmatrix} 36 & 3L & -36 & 3L \\ 3L & 4L^2 & -3L & -L^2 \\ -36 & -3L & 36 & -3L \\ 3L & -L^2 & -3L & 4L^2 \end{bmatrix} \begin{bmatrix} \eta_i \\ \phi_i \\ \eta_j \\ \phi_j \end{bmatrix}$$

If the axial force is constant, the stiffness matrix of the member may be formed by adding the two matrices. When the axial force varies, the stiffness matrix depends on the loads, and the problem becomes nonlinear and is solved by iteration (CSI, 2012). The correction converges rapidly, and in many cases it is sufficient to analyse all loads starting from the stiffness matrix written for the primary load condition (for most machines, self-weight and payload). The results of the analysis may thus be superimposed for the purposes of design.

10.5. Buckling analysis

If the compressive force in a member is large enough, the transverse stiffness goes to zero, the stiffness matrix becomes singular, and the member is said to have 'buckled'. Linear buckling analysis seeks the instability modes of a structure due to the P-delta effect under a specified set of loads through the solution of the generalised eigenvalue problem. Each eigenvalue–eigenvector pair is called a 'buckling mode' of the structure. The eigenvalue is called the 'buckling factor' and it is the scale factor that must multiply the loads to cause buckling in the given mode.

The buckling modes of a structure depend on the load. There is no set of natural buckling modes as there is for the natural vibration modes. As most types of bridge construction equipment involve the additional complication of movement, buckling is analysed for different load cases and different configurations of the machine. Inertial effects and load oscillations also affect the buckling factor during load handling.

Structural analysis programmes identify the buckling modes in a sequence with increasing buckling factors. The first modes are typically out-of-plane modes and the analysis progresses toward more and more localised modes with shorter effective length. Underdesigned members often buckle locally within a sequence of out-of-plane modes.

Figures 10.3 and 10.4 show the first buckling modes of an overhead gantry for span-by-span macro-segmental construction at lifting of the pier-head segment from the casting cell and at segment lowering on the leading pier cap. In the last phases of handling, the segment is shifted rightward with the two hoist crabs: this increases the load applied to the right truss and reduces its buckling factor. As the winch trolleys exert a lateral restraint action between the two trusses, they are included in the model. The rear pendular frame is lifted prior to traversing the gantry; deflections of the right cantilever of the front portal frame further reduce the buckling factor of the right truss.

Buckling is analysed using the stiffness matrix written for the deformed geometry of the machine under the loads that generate buckling. Geometry imperfections may be surveyed and applied to the model with equivalent lateral loads or by modifying the node coordinates. The buckling factors of trusses and tall pier towers are sensitive to geometry imperfections because, in one of the many load and support conditions of the machine, the imperfections may have the same shape as a buckling mode.

Figure 10.3. First buckling mode at lifting of the macro-segment

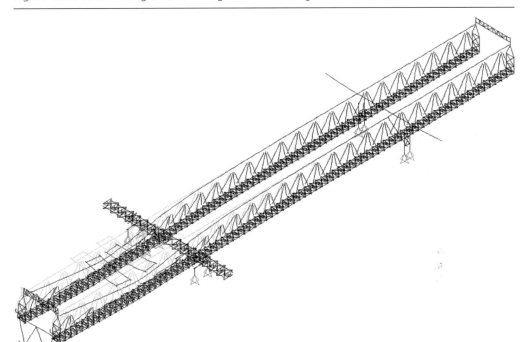

Because of the structural complexity of these machines, approximate values for the critical load for out-of-plane bucking are used only during preliminary design or to quickly assess site assembly operations. Setting L as the span length or the distance between the lifting points, the critical load for out-of-plane buckling may be expressed as

$$q_{cr} = k\frac{\sqrt{EI_2 GJ}}{L^3}$$

where I_2 is the moment of inertia of the truss about the vertical axis, J is the constant of torsion, the shear modulus is $G \approx 0.4E$ and k is a coefficient that depends on the restraint conditions at the lifting points. The trusses are typically assembled on the ground and lifted on the launch saddles in sections that are as long as possible. During handling the trusses are subjected to self-weight only, and therefore, setting q as the distributed weight of the truss, it must be that

$$q < \frac{q_{cr}}{2.5}$$

Excessive confidence with a machine resulting from having already handled similar loads in the past may be a serious mistake (Rosignoli, 2007). The stability of these machines depends not only on the entity of the load but also on how the load is applied to the machine. When a gantry handles a long precast beam, the winch trolleys are at the ends of the beam. When the winch trolleys work close to each other (e.g. for simultaneous application of the starter segments of the hammer) the total load may be similar but the loads in the diagonals, verticals, launch bogies and crossbeams may be more than double.

Figure 10.4. First buckling mode at insertion of the pier-head segment within the front portal frame

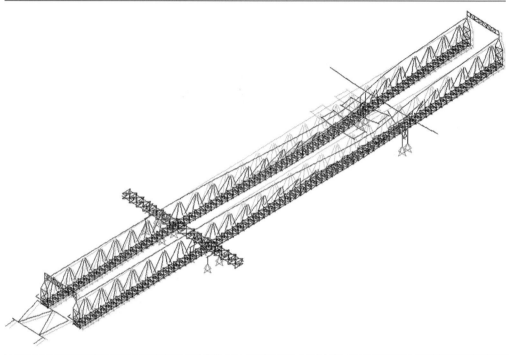

Neither is launching less demanding: Figure 10.5 shows the first buckling mode of an 86 m cantilever. The winch trolleys suspend counterweights at the rear end of the gantry to control overturning, and most of the weight of the gantry is applied to the front support frame.

Overloaded members may buckle suddenly and without warning, and careful inspections are therefore necessary during site assembly and at regular intervals during operations. Damaged members must be reinforced or replaced, even when they are not in critical locations, because they could trigger collapse under different load or support conditions in the future. The cross-diaphragms are also inspected frequently and the reasons for out-of-flat investigated, as this makes the members susceptible to local buckling.

10.6. Robustness of trusses

The robustness of bridge construction equipment depends on the margin to local failure of the component members and systems and on how the machine responds to local failure (Rosignoli, 2007). Local buckling of primary load-carrying members is often critical in the pier towers, while stable alternative load paths often exist in such redundant trusses; however, local buckling can trigger a chain reaction of local failures, causing progressive collapse.

One of the most reliable ways to ensure robustness and to reduce the risk of progressive collapse (Starossek, 2009) is to require insensitivity to local failure – that is, local buckling of a primary load-carrying member must not cause collapse of the machine or of a major part of it. Clear analysis procedures are then specified to ensure the desired level of robustness.

Figure 10.5. First buckling mode of the 85.9 m front cantilever at landing

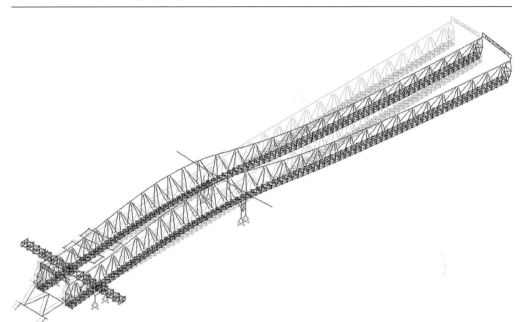

Although the structural damage caused by local buckling is often localised, the sudden load redistribution due to the loss of carrying capacity of a member is a highly dynamic process that requires analysis in the time domain. The effects can be analysed using the following approach (Rosignoli, 2007).

1 Start from an accurate numerical model of the machine.
2 Calculate the forces and moments at the end nodes of the buckling member under the loads that generate buckling.
3 Remove the buckled member from the model and apply end forces and moments to the end nodes with a specific load case.
4 Define a loading time function that increases linearly from 0 to 1 and then holds constant. Lengthen the load ramp or the end plateau and/or increase damping to damp vibrations.
5 Define an unloading time function that abruptly jumps from 1 to 0 and remains constant at 0 for the duration of the dynamic analysis.
6 Load the model by applying buckling loads and nodal loads using the ramp function.
7 Analyse the dynamic response to member buckling by removing the end nodal loads using the unloading function.

With regard to the time-integration technique, modal superposition in the time domain provides a highly efficient procedure that has a shorter runtime than direct integration (CSI, 2012). Closed-form integration of modal equations is used to compute the response, numerical instability problems are never encountered, and the time step may be any sampling value that is deemed fine enough to capture the peak response values. Direct integration involves a longer runtime and the results are sensitive to time-step size; however, impact problems that excite a large number of modes are usually solved more efficiently by direct integration.

Figure 10.6. Axial force time history in the adjacent compression diagonal

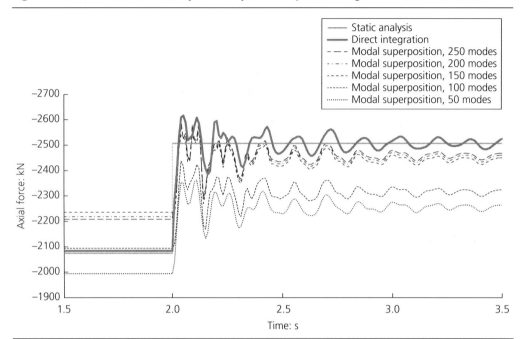

Figure 10.6 illustrates the time history of the axial force in the left diagonal at the front support of the gantry shown in Figure 4.6 when the right diagonal buckles. Direct integration captures the initial and final static forces accurately, while modal superposition converges slowly and only by including so many modes that runtime savings become insignificant. Static analysis prior to and after buckling is recommended for assessment of results.

The resulting stresses are checked like any other ultimate limit state condition. Dynamic amplification is typically low due to the flexibility of these trusses. The launch bogies are long in heavy machines, and several pairs of diagonals are therefore involved in the load paths at supports, which further reduces the dynamic effects of load redistribution.

If the analysis detects overloading of another member, the same approach can be followed to investigate the entire buckling sequence until either load stabilisation or collapse. The analysis can be automated by inserting plastic hinges in the buckling members that drop axial load, bending, shear and torsion upon reaching the local buckling load, although nonlinear direct integration lengthens the runtime significantly when using such large numerical models.

10.7. Instability of I- and box girders

An I-girder subjected to mainly flexural stresses is affected by three modes of instability (Rosignoli, 2002).

1 The lateral–torsional instability of the compression flange, where a girder loaded in the vertical web plane suddenly becomes unstable and twists laterally under a load smaller than the flexural strength.

2   The torsion instability of the compression flange, which is partially related to the previous mode and controlled by vertical web stiffeners and lateral bracing.
3   Local instability of web panels along horizontal lines in the case of pure bending and along inclined lines in the presence of shear forces.

The first two forms of instability are controlled with diaphragms and lateral bracing. The diaphragms avoid relative movements between the flanges, balance the vertical load between the girders, and oppose distortion. The diaphragms are designed as crosses or flexural frames with mid-depth crossbeams and flexural splices to web stiffeners. Lateral bracing stiffens the girders, resists wind loads and transverse forces, resists torsion when applied to both flanges, and distributes the stiffening action of the diaphragms.

The I-girders should be restrained at the supports against rotations about the longitudinal axis. Direct restraint with diaphragms is impossible during launching, and indirect restraint is therefore obtained using lateral guides at the launch bogies and oversized bracing. BS 5400: Part 3: 1982 (BSI, 1991) provides instructions for optimum sizing of bracing and crossbeams. BS EN 1993-1-1: 2005 (BSI, 2005) takes the imperfections of welded girders into account and provides information on lateral–torsional instability.

Most design standards for steel structures provide instructions for checking buckling of web panels under different load conditions, such as uniform and non-uniform compression fields, pure bending, bending with prevailing tensile stress, and pure shear. When the critical buckling stress exceeds the yield strength of steel, some standards provide reduction factors for the critical stress.

The web plates are designed conservatively to allow for the uncertain load distribution between the webs and to distribute the vertical stiffeners according to bracing requirements. Based on the geometry of lateral bracing, plate thickness is chosen to control buckling of web panels, and it is rarely less than 10–12 mm to facilitate workshop handling. The length and flexibility of launch bogies and PTFE skids control web crippling over the supports. The support saddles may be placed on interconnected hydraulic cylinders to create torsional hinges that equalise the load in the webs.

A non-stiffened web panel subjected to a patch load applied through the bottom flange is affected by three failure modes that depend on the load intensity and the slenderness of the web panel (Rosignoli, 2002). A first mode is caused by web yielding immediately above the load followed by plastic deformations in the bottom flange. A second mode is triggered by localised buckling in the lower part of the web panel followed by web yielding and onset of a plastic mechanism in the bottom flange. A third mode is general buckling of the web panel extended to most of its depth.

BS EN 1993-1-1: 2005 (BSI, 2005) provides check instructions for the three failure modes. In the absence of torsional hinges, the design value for the support reactions is calculated by allowing for misalignment of the launch saddles and geometry irregularities in the bottom flanges due to fabrication and assembly tolerances (Rosignoli, 1998). A 30% increase in the theoretical value is often sufficient, provided that stringent geometry tolerances are prescribed and actually respected.

The flexural capacity of an I-girder is determined as the lower value of the plastic moment of the cross-section and the critical moment for web-panel buckling. BS EN 1993-1-1: 2005 (BSI, 2005) specifies an additional stability criterion to account for the combined risk of the two failure modes.

The flexural capacity depends on the type of girder. For an I-girder, the stress–strain relationship is bilinear. In the absence of buckling of the web panels – Class 1 and Class 2 sections according to BS EN 1993-1-1: 2005 (BSI, 2005) – the cross-sectional capacity is reached when the applied moment reaches the plastic moment capacity of the section. For Class 1 and Class 2 sections the ratio of the plastic moment capacity to the first-yield moment is approximately 1.15, for Class 3 sections it is 1.00 (instability prevents taking advantage of the post-yielding domain), while the flexural strength of a Class 4 section is governed by instability.

The behaviour of a box girder subjected to negative bending and shear is more complex. Simple beam theory cannot predict the stress distribution in the flange because plane sections do not remain plane. If the webs are stocky and the bottom flange is slender, first yield occurs in the outer flange panels where shear lag maximises axial and tangential stresses. As the outer flange panel is poorly stiffened, local buckling maximises the stresses at the plate surface, and both the first-yield moment and the plastic moment capacity may be lower than the linear-elastic values. Two assumptions may be made (Rosignoli, 2002).

1   The flexural capacity is reached when first yielding occurs at the outer panel. This
    assumption is on the conservative side.
2   The bottom flange has the capacity to redistribute the compressive stress across its breadth
    so as to reach yielding also in the less stressed central panel. In reality, the flexural capacity
    depends on whether the outer flange panels continue to carry compression load as
    redistribution occurs, or they rapidly shed load due to local buckling.

The behaviour of the bottom flange of a box girder when it moves over the launch supports depends on the behaviour of each component of the cross-section and on the effects of the boundary conditions. Analysis includes the effects of shear lag, restrained warping, cross-sectional distortion, and imperfections. Several design standards provide instructions for analysis of shear lag; the governing parameters are the flange orthotropy, the ratio of flange area to web area, the cross-sectional shape, the aspect ratio of the flanges, the location of point loads, the load combinations and the support conditions.

With the exception of circular and square thin-walled sections, torsion is accompanied by warping of the cross-section. If warping is resisted, restrained warping stresses are induced in the section. Some design standards propose equations to predict peak warping stresses in box sections. Eccentric loads also deflect, twist and distort the cross-section between diaphragms. Although distortion is small in terms of displacement, it may cause significant transverse bending and longitudinal warping stresses. Several design standards provide equations for the analysis of these stresses.

The behaviour of the bottom flange of a box girder over the launch supports can be predicted using elastoplastic analysis. The effects of residual stresses and the initial lack of flatness can be allowed for in the analysis, although it is difficult to measure these levels of geometric imperfections. Residual stresses tend to weaken flange plates in compression. Analyses have shown that geometric imperfections can decrease the plate strength significantly, while cylindrical out-of-plane distortion has little weakening effect. Tests have shown that transverse fillet welds of diaphragms to flange plates do not reduce plate strength, while steps at butt welds reduce the strength, particularly where a thinner plate is connected.

When rectangular launch rails or outer lateral guides keep the support reactions aligned with the webs, transverse bending is low and the longitudinal axial stress due to flexure and the vertical axial stress due to dispersal of the support reaction within the web have the same sign and decrease along the web depth. Although the tangential stress is lower at the web edges than at mid-height, the ideal stress typically peaks near the main weld between the web and the bottom flange, which must be checked accurately during fabrication.

Articulated assemblies of cast-iron wheels on equalising beams apply the support reaction to rectangular launch rails. The flanges of the wheels are guided by the sides of the rail for transfer of lateral loads. Rigid rollers mounted on tilting arms often use outer lateral guides. PTFE skids and friction launchers also use lateral guides; both types of launch support load a wide strip of the bottom flange, and transverse bending may become a major stress component in the bottom flange, the base region of the web, and the weld of web to flange.

Transverse bending is calculated under the assumption that the patch load is centred with the web and the support reaction is equally distributed between the two girders, and then it is increased to allow for lateral eccentricity. Closely spaced vertical stiffeners are often applied to the webs in the operational support regions of the machine to create additional support lines in the bottom flange and to improve dispersal of the support reaction within the webs.

10.8. Instability of support members

Vertical and horizontal loads are transferred to the bridge foundations through complex load paths that typically include geometry adjustment systems. The auxiliary pendular legs are among the most delicate components of bridge construction equipment because their length must be varied to take support on the deck and pier caps, and their transverse geometry must also be varied in curved bridges.

Starting from a short configuration for support on the deck, the leg is lengthened by inserting one or two articulated bottom extensions into the load path, and multiple aligned articulations make the leg prone to kinematic instability when fully extended. The leg is typically pinned at the top for vertical setting when the truss follows the bridge gradient, and the tilt saddles of the support cylinders also work as hinges. Even in the absence of kinematic instability, therefore, the effective length of the assembly cannot be shorter than the actual length of the leg. The front pendular legs of self-launching gantries for balanced cantilever erection of precast segments are particularly long due to the depth of the hammerhead segments of long-span bridges, and they are also heavily loaded during placement of the hammerhead segment and of the support crossbeam onto it.

Overhead gantries and MSSs apply eccentric loads also to the tower–crossbeam assemblies. As cantilevers partially fixed at the base, the towers are sensitive to couples and horizontal forces applied at the top. Diaphragms and lateral bracing govern the stability and load capacity of compression columns. When the slenderness ratio is low, the deflections may be disregarded and the towers are checked using the moment magnifiers specified by the design standards.

Figure 10.7. Pier cap forces for a launch gradient smaller than the breakaway friction

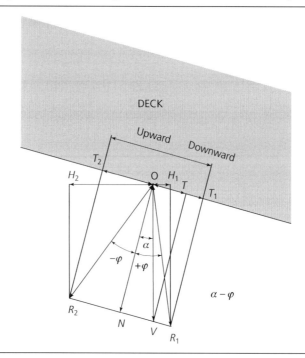

In most cases, however, the towers are flexible and are checked by accounting for the P-delta effect.

The load applied to the towers is eccentric in the longitudinal and transverse directions. Thermal deformations, launch friction and the launch gradient cause longitudinal load eccentricity, lateral wind is the main cause of transverse eccentricity, and tolerances of verticality cause load eccentricity in both directions. Starting from the breakaway friction coefficient μ of the launch saddles, the friction angle φ is such that:

$$tg\varphi = \mu$$

Setting α as the launch gradient at the tower, Figures 10.7 and 10.8 define the vectorial combinations for $\alpha < \varphi$ and $\alpha > \varphi$, respectively (Rosignoli, 2002). The vertical load applied to the tower is the vector $\overline{OV}$. Its component orthogonal to the launch plane is

$$\overline{ON} = \overline{OV} \cos \alpha$$

and the tangential component is

$$\overline{OT} = \overline{OV} \sin \alpha$$

When the truss is pushed forward, the resultant vector rotates, and the movement starts when its inclination exceeds the equilibrium angle $R_2OR_1 = 2\varphi$. The limit equilibrium resultants are, therefore,

$$\overline{OR_1} = \overline{OR_2} = \overline{OV}\frac{\cos \alpha}{\cos \varphi}$$

Figure 10.8. Pier cap forces for a launch gradient greater than the breakaway friction

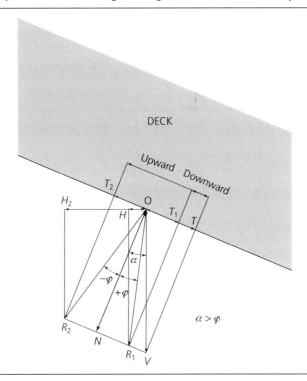

When the truss is launched uphill, rotating the resultant from $\overline{OV}$ to $\overline{OR_2}$ requires a launch force F equal to

$$F = \overline{TT_2} = \overline{OT_2} - \overline{OT} = -\overline{OV}(\cos \alpha \, tg\varphi + \sin \alpha)$$

and the longitudinal force applied to the tower is

$$H = \overline{OH_2} = -\overline{OV}\,\frac{\cos \alpha}{\cos \varphi}\sin(\varphi + \alpha)$$

When launching occurs downhill, the launch gradient may be lower or higher than the breakaway friction angle. In the first case ($\alpha < \varphi$) it is necessary to push the truss until winning the breakaway friction, and the resultant rotates from $\overline{OV}$ to $\overline{OR_1}$ under a force F equal to

$$F = \overline{TT_1} = \overline{OT_1} - \overline{OT} = \overline{OV}(\cos \alpha \, tg\varphi - \sin \alpha)$$

If the gradient is higher than friction ($\alpha > \varphi$) the truss slides freely and must be braked. In both cases the longitudinal load applied to the tower is

$$H = \overline{OH_1} = \overline{OV}\,\frac{\cos \alpha}{\cos \varphi}\sin(\varphi - \alpha)$$

Gantries and MSSs may apply large longitudinal loads to the piers. The MSSs for long spans are particularly demanding due to their weight and the frequent use of PTFE skids to transfer large vertical forces. The breakaway friction of painted steel surfaces on greased PTFE skids may amply exceed 10%, and the bridge gradient further increases the longitudinal load when launching uphill. Longitudinal bending may be particularly demanding at the landing pier due to the low axial load.

PTFE skids are thin and can rarely accommodate support cylinders for the trusses. The support cylinders are therefore inserted between tower and the crossbeams, or between through girders and W-frames, and in both cases the PTFE skids are designed for the entire load of span and erection equipment. The load is applied to the piers through inclined supports, and the longitudinal load increases further. In the most delicate cases, the bridge piers must be realigned after repositioning the MSS or be stiffened with stay cables. Stay cables are frequently used in the temporary steel piers.

The slenderness ratio of the columns of a pier tower is

$$\lambda = \frac{L_E}{r}$$

where L_E is the effective length between counter-flexure points and r is the cross-section radius of gyration in the plane under consideration. The stability against local buckling is checked using load magnifiers related to the grade of steel, shape of cross-section, slenderness ratio and Euler load of the member. Bending typically varies linearly along the member, and the checks are performed using an equivalent moment determined starting from the moments at the two restrained ends of member.

Most design standards specify limits for the slenderness ratio for the main and secondary members. CNR 10011 (CNR, 1988) requires $\lambda < 200$ for main members, $\lambda < 250$ for secondary members, and $\lambda < 150$ and $\lambda < 200$, respectively, in the presence of significant dynamic actions. If the minimum size members of lateral bracing are selected to satisfy the slenderness ratio, the demand calculated using the load magnifiers is typically below capacity. CNR 10011 (CNR, 1988) also requires that lateral bracing be able to carry a notional compression load:

$$P = \frac{\omega P_L}{100}$$

where P_L is the compression load in the column and ω is the load magnifier. In the presence of dynamic actions, the notional design load is increased by 25%.

10.9. Material-related failures

Steel plates and shapes are manufactured under controlled conditions and material defects are not frequent. The root causes of most material-related failures in bridge construction equipment are improper design, detailing, fabrication, assembly or operation. Improper fabrication of connections (in particular, poor-quality welding and plate misalignment) is a major source of material-related failures.

Material-related failures include ductile fracture, brittle fracture, fatigue cracking, general atmospheric corrosion, and stress corrosion. Failure may be sudden or may be preceded by crack growth, plastic deformations or corrosion deterioration. Defects and damage mechanisms may be related to fabrication or may develop during the service life of equipment.

The expected service life influences several aspects of design. The longer the expected service life, the more the requirements of handling, shipping, assembly, inspection, maintenance, repair and replacement will prevail over the requirements of fabrication. This is particularly true for components or systems that have shorter service life than the structural components of the machine or accelerated obsolescence.

10.9.1 Ductile fracture

Yielding is not infrequent in bridge construction equipment and is typically related to overloading or impacts. Ductile fracture is characterised by increasing plastic deformations. Fracture occurs when the principal strain reaches the ultimate deformation capacity of steel.

Large plastic deformations dissipate energy through inelastic work, relax peaks of local stresses, and redistribute the load effects within the structure. The ductility parameters of steel (total elongation and necking) play a fundamental role in designing structures operating in a ductile regimen. All welded structures exhibit local strain peaks that can exceed the yield point of steel. The ability to deform plastically so as to relax and redistribute these stress peaks is essential to the practical use of the measured yield and tensile strengths.

Ductile fracture often occurs at the field splices where the section is weakened by holes and geometry changes. In many cases ductile fracture is the consequence, not the root cause, of failure. Cracks in the paint in bands orthogonal to the direction of the principal stress signal large plastic deformations during inspection. Ductile fracture is irregular and exhibits localised necking in the plates; scanning electron micrographs show micro-void coalescence.

Temperature, load rate and level of constraint may affect the ultimate strain capacity of steel. As the temperature decreases or the load rate and level of constraint increase, the yield strength increases, the ultimate strain capacity diminishes, and failure may transition from ductile to brittle.

10.9.2 Brittle fracture

Brittle fracture occurs with little deformation, often at stress levels that are below the yield point of steel and sometimes even below the SLS stress limits. Fracture propagates at high speed and gives minimal or no warning; brittle fracture may therefore cause significant or catastrophic structural damage. Brittle, low-energy failure is characterised by flat fracture surfaces with little or no plasticity and chevron marks that identify the direction of propagation and hence the point of fracture initiation. Scanning electron micrographs of the fracture surface show cleavage in the steel microstructure (Ratay, 2009).

Brittle fracture typically initiates from stress concentrations such as weld defects. When the local constraint is high (e.g. an accumulation of orthogonal welds in thick plates) brittle-like fracture can also occur in steels with good ductility and toughness, because tri-axial states of stress limit the strain capacity.

Structural analysis does not provide any safety against brittle fracture. Failure is prevented by the quality of the steel, proper design of welded connections, and quality control during fabrication to prevent defects and discontinuities that may cause crack initiation and growth. The quality of steel is specified in terms of strength, chemical composition and weldability.

FEM 1.001 (FEM, 1987) recognises three causes of brittle fracture: the concomitance of shrinkage strains and longitudinal tensile strains in the welds of web to flange, the thickness of the plates, and the influence of low temeperature. The influence of these three factors is measured using evaluation indexes, and the required quality of steel depends on the sum of the three indexes.

Let σ_{all} be the allowed SLS tensile stress of steel for load condition I, and σ_{con} be the longitudinal tensile stress due to constant loads. The risk of brittle fracture is increased by stress concentrations, and is particularly high in the presence of tri-axial tensile stresses resulting from weld accumulation. The first evaluation index in the absence of longitudinal welds or when thermal-stress-relieving treatments (600–650°C) are applied to welds of any type is

$$Z_1 = \frac{\sigma_{\text{con}}}{0.5\sigma_{\text{all}}} - 1$$

which applies only for $\sigma_{\text{con}} \geq 0.5\sigma_{\text{all}}$. In the presence of longitudinal welds of the web to the flange, if the welds of the vertical stiffeners to the web and the flange do not cross the main longitudinal weld, the evaluation index is

$$Z_1 = \frac{\sigma_{\text{con}}}{0.5\sigma_{\text{all}}}$$

In the case of accumulation of orthogonal welds, the evaluation index is

$$Z_1 = \frac{\sigma_{\text{con}}}{0.5\sigma_{\text{all}}} + 1$$

For plate thickness in the range $5 \leq t \leq 20$ mm, the second evaluation index is

$$Z_2 = \frac{t^2}{278}$$

For thickness $20 \leq t \leq 100$ mm, it is

$$Z_2 = 0.65\sqrt{t - 14.81} - 0.05$$

A notional thickness t' is used for rolled shapes. For round bars with diameter d it is $t' = d/1.8$, and for square and rectangular bars with breadth $b > t$ it is $t' = b/1.8 \leq t$. The third evaluation index is determined in relation to the minimum temperature T_{min} at the utilisation site of the machine. For temperatures in the range $-55°C < T_{\text{min}} < -30°C$ it is

$$Z_3 = \frac{-2.25 T_{\text{min}} - 33.75}{10}$$

while in the range $-30°C < T_{\text{min}} < 0°C$ it is

$$Z_3 = \frac{T_{\text{min}}^2}{267}$$

The total evaluation index is

$$Z_{\text{tot}} = Z_1 + Z_2 + Z_3$$

The steels are divided into four classes of quality in relation to their resilience, which is measured according to Euronorm 45-63 and ISO R148 at a given test temperature. Steels of different classes can be welded together.

- Steels of Quality Class 1 (Fe360-A and Fe430-A according to Euronorm 25, St37-2 and St44-2 according to DIN 17100, E24-1 according to NF A 35-501, and 43A and 50B according to BS 4360: 1990 (BSI, 1990)) devoid of resilience tests can be used only for $Z_{\text{tot}} \leq 2$.

- Steels of Quality Class 2 (Fe360-B, Fe430-B and Fe510-B according to Euronorm 25, RSt37-2 and RSt44-2 according to DIN 17100, E24-2, E26-2 and E36-2 according to NF A 35-501, and 40B and 43B according to BS 4360: 1990 (BSI, 1990)) with a resilience of 35 N/cm^2 at $+20°C$ test temperature can be used only for $Z_{tot} \leq 4$.
- Steels of Quality Class 3 (Fe360-C, Fe430-C and Fe510-C according to Euronorm 25, St37-3U, St44-3U and St52-3U according to DIN 17100, E24-3, E26-3 and E36-3 according to NF A 35-501, and 40C, 43C, 50C and 55C according to BS 4360: 1990 (BSI, 1990)) with resilience 35 N/cm^2 at $0°C$ test temperature can be used only for $Z_{tot} \leq 8$.
- Steels of Quality Class 4 (Fe360-D, Fe430-D and Fe510-D according to Euronorm 25, St37-3N, St44-3N and St52-3N according to DIN 17100, E24-4, E26-4 and E36-4 according to NF A 35-501, and 40D, 43D, 50D and 55E according to BS 4360: 1990 (BSI, 1990)) with a resilience of 35 N/cm^2 at $-20°C$ test temperature can be used for $Z_{tot} \leq 16$.

According to FEM 1.001 (FEM, 1987), steels of Quality Class 1 can only be used in rolled shapes with thickness $t \leq 6$ mm. Plates thicker than 50 mm can be used in welded connections only upon prequalification of the manufacturer and the welding process. Prior to welding, plates thicker than 40 mm should be checked ultrasonically for imperfections and delamination.

Fracture toughness may also be determined using the Charpy V-notch impact test per ASTM E23 on specimens whose size and location are defined by ASTM A673 (ASTM, 2012a, 2012b). Although different testing standards identify different classes of resilience, the physical principles underlying the classification are the same and the classes of steels are therefore more or less equivalent.

Brittle fracture is often associated with welds, which are a composite of microstructures and materials. Fracture toughness is highly desirable in weld materials, as they are more likely to experience defects; weld metals, however, for a variety of reasons, rarely have the same composition and microstructure as the plates they join (Ratay, 2009). The toughness of the weld is also influenced by consumables (flux, gas covering) and by the depth and microstructure of the coarse-grained heat-affected zone. Separate fracture toughness tests may be performed on the base metal, the weld metal and the heat-affected zone.

Brittle fracture due to metallurgical defects may also occur in the high-strength bars used in the frontal splices and as tie-downs. Robustness and redundancy of field splices and anchor systems are prime concerns with most types of bridge construction equipment, and are critical issues in purely cantilever systems such as form travellers and lifting frames. In addition to duplicating the anchorages for redundancy, the tie-down bars should be tested at full load prior to their first use. Proof testing on site increases the confidence of workers. Bar stressing at 50–60% of the tensile strength is typically specified for the operations of erection equipment, along with a limit of 20–30 reuses of bars and anchor nuts prior to disposal.

10.9.3 Fatigue cracking

Ductile and brittle fractures are related to a single load event, while fatigue fracture is due to the progressive and localised structural damage that occurs when a material is subjected to repeated loading and unloading. Failure is characterised by the initiation and growth of one or more cracks. Cyclic loads generate microscopic inelastic damage at regions of local stress concentration such as fillet weld toes, lack-of-fusion flaws, flame-cut penetration zones and bolt holes. If sufficient inelastic damage accumulates, a small crack develops and propagates through the member perpendicular to the direction of the principal tensile stress.

In metals and alloys, fatigue failure starts with crystallographic defects and dislocations that eventually form persistent slip bands that initiate short cracks (Ratay, 2009). Damage is cumulative and the material does not recover when rested. The fracture surface is flat and exhibits crack growth bands, markings (ratchet marks) that identify the zone of crack initiation, and minimal plastic deformation. The crack grows until reaching a critical size that causes member failure. As fatigue is a progressive damage mechanism, the cracks can often be identified and repaired before significant damage occurs. Some second-hand gantries have been repaired multiple times and still perform satisfactorily.

As fatigue life does not increase proportionally with the tensile strength of steel, the high-grade steels commonly used in bridge construction equipment are more prone to fatigue as they are subjected to wider stress ranges. Fatigue failure mostly occurs at welded connections and is mostly due to inadequate design rather than inadequate welding. Non-welded connections and field splices may cause other types of structural distress and failure, but fatigue failures from these sources are rare.

Members subjected to cyclic loading should have connections designed to avoid displacement-induced fatigue, proper detailing of welds, and mandatory values of Charpy V-notch energy to ensure that fatigue cracking does not result in premature brittle fracture. Quality control prevents weld defects and discontinuities that are susceptible to crack initiation and growth, and frequent inspections during operations permit repairs when the fatigue cracks are still small.

Fatigue tests of steels use smooth or notched rotating or axial bar specimens to create S–N Wöhler curves that relate the stress range to fracture (S) to the number of load cycles (N). Fracture results from initiation and growth of fatigue cracks in the specimen. Actual structures are rarely flawless, and welding creates residual strains and microscopic flaws. This eliminates the crack initiation portion of the S–N curve, and material testing is therefore not very helpful in assessing the fatigue behaviour of complex welded structures.

Analysis of fatigue failure in steel structures with planar flaws due to fatigue cracks or weld discontinuities requires failure assessment diagram (FAD) analysis according to the principles of fracture mechanics. Weld porosity and slag inclusions are considered volumetric discontinuities, and are typically less severe on fracture initiation. FAD analysis is particularly useful in the evaluation of structures that can transition from ductile to brittle failure. FAD analysis is rarely performed in the design of bridge construction equipment, even though the cost of some of these machines could warrant design optimisation of welded connections.

Equipment that is properly designed, detailed, fabricated, assembled, operated and maintained is unlikely to sustain fatigue damage during construction of the first bridge. Assessing a second-hand machine is complicated by the absence of a documented working history in most cases. MSSs are rarely affected by fatigue issues due to the inertia provided by the casting cell and the stiffening action of the concrete span. Long and light trusses for balanced cantilever erection of precast segments may be affected by several types of fatigue distress, and not necessarily in relation to the demanding load spectrum used for the structural classification of the machine.

Wear and tear of the wheels and rails of travelling cranes may cause fatigue cracking at the fillet welds between the web and the top flange in the runway beams due to transverse bending. Fatigue cracking is also not unusual around vertical steps in the runways at the field splices due to vibration generated by wheel impacts in such poorly damped structures. If a winch trolley has two four-wheel bogies on either side and every impact can be equated to 10 full vibration cycles, one passage of the trolley over the splice generates 80 complete load cycles. Lateral bracing is also susceptible to a large number of load cycles due to wind-induced vibration.

The typical procedure used to assess fatigue distress is to make a full visual check of the machine and 5–10% non-destructive checks on the welds at the locations that, from experience, are considered to be the most likely sources of fatigue cracking. Such checks offer only a statistical level of confidence, which is not the same as a guarantee.

Weld hammering methods such as HiFIT have proven to be effective and inexpensive treatment means of increasing the service life of dynamically loaded welded steel structures. HiFIT can be applied to new and existing machines. A hardened pin hammers the weld toe with an adjustable frequency of 160–300 Hz. The weld toe is deformed plastically, and the induced residual compressive stress prevents crack propagation on the surface.

The initial low-cycle portion of the S–N curve is related to load conditions characterised by large strain variations and a small number of load cycles. The research around the Coffin–Manson relation and the fatigue ductility coefficient of steels is addressed to plastic strain ranges. Hangers for precast segments and casting cells, anchor bars of spreader beams for precast segments and spans, tie-downs of form travellers and lifting frames, the hoisting strand of strand jacks, and other such types of highly stressed tension members of bridge erection equipment may also experience local plasticity and low-cycle fatigue.

In the absence of guidance from the design standards on the design for low-cycle fatigue, it is common industry practice to limit the load in these devices to 50% of the tensile strength and the number of reuses to a few tens. The hoisting strand is reused only a few times because of the local damage generated by the anchor wedges.

10.9.4 Atmospheric corrosion

Damage due to oxidation and atmospheric corrosion is the most common failure mechanism in steel structures. Corrosion is an electrochemical process characterised by rusting and metal loss. Corrosion may lead to structural failure by section loss, and can promote the initiation and growth of fatigue cracks through section loss or pitting and grooving.

Corrosion often develops where dirt and debris are allowed to accumulate and cause active corrosion cells due to the presence of water. Salts, bird guano and other contaminants contribute to the electrochemical cell activity. Most corrosion conditions are time-dependent and can be controlled by inspection and corrected.

Rust is the product of reactions of iron and steel with oxygen and moisture. The typical remedy is to create a barrier between the corroding solution and the steel surface, usually by painting or galvanisation. Corrosion occurs when the protective barrier fails or is removed. The integrity of paint and galvanisation is controlled by visual inspection.

Accelerated and more aggressive corrosion processes may occur at the field splices due to water penetration between faying surfaces. If not sealed, field splices may collect and retain moisture. Electrochemical cells change water chemistry and promote the corrosion of steel plates and, in some cases, stress corrosion of high-strength bolts and stressed bars. Sometimes the corrosion products that form in these confined spaces exert considerable pressure, which may separate members, crack bolts and induce mechanical failure.

The field splices of bridge construction equipment are mostly designed as slip-critical connections. The faying surfaces can be sandblasted bare steel or protected with approved paints as designated by the RCSC (2009) or similar specifications. AISC Class A splices have a slip coefficient of 0.33 or higher and Class B splices have a slip coefficient of 0.50 or higher (AISC, 2006). This rating is the result of a test of the slip resistance of the bolted surface and a 30-day creep test.

Many organic coatings meet the Class A standard. Class B connections require special inorganic or organic zinc-rich primers or hot-dip galvanisation as per ASTM A123 (ASTM, 2012c). Both solutions provide a high slip coefficient and long-term corrosion protection, and solve the problems of rust in the connection and concerns about deterioration of the slip resistance.

The faying surface is abrasive blasted to the specified finish and primed with a single coat at the recommended dry film thickness. The surface is then masked to avoid application of the top-coat. After bolting the connection, the outer surfaces are cleaned of residues or bolting lubricants, touched up with a surface tolerant primer, and top-coated if desired. Finally, the perimeters of the faying surfaces are sealed with paint.

Improper storage of bridge construction equipment when not in use may also cause or accelerate corrosion. If stored outside, steel members should be kept well above the ground and should not collect water.

10.9.5 Stress corrosion

Stress corrosion is a crack-growth mechanism that occurs in high-strength steels subjected to high static tensile stress in the presence of a moist environment where hydrogen can be generated. Stress corrosion is characterised by cracking phenomena rather than section loss; it often occurs without warning and can result in catastrophic failures.

Stress corrosion typically occurs in high-grade alloy steels with $f_y \geq 700$ N/mm^2, but it may also occur in carbon and stainless steels with lower yield points. High-strength bolts and stressed bars of frontal splices may be affected if yield point or hardness is too high. The ASTM specifications for high-strength bolts limit hardness to avoid such failures.

Sealing the perimeter of lap plates and frontal splices with paint is recommended to avoid gathering of moisture and the formation of electrochemical cells. The exposed surfaces of bolts, nuts and stressed bars should also be painted. Galvanisation protects bolts and bars with a more reactive material such as zinc or magnesium, which corrodes and generates microscopic electric currents that protect the steel surfaces.

10.10. Permanent connections

Material-related failures of steel structures mostly occur at welded connections and field splices, and all types of bridge construction equipment include multiple connections of both types.

Subassemblies of transportable dimensions are often welded or assembled in the workshop to simplify and accelerate site assembly. Diagonals and verticals of triangular trusses are shipped in closed 40 ft containers and spliced to long chord modules. Workshop-welded braced A-frames with pinned end splices are often used for the diagonals, in combination with modules of paired bottom chords connected by welded braces. The A-frames typically have the same design throughout the truss for faster and simpler assembly.

Complex mechanical components such as articulated launch saddles, hoist crabs and friction launchers are assembled and tested in the workshop and shipped ready for operations. PPU, hydraulic pumps, heat exchangers and control systems are mounted on modular platforms that are commissioned in specialised facilities, shipped to the site, and applied to the machine in one lift.

The chords of trusses comprise H-shapes or box sections made with side channels, beams or plates. Open sections facilitate inspection, maintenance and field splicing. Wide flange shapes are used frequently, even though the difference in back-to-back dimensions for the same nominal size members with different weights requires fills at the field splices. Steel plates may be welded together to make I-girders, box sections and other arrangements necessary to achieve the required stiffness and cross-sectional area. High-grade steels provide a more favourable strength-to-density ratio and smaller welds. Hybrid steel construction is rarely used, as it may complicate identification and cause assembly errors.

H-shapes, square tubes and round tubes are used for the diagonals and verticals of triangular and square trusses and for the self-launching frames of form travellers and forming trolleys. First-generation narrow rectangular trusses use L- or I-shapes welded into framed subassemblies to simplify field splicing. Lateral bracing includes crosses or K-frames, connections designed to minimise displacement-induced fatigue, and sufficient flexural stiffness to resist vibration stresses. Diaphragms connected to the chords at the same locations of lateral bracing distribute torsion, control distortion and provide transverse stiffness to the truss.

Permanent connections are created during fabrication and are almost always welded. Field splices are based on the use of bolts, pins and stressed bars. Both types of connections are designed with the maximum possible symmetry to streamline the load path and stress distribution throughout the connection. Design should be as simple as possible, and should take onto consideration the requirements of fabrication, inspection and maintenance. The same type of connection should be used throughout the truss, instead of combining different schemes.

Distortion due to overloading, buckling of members or accidents is primarily resisted by the connections. Therefore, when possible, the connections are designed to develop the full capacity of the connecting members. This also facilitates reusing the machine in different assembly configurations.

Welded connections may be affected by structural discontinuities, residual stresses and load eccentricity due to geometry imperfections. As local bending peaks at the connections, these are the zones where a steel truss is most prone to failure. Most fabrication standards recognise this fact by placing more emphasis on fabrication and inspection of welded connections than on other parts of the structure.

Welded connections typically fail because the welds contain discontinuities that initiate brittle fracture or fatigue cracking. Rounded volumetric discontinuities such as inclusions or porosity rarely initiate failure. Linear discontinuities such as cold cracks and lack of fusion are generally more severe.

Hydrogen-induced cold cracks are the most common type of linear discontinuity in structural welds. Cold cracks form hours after welding in the heat-affected zone of the base plate and are, therefore, difficult to detect. The susceptibility of steel to hydrogen-induced cold cracking is expressed in terms of carbon equivalent C_{eq}:

$$C_{eq} = C + \frac{Mn}{6} + \frac{Cr + Mo + V}{5} + \frac{Ni + Cu}{15}$$

where C, Mn, Cr, etc. are the weight percentages of the alloy elements in the steel. A steel with $C_{eq} \geq 0.45\%$ is susceptible to cold cracking and requires preheat prior to welding. Preheat starts at 60–70°C, but may exceed 150°C for thick plates and steels with $C_{eq} \geq 0.70\%$. AWS D1.1 (AWS, 2010) lists the recommended minimum preheating for most structural steels. Joints free of moisture and grease, low-hydrogen electrodes and fluxes, and proper care of electrodes and consumables are indispensable preventive measures.

Another possible cause of brittle fracture and fatigue cracking is lack of fusion, which leaves a crack-like non-fused linear discontinuity in the joint. Lack of fusion is often due to the use of backing bars to form the root of the weld in complete penetration joints welded from above, especially if the bars are not removed and the root repaired. Ceramic backing plates fixed with magnetic clamps have been used successfully for control of lack of fusion. Geometry irregularities on the surface of the weld seam may also initiate fatigue cracking over time.

Elastoplastic finite-element analysis is sometimes used to investigate the failure mechanisms of connections. Factors such as flaws and discontinuities in the welds, residual stresses and their effects on the mechanical and metallurgical properties of the base material, geometry imperfections due to thermal distortion, thermal shrinkage, chamfering tolerances, surface preparation and misalignment of connecting members, and residual stresses and deformations due to the use of the machine, are difficult or impossible to model and may have a marked influence on the ultimate behaviour of the connection.

In relation to the welding process, connections may comprise fillet welds, partial penetration welds, and complete penetration welds when the full capacity of the connected members must be attained. The cost of welding increases in the same order. Combining different types of welds is not recommended due to the different stiffness of welds.

The technical specifications of the machine define the level of control of welding. Full-length control is often specified for field welds and critical welds the failure of which can lead to collapse. Radiography and ultrasound imaging have pros and cons; ultrasound imaging is becoming the most commonly used analysis as the technology progresses. Fillet welds cannot be radiographed or ultrasounded, and are checked using magnetic particle or dye-penetration test procedures, or certified by the quality control procedures of the manufacturer and the qualifications of the welders.

10.11. Field splices

The design of the field splices influences the structural performance of bridge construction equipment and the cost of site assembly and decommissioning.

Figure 10.9. Welded A-frames with pinned field splices to long chord modules (Photo: HNTB)

The splices in the chords of triangular or square trusses may be designed with through pins to simplify site assembly. Some design standards discourage the use of pins in the presence of stress reversal, while others specify the minimum thickness and the dimensions of the support plates in relation to the load range and contact pressure. The chords are primary load-carrying members of a truss, and mid-panel rotational releases of degrees of freedom cause flexural peaks at the adjacent nodes. Local bending generated by the launch bogies is higher than the local bending generated by travelling cranes, and the mid-panel pins are therefore located in the top chords (see Figure 5.13). The pinned splices are designed with $2+1$ or $3+2$ support plates to minimise bending in the pins. When the truss modules are of transportable dimensions, the diagonals may be workshop welded to the chords.

When the field splices of diagonals to continuous chords are pinned, the diagonals of triangular trusses are paired into welded A-frames that require only two pins at the bottom and one at the top for assembly (Figure 10.9). Welded lateral braces and cross diaphragms are used to frame long bottom chord modules for fast assembly with end pins or frontal splices. Maximised workshop welding simplifies and accelerates site assembly, decreases the final cost of equipment and facilitates future reuses. All pins are equipped with split pins that prevent expulsion due to vibration.

Pinned splices are more flexible than slip-critical connections due to minor transfer of bending throughout the joint and the need for the pin to reach a new contact at every load reversal. Local bending generated by the eccentricity of the pin from the geometric nodes of the panel can demand heavier chord members, and the local stresses can reduce the fatigue life of the connection. Some gantries have indeed been scrapped because fatigue cracking was too extensive to make repair economical.

The pins are typically turned from high-grade steels. According to FEM 1.001 (FEM, 1987) the allowed SLS axial stress for special steels with $f_y/f_t > 0.7$ is

$$\sigma_{all} = \frac{f_y + f_t}{f_{y.510} + f_{t.510}} \sigma_{all.510}$$

where $\sigma_{all.510}$ is the allowed SLS axial stress for Fe510 steel ($f_{y.510} = 355$ N/mm^2 and $f_{t.510} = 510$ N/mm^2). For example, the allowed axial stress for 39NiCrMo3 steel ($f_y = 650$ N/mm^2 and $f_t = 850$ N/mm^2) is $\sigma_{all} = 417$ N/mm^2, it being $\sigma_{all.510} = 240$ N/mm^2. The allowed stress is 67.6% of the yield point for Fe510 steel and 64.2% for 39NiCrMo3 steel, to reflect the lower ductility of high-grade steels in case of accidental overloading. The SLS limit stress in the mechanical components is calculated by applying resistance factors to the tensile strength of steel rather than to the yield point. The SLS resistance factors are $\phi_{LC1,2} = 0.455$ for load conditions I and II and $\phi_{LC3} = 0.556$ for load condition III.

Primary bolted splices subjected to load reversal are typically designed for friction. Noise and vibration due to joint slippage cause concerns in the operating crews and impacts on highly stressed machines. Calibrated bolts for bearing splices require tight tolerances that complicate fabrication and site assembly, joint slippage complicates geometry adjustment during operations, and progressive joint deterioration discourages from multiple reuses of the machine. The geometric effects of joint slippage may be calculated using nonlinear analysis, although the reliability of results depends on the entity of loading.

Slip-critical connections with lap plates at webs and flanges are frequently used in the box girders of portal carriers, span launchers and underbridges. These connections were also used for the trusses in the past, but have been progressively abandoned due to the labour demand of site assembly and decommissioning. The stiffness of slip-critical connections is similar to the stiffness of welded connections, and these field splices are therefore suitable for load reversal, impacts, vibrations and applications that require a reliable geometry of the machine. Compared with pinned connections, the slip-critical connections also have an improved fatigue life.

Bolts with strength higher than 8.8 or 10.9 are rarely used due to their fragility. Oversized holes may be specified to facilitate site assembly. The use of bolts with different diameter in the same splice is not recommended, and the use of direct tension indicators is recommended to facilitate inspection and control.

Most design standards specify the shear capacity of a controlled-tightening friction bolt in relation to the number of friction planes in the splice, the tightening force in the bolt and the friction coefficient of the faying surfaces, which is related to their preparation. According to FEM 1.001 (FEM, 1987) the resistance factors for slip-critical connections are $\phi_{LC1} = 0.667$ for load condition I, $\phi_{LC2} = 0.752$ for load condition II, and $\phi_{LC3} = 0.909$ for load condition III.

Bolted splices are critical elements in machines that are assembled and decommissioned many times. Repairs are easier in the workshop, and the field splices of second-hand machines are therefore inspected prior to shipping. Buckled plates are reinforced or replaced, and the faying

Figure 10.10. Slip-critical frontal connection with stressed bars for bending

surfaces are re-sandblasted. Oval-shaped holes indicate joint slippage in previous usage. Splice inspection can reveal old defects in many freshly painted second-hand machines.

The labour demand of a great number of field splices, repeated load reversals, and the possibility of impacts and vibrations suggest that the primary field splices of bridge construction equipment be designed as mechanical connections rather than conventional slip-critical or pinned connections. The most reliable and cost-effective splice arrangement for I-girders is showed in Figure 10.10.

Controlled-tightening bolts at the end plates transfer shear by friction, and longitudinal stressed bars anchored to the flanges transfer bending. The bars are housed in boxes welded to the web and flange, and the absence of lateral contact between the bar and containment box avoids the transfer of shear forces. The welds of the box to the flange and web transfer longitudinal axial force from the flange to the anchor nut of the bar.

After tightening the shear bolts, the bars are inserted into the anchor boxes for symmetrical tensioning. The absence of bolts and lap plates at the bottom flange is particularly advantageous when friction launchers or PTFE skids are used for launching. The end plate acts as a vertical web stiffener, longitudinal web stiffeners are welded to the end plate and spliced without bolts, and the shear force applied to each bolt is independent of its distance from the centre of rotation of the splice. The stiffness of this splice arrangement is similar to the stiffness of a welded connection.

Frontal splices are also used in the columns of the pier towers. The behaviour of these connections depends on the tension-to-compression ratio. When the column is always compressed and the contact with the end plate is milled to bear, the end weld and the frontal bolts are subjected to minimal stresses. In case of load reversal, the net tension is low in most cases, the milled contact is designed for transfer of compression, and the weld and frontal bolts are designed for shear, tension and prying action. However, if the contact is not uniform and the weld cracks in compression, the column can detach from the end plate at load reversal. The robustness of this type of connection depends on the reliability of the quality-control processes implemented during fabrication and on the tension-to-compression ratio: if tension is high and the reliability of quality control is uncertain, vertical stressed bars housed within boxes welded to the columns are used to transfer tension from column to column bypassing the end plates, and the weld and frontal bolts are designed for shear.

Not many standards designate the design of frontal splices that combine large-diameter stressed bars and shear bolts. The bars are stressed by torque or with hollow plunger cylinders; the use of aluminium cylinders is recommended for easier handling, and double-acting operation permits fast retraction. For torque tensioning, FEM 1.001 (FEM, 1987) defines the tangential stress τ_B generated by torque as

$$\tau_B = \frac{2d_{B.min}\sigma_B}{d_B}\left(\frac{p_T}{\pi d_{B.min}} + 1.155\mu_T\right)$$

where d_B is the diameter of the bar, $d_{B.min}$ is the minimum diameter, σ_B is the axial stress generated by tightening, p_T is the pitch of the thread, and μ_T is the friction coefficient of the thread. The ideal stress in the bar at tightening is assessed using the following equation:

$$\sigma_{id} = \sqrt{\sigma_B^2 + 3\tau_B^2} \leq 0.8f_y$$

Initial bar stressing must be accurate, the bars can be re-tensioned in case of need, and a load tolerance of $\pm 10\%$ can be easily maintained over time. A resistance factor $\phi = 0.9$ may therefore be used for the design of the splice. When the bars are tensioned using a hollow plunger cylinder and no torque is applied, the assessment condition becomes

$$\sigma_B \leq 0.8f_y$$

The dimensions of stressing cylinders and anchor nuts govern the geometry of the connection. The anchorages of multiple bars may be staggered longitudinally to reduce the stress concentration and to facilitate transfer of axial force from the flange plate to the anchor boxes. The prestressing force in the bars is designed to avoid joint decompression. FEM 1.001 (FEM, 1987) specifies not exceeding the yield strength of steel with load factors γ_1 and γ_2, and avoidance of joint decompression with load factors γ_2 and γ_3 and resistance factor $\phi = 0.9$. The first load factor depends on the conditions of the faying surfaces and for flattened end plates it can be taken as $\gamma_1 = 1$. The second load factor corresponds to the margin from yielding, and is $\gamma_2 = 1.50$ for load condition I, $\gamma_2 = 1.33$ for load condition II and $\gamma_2 = 1.10$ for load condition III. The third factor determines the margin from joint decompression, and is $\gamma_3 = 1.30$ for load condition I and $\gamma_3 = 1.00$ for load conditions II and III.

The load factors γ_2 and γ_3 are applied to the least favourable condition in relation to the time-dependent relaxation of tension steel and conservative load distribution models between adjacent girders. For the yield checks, an equivalent cross-sectional area of the assembly is determined using the following equation:

$$A_{A.eq} = \frac{\pi}{4}\left[\left(d_{CS} + \frac{L_{tot}}{10}\right)^2 - d_H^2\right]$$

where d_{CS} is the diameter of the contact surface of the bar anchor nut, L_{tot} is the total nut-to-nut length of the assembly, and d_H is the diameter of the hole lodging the bar. The equivalent cross-sectional area $A_{A.eq}$ is used to compute the shortening ΔL_A of the assembly under the tightening force. Bar elongation ΔL_B at tensioning is calculated taking into account the threaded end portions of the bar (net area) and the central non-threaded portion. It is then:

$$\delta_B = \frac{\Delta L_A}{\Delta L_A + \Delta L_B}$$

The net area A_B of the bar is determined for $d_{B.min}$ and the SLS axial stress limit in the bar is

$$\sigma_{B.max} = \sqrt{f_y^2 - 3\tau_B^2}$$

Setting F_B as the longitudinal tensile force that the bending moment in the splice applies to the bar, yielding is checked using the equation

$$\frac{F_B}{A_B} \leq \frac{\sigma_{B.max} - \sigma_B}{\gamma_1 \gamma_2 \delta_B}$$

and detachment is checked using

$$\frac{F_B}{A_B} \leq \frac{\phi\sigma_B}{\gamma_2 \gamma_3 (1 - \delta_B)}$$

Both checks are performed for every load condition.

Stressed bars are also used for splicing of truss chords. Transfer of axial tension governs the design of chord splices. Bar anchor pipes are welded symmetrically to the flanges and web in order to streamline the stress flow to the bars and to increase the robustness and redundancy of the splice (Figure 10.11). The use of several bars with staggered anchor nuts simplifies welding and often permits the use of commercially available prestressing bars for the connection. Local bending and shear due to travelling cranes and transfer of support reactions from the launch bogies are checked using the criteria discussed above. The end plates are often equipped with spindles that act as shear-keys and help keep the chords aligned during assembly to avoid involuntary eccentricities. Shear bolts are rarely used at the end plates in the chords.

Figure 10.11. Frontal splice with staggered stressed bars (Photo: VSL)

Frontal splices with stressed bars and shear bolts or spindles are structurally efficient and quick to assemble. Because of the high capacity and the high resistance factor of the bars, however, these splices must be designed and tightened with care.

1   The distribution of bending and shear between chords and braced I-girders is often uneven due to geometry tolerances. The splices are therefore designed allowing for the possibility of accidental overloading.

2   The use of two bars at every flange is a non-redundant detail. Failure of one bar may lead to collapse due to local bending generated by load eccentricity and the absence of alternative load paths and stress redistribution. Bars, anchor boxes and welds are oversized to allow plasticity and local damage at bar failure but to avoid collapse. Wide flange plates can accommodate two anchor boxes on either side of the web for enhanced redundancy. Narrow flanges accommodate only one box on either side, but a second bar can be housed within a round pipe welded to the edge of the flange or a lower anchor box welded to the web.

3   Stiff anchor boxes or pipes are used to protect the bars from impacts. Thick anchor plates and short-pitch threads reduce the prestress losses due to anchor set. The central portion of the bars is not threaded to save machining costs and to increase the shear capacity of the bar in case of accidental shear transfer due to splice slippage. Commercially available prestressing bars are often adequate for chord splicing.

4   The use of calibrated bolts for bearing shear transfer is not recommended, as splice slippage at load reversal could transfer shear forces to the bars. Symmetrical hole patterns, the use of large-diameter bolts, the conservative friction coefficients specified by the design standards, and the additional compression generated by the bars typically lead to field

splices with a low D/C ratio for shear and large margins from slippage. The bolts are tightened concentrically from the centre of the splice, and the bars are inserted into the anchor boxes after completion of bolt tightening.

5   Accurate flattening and sandblasting of the faying surfaces are recommended, along with the use of direct tension indicators on the bolts. Sealing the perimeter of the splice with paint at the end of tightening is also recommended, as corrosion of contact surfaces can reduce friction.

Frontal splices are also used in the box girders. The bars are anchored to the central flange panels between webs, and the end shear plates are located within the box cell for easier assembly, inspection and final dismantling. Inner end plates are easily braced to stiffen the cross-section and to control distortion. The bars may be anchored to both sides of the flanges because the anchor boxes do not interfere with the launch saddles.

10.12. Repairs during operations

Repairing bridge construction equipment during operations is a potential cause of structural failure. Nevertheless, repairs may be necessary to correct corrosion or fatigue cracking, to replace damaged or deformed members, to improve access, or to solve geometry conflicts.

Modifications may also be necessary for multiple reasons. Hangers for pipes or cables and weld plugging of misdrilled holes are typical examples. Often, these modifications are not subjected to any design process, are made at incorrect locations, are executed carelessly, or are of insufficient strength. Replacement bolts are rarely tightened properly.

Repair welds made to a stiff assembled structure require greater care and skill than workshop welds made on flexible members that can be positioned so as to facilitate welding. They also require higher preheating, which typically damages paint. In many cases, poorly executed repair welds cause fatigue cracking due to stress concentration or weld cracking.

Repairs should preserve useful information about the damaged element. Welding to fracture surfaces destroys the surface and prevents assessment of the cause of failure. Bolted plates are often applied to strengthen cracked members; they are easy to install and the drill geometry can be selected to maximise the net area.

Small fatigue cracks in the flange and web plates are arrested by drilling holes that intercept the tip of the crack. A part of the fracture surface may be removed to investigate the cause of cracking. Prior to removal, the size, location and orientation of the crack should be registered and the crack photographed. During removal it is necessary to ensure that the crack surfaces do not come into contact, as this would damage the fracture surface. Major fatigue cracks are repaired by grinding, welding and dressing flush. When fatigue cracks are discovered in a machine in operation, additional checks of other similar locations on the machine are strongly recommended.

Because of the high structural efficiency of most types of bridge construction equipment, removal of payload is an effective means to reduce the stresses in the regions to be repaired. Additional stress reduction may be achieved by suspending loads at proper locations or by shifting the trusses to locations that minimise bending or shear as needed. Replacement of buckled members may require temporary shoring or bracing to prevent further damage.

10.13. Programmable control systems

A PLC is a digital computer used for automation of electromechanical processes. PLCs are used in many industries and machines. Unlike general-purpose computers, PLCs are designed for multiple input and output (I/O) arrangements, extended temperature ranges, immunity to electrical noise, and resistance to dust, moisture, vibrations and impact.

The functionality of new-generation PLCs includes sequential relay control, motion control, process control, distributed control systems and networking. The data handling, storage, processing power and communication capabilities are similar to those of desktop computers. The operating systems for PLCs are designed for deterministic logic execution and are more stable than the operating systems for personal computers in order to respond to changes in logic state or input status with extreme consistency in timing.

Programmable logic relays (PLRs) are branded by the manufacturers of larger PLCs to fill out their low-end product range. PLRs have limited networking capability, but their cost is much lower than a PLC and they still offer robust design and deterministic logic execution. PLRs are rarely used for the process controls of bridge erection machines due to the small number of I/O points: 8–12 digital inputs, 0–2 analogic inputs and 4–8 digital outputs, which is rarely enough.

A PLC panel consists of a power supply, a controller and an expandable network of modular I/O units that connect the PLC to sensors (input) and actuators (output). A PLC reads digital signals and analogic process variables. Digital signals from push buttons and limit switches behave as binary switches and are sent by means of ranges of voltage or current. Analogic signals from displacement, pressure, flow and temperature transducers have a range of values between zero and full scale, and these are counted by the PLC. Count accuracy depends on the sensitivity of the sensors and the number of bits available to store the data; as PLCs typically use 16-bit processors, the range of integer values is 65.5k. Analogic signals use voltage or current with a magnitude proportional to the value of the process signal. Current inputs are less sensitive to electrical noise (i.e. from welders or electric motor starts) than are voltage inputs. On the actuator side, a PLC operates electric motors, hydraulic cylinders, magnetic relays, solenoid valves or analogic outputs.

New-generation PLCs can be programed in a variety of ways, ranging from relay-derived ladder logic and special dialects of BASIC and C++ to Windows applications on personal computers. Most software provides debugging functions that highlight portions of the logic to show current status during operations or to simulate the process. The program is transferred to the PLC via Ethernet and stored in a memory that is backed up by a battery or in a non-volatile memory.

A PLC programme is executed repeatedly as long as the controlled system is running. The status of physical input points is copied to an area of memory accessible to the processor, sometimes called the 'I/O image table'. The program is then run from its first instruction down to the last, and the I/O image table is updated with the status of outputs. The scan time of fast processors is of the order of a few milliseconds, which ensures immediate response of the PLC to different input conditions.

REFERENCES

AISC (American Institute of Steel Construction) (2006) *Base Plate and Anchor Rod Design*. AISC, Chicago, IL, USA.

ASTM (American Society for Testing and Materials) (2012a) ASTM E23 – 12c Standard test methods for notched bar impact testing of metallic materials. ASTM International, West Conshohocken, PA, USA.

ASTM (2012b) A673 – 07 Standard specification for sampling procedure for impact. Testing of structural steel. ASTM International, West Conshohocken, PA, USA.

ASTM (2012c) ASTM A123/A123M – 12 Standard specification for zinc (hot-dip galvanized) coatings on iron and steel products. ASTM International, West Conshohocken, PA, USA.

AWS (American Welding Society) (2010) AWS D1.1 Structural welding code – steel. AWS, Miami, FL, USA.

BSI (1990) BS 4360: 1990 Specification for weldable structural steels. BSI, London.

BSI (1991) BS 5400: Part 3: 1982 Steel, concrete and composite bridges. Part 3. Code of practice for design of steel bridges. BSI, London.

BSI (2005) BS EN 1993-1-1: 2005. Eurocode 3. Design of steel structures – Part 1.1: General rules and rules for buildings. BSI, London.

CNR (Consiglio Nazionale delle Ricerche) (1988) CNR 10011. Strutture di acciaio. Istruzioni per il calcolo, l'esecuzione, il collaudo e la manutenzione. CNR, Rome, Italy.

CSI (2012) *SAP 2000 version 15. Analysis Reference Manual*. CSI, Berkeley, CA, USA.

FEM (Fédération Européenne de la Manutention/European Federation of Materials Handling) (1987) FEM 1.001. Règles pour le calcul des appareils de levage. FEM, Brussels, Belgium.

Ratay R (2009) *Forensic Structural Engineering Handbook*, 2nd edn. McGraw Hill Professional, New York, NY, USA.

RCSC (Research Council on Structural Connections) (2009) *Specifications for Structural Joints Using High-strength Bolts*. RCSC. Chicago, IL, USA.

Rosignoli M (1998) Misplacement of launching bearings in PC launched bridges. *ASCE Journal of Bridge Engineering* **3(4)**: 170–176.

Rosignoli M (2002) *Bridge Launching*. Thomas Telford, London.

Rosignoli M (2007) Robustness and stability of launching gantries and movable shuttering systems – lessons learned. *Structural Engineering International* **17(2)**: 133–140.

Rosignoli M (2010) Self-launching erection machines for precast concrete bridges. *PCI Journal* **2010(Winter)**: 36–57.

Rosignoli M (2011) Bridge erection machines. In *Encyclopedia of Life Support Systems*, Chapter 6.37.40, 1–56. United Nations Educational, Scientific and Cultural Organization (UNESCO), Paris, France.

Starossek U (2006) Progressive collapse of structures: nomenclature and procedures. *Structural Engineering International* **16**: 113–117.

Starossek U (2009) *Progressive Collapse of Structures*. Thomas Telford, London.

Bridge Construction Equipment
ISBN 978-0-7277-5808-8

ICE Publishing: All rights reserved
http://dx.doi.org/10.1680/bce.58088.377

Chapter 11
From procurement to safe operations

11.1. Contractual environment

Several engineering professionals interact at different levels and with different responsibilities during the design and construction of bridges.

- The engineer of record (EOR), also referred to as the designer or the design engineer, is an engineering professional engaged by the owner for the design of the project. The EOR coordinates the activity of multiple engineering disciplines, such as highway, structures, geotechnical work and maintenance of traffic. The EOR maintains close contact with the owner during project development, and represents the interests of the owner.
- The erection engineer is an engineering professional engaged by the bridge contractor for the design and plan production of temporary portions or special items of the project that are not fully detailed in the plans. With regard to the use of bridge construction equipment, the erection engineer is responsible for the verification of the final structure in relation to the construction method selected by the bridge contractor, and for the design of required changes to the final structure. This applies to the final structure itself, not to the bridge construction equipment. The above changes include changes of support locations or cross-section member sizes, relocation of construction joints, changes in sequence and application of prestressing, and changes of conditions during casting, lifting, delivery and placement of structural elements. In the event of cracking or damage, the erection engineer is responsible for inspection, development of repair proposals, and verification of the repaired structure. On many design-build projects the erection engineer and the EOR are the same person, and the EOR is removed from the role of advising, working for and protecting the interests of the owner. In this case the owner may assign review of submittals and technical advice to a general engineering consultant (GEC) firm.
- The equipment designer is an engineering professional engaged to undertake the design of bridge construction equipment and to write operation manuals and checklists for its operations. In the past, the equipment designer was engaged by the bridge contractor, and the equipment was fabricated by local workshops with some in-house experience of mechanical systems. Nowadays the equipment designer is typically engaged by the equipment manufacturer to facilitate the fabrication and reuse of equipment within the manufacturer's organisation, and the equipment is returned to the manufacturer on completion of the project. As far as the owner and EOR view the use of construction equipment, the equipment designer is responsible for everything related to equipment operation during the bridge construction. The equipment designer is rarely the erection engineer; he or she is generally a mechanical engineer or a specialised structural engineer, perhaps with little knowledge of the bridge and its serviceability requirements.

- The technology consultant is an engineering professional engaged by the bridge contractor to supervise the construction means and methods to be used in the project. The technology consultant maintains close contact with the bridge contractor throughout the project, represents the interests of the bridge contractor, and provides advice throughout the process. Duties include close contact and liaison with the bridge contractor; establishing performance requirements, technical specifications, design criteria and quality standards for design, fabrication, assembly, commissioning and operation of construction equipment; assisting the bridge contractor during the contract negotiation with the manufacturer; reviewing and approving the manufacturer's proposals and submittals; responding to the manufacturer's requests for information; coordinating multiple equipment manufacturers used on the project on behalf of the bridge contractor; performing independent checking of equipment design; supervising fabrication, assembly, commissioning and load testing of equipment; and advising the bridge contractor on matters of technical compliance and QA/QC.
- The construction engineers and inspectors (CEI) team, also referred to as the resident engineer or site representative, is an engineering firm engaged by the owner to represent the owner's interests at the project site and to ensure compliance with the requirements of the construction contract. Duties and responsibilities of the CEI are representing the owner, ensuring contract compliance, establishing and performing QA to check the contractor's QC, measuring the work done, monitoring compliance with incentives and disincentives, providing the owner with regular progress reports, approving progress payments, resolving issues, responding to submittals, expediting requests for information, issuing and monitoring change orders or variations, and keeping diaries and records.

The owner communicates with the CEI, the EOR and the prime contractor. The CEI is a vital element of communication between the parties. Although the prime contractor is ultimately responsible for coordinating construction, the CEI is often crucial to the success of the project – for instance, by anticipating forthcoming needs, activities, events and making sure that parties not legally bound under the terms of the contract communicate and are informed as necessary.

The prime contractor communicates with the bridge contractor, who is supported by the erection engineer, the technology consultant and multiple equipment manufacturers with their own equipment designers. The erection engineer analyses the temporary conditions of the bridge during construction and explores alternative erection scenarios and means and methods. The erection engineer may also provide the engineering for fabrication and submittal requirements associated with bridge construction. The design changes proposed by the erection engineer are typically subjected to the approval of the EOR or GEC.

Mechanised bridge construction brings new engineering disciplines into the design of transportation infrastructure. For this reason, the bridge contractor and erection engineer often need guidance in dealing with bridge erection machines. These coordination activities are the responsibility of the technology consultant. The erection engineer and technology consultant communicate with the EOR or GEC, using shop drawings and erection manuals to document the stress history of the bridge during construction and to fulfil the contractual requirements of having licensed professionals responsible for providing a safe workplace.

11.2.   Request for Proposal

Proper selection of bridge construction equipment is a key factor in planning a project and estimating a tender (Homsi, 2012). The decision is driven by a large number of variables that are often unique to the project.

- Project schedule and construction duration often drive the choice of the construction method. If time were the only constraint, however, bridge contractors would move from simple, low investment, labour intensive solutions toward prefabrication and fast erection with reduced on-site labour. The productivity of traditional scaffolding solutions, incremental launching, span-by-span casting with an MSS, precast segmental construction and full-span precasting increases from some weeks to 1 day per span. The fact that all these methods are used indicates that the decisions about the means and methods to use are rarely made based on time alone.
- Bridge geometry has a major impact on the construction method. It includes bridge length, span length, pier height, plan curvature, longitudinal gradient, crossfall, and consistency and repeatability of these characteristics throughout the project. Construction methods and break-even points between different alternatives are also different for one bridge or two adjacent bridges.
- Access and site conditions are among the most important factors. When working over land, environmental issues and interference with traffic, railways, utilities and businesses are major concerns. When working over water, water conditions (depth, tides, ice), permits, feasibility and cost of access trestles, and availability of barges, tugboats and docking facilities lead the decision. For in-place casting, additional factors include soil conditions, crane access, availability of ready-mix concrete, weather conditions and climate. The cost-effectiveness of segmental precasting is influenced by the availability of an existing precasting facility or suitable land to install a new one, the segment storage area, the distance of the precasting facility from the erection site, and the need for double handling of segments. The delivery route often dictates the maximum size and weight of the segments.
- Experience of the project team is another major factor. The more experienced the bridge contractor, the more open the evaluation of sophisticated or innovative construction methods. Availability of skilled labour at the project location is another factor to be considered prior to selecting complex construction methods.
- Availability of second-hand equipment and competition between manufacturers influence the cost and delivery schedule of special equipment and, ultimately, the selection process. Similarly, if a bridge contractor has retained equipment from previous projects, reusing that equipment will often be more efficient than using different means and methods.

The procurement process for bridge construction equipment is initiated by the bridge contractor. The bridge contractor collects the information necessary for the manufacturers to prepare a proposal as a set of documents called the Request for Proposal (RFP). The contractor sends the RFP to multiple manufacturers and, eventually, receives their proposals. Meetings are held between the parties during the preparation of the proposals to clarify issues and acquire additional information.

The RFP typically requires qualification of bidders – that is, documentation on the prospective bidder's financial capability, performance bonding ability, insurance coverage, a list of similar project experience, adequate supervision and the ability to meet the project schedule (CNC, 2007).

The RFP should include all the information necessary for the manufacturers to prepare their proposals, as this would minimise the requests for information and result in prompt, uniform and comparable offers. In reality the bridge contractor often releases additional information during the preparation of the proposals in response to requests for information from the manufacturers. After receiving and scoring the proposals, the contractor selects the equipment manufacturer. The contractor and manufacturer negotiate the details of the contract, make the necessary verifications with the EOR or GEC, and execute the contract.

The RFP is written by personnel of the bridge contractor, often with the support of the technology consultant. Several contractors have well-staffed technical offices, specialise in a few construction methods, and do not need external consultants. General contractors typically hire a technology consultant based on the technical features of the project so as to minimise the long-term overhead costs and to get more specialised assistance.

The RFP describes the project and the technical specifications (TS) for special erection equipment as resulting from the contractual obligations and the quality management policy of the bridge contractor. The RFP also includes the performance requirements (PR) identified for the special equipment in the pre-award bidding phase. In many cases, the PR and TS are so tightly interrelated that they are collected into a single document called the performance requirements and technical specifications (PR&TS). General features of the project and PR&TS for the special erection equipment are typically sufficient for a manufacturer to prepare a proposal.

The manufacturers may provide unsolicited information in the proposal and may suggest modifications to the bridge design to enhance compatibility with second-hand equipment. Recycling time-tested equipment and using the deck design from a previous project offer several advantages, and all suggestions should be encouraged and appreciated. Critical skill and deep knowledge of the industry, however, are indispensable for a successful negotiation.

Many manufacturers have high reputation, financial solidity and a well-documented background, but backlog and market conditions change day by day, and financial capability, performance bonding ability and insurance coverage should be scrutinised prior to moving a negotiation forward. Some manufacturers are financially connected to bridge contractors or are subcontractors themselves, and will probably try to widen the scope of work. This is part of the business and may bring advantages to both parties; however, losses on equipment lease or favourable buy-back conditions may be recovered with gains on the erection cost. A low offer for the equipment may be part of the strategy and should not unbalance competition with manufacturers that do not provide construction services. In most cases, in other words, information provided by the manufacturers of special equipment will be a blend of marketing and engineering.

There are many manufacturers and subcontractors on the market, and specific situations may bring unexpected opportunities that become part of the negotiation. Machines may have been purchased, never moved from the storage yard, and ultimately returned to the manufacturer for the owner to claim immobility costs. Machines may have fallen from the deck after casting the last span, and the cost reimbursed by the insurance carrier. Mixing new and used components prior to repainting may be as tempting as poor adherence to QC procedures. Negotiation, in other words, is sometimes flexible.

Negotiation is carried out between two parties. The bridge contractor and technology consultant sit on one side of the table with the manufacturer and equipment designer on the other, and hands are supposedly shaken in between. Both parties provide information during the progress of negotiation. The technology consultant facilitates the discussion and registers the technical agreements made between the parties on advanced drafts of the PR&TS, which are distributed to the parties in preparation of another meeting. The technology consultant and erection engineer may need to consult third parties, the EOR or the GEC during the process.

At the end of the negotiation, the PR&TS describes the aspects of the bridge construction equipment that have been discussed and agreed upon by the parties. Components not included in the PR&TS will often be considered as change orders or additional orders. The language may also be distorted by the more experienced party to prepare the ground for change orders or claims for extra costs in the future. Business is business.

The final version of the PR&TS includes schematic drawings of the machine and of the bridge, and becomes part of the contractual documentation of the machine. Basically, the contract between the bridge contractor and the equipment manufacturer includes two documents: the final version of the PR&TS and the Legal and Payment Conditions.

11.2.1 General description of the bridge project

Several features of the bridge project affect the design and operations of construction equipment. This influence is obvious in the machines for in-place casting, but launching gantries, lifting frames, portal carriers, tyre trolleys and span launches are also affected in many ways. This section of the RFP includes contractual and technical information on the project, such as the following:

- Time-schedule.
- Bridge geometry: plan and vertical alignment, minimum plan and vertical radii, maximum gradient and crossfall, span sequence, irregular geometries (varying-width areas, bifurcations, ramps), irregular supports (C-piers, L-piers, straddle bents, hammerheads of LRT aerial stations), vertical clearance above and below the deck, horizontal clearance from fixed obstacles.
- Deck articulation and erection direction: piers with sliding bearings, continuous piers, expansion joints at the end of continuous frames, saddle joints in the spans, longitudinal fixity points, thermal displacements during erection.
- Longitudinal section and cross-sections of typical and atypical spans. In an MSS, bridge geometry governs shape and modularity of inner and outer forms, the geometry of hangers and bottom frame of the casting cell, the location of the main girders, and the design of pier brackets and support crossbeams. In a precast segmental bridge, segment geometry influences casting cells, manipulators, lifting beams, spreader beams, suspension frames and the hoist capacity of the winch trolleys. In both cases the weight and length of the span are the main design parameters for the self-launching frame of the machine. The influence of deck geometry becomes more and more pronounced with the increasing level of specialisation of equipment. Heavy gantries for macro-segmental construction, portal carriers with underbridge, tyre trolleys and span launchers are custom designs in most projects.
- Longitudinal section and cross-sections of the pier diaphragms. In an MSS for one-phase casting of box girders, the geometry of the pier diaphragms governs the design and

kinematics of the inner tunnel form. In an overhead MSS for continuous spans, the pier diaphragms influence the location and bracing of the support legs of tower-crossbeam assemblies. The weight and geometry of precast hammerhead segments may govern the design of straddle carriers, lifting frames and winch trolleys of self-launching gantries.

■ Geometry of longitudinal and transverse prestressing. Tie-downs, recesses and embedded items for bridge construction equipment often interfere with internal tendons. In an MSS, the anchor blisters of post-tensioning may require interchangeable inner form modules, easy access for removal of shutters and positioning of stressing jacks, and working clearances within the inner tunnel form for tendon fabrication and tensioning. External post-tensioning is rarely used in combination with in-place casting due to the longer curing time and the need for extracting the inner tunnel form prior to tendon fabrication.

■ Upward span deflection at the application of prestress and allowed residual support action prior to span release. Deflection compatibility of span and self-launching frame governs the residual support action exerted by the erection equipment for span-by-span construction prior to span release. Control of concrete cracking at short curing may result in a lower-bound stiffness for the self-launching frame or in staged span release combined with staged application of prestressing.

■ Dimensional tolerances. Form flexibility under the pressure of fluid concrete and geometry segmentation in curved spans increase the weight of the structural members and the load applied to the machine. Load increase may be significant when the cross-section has thin plates.

■ Geometry of piers and pier caps. Pier geometry governs the opening kinematics of the outer form in most types of MSS; design of pier brackets, W-frames on through girders and tower–crossbeam assemblies also depends on pier geometry. Most overhead gantries and MSSs take support on the leading pier of the span to be erected, and the support systems are designed to avoid interference with the bridge bearings and to anchor the machine in service and out-of-service wind conditions. Slender piers may require front brackets or temporary towers that increase the cost of the machine and the complexity of operations. Anchorages and tie-downs on narrow pier caps are particularly complex with tall machines such as telescopic overhead gantries and span launchers.

■ Abutment geometry. The abutments of span-by-span bridges erected with underslung gantries and MSSs are often cast in two phases (the pier-like central section first, and then the wing walls) to simplify construction of the first and last span of the bridge with a completely assembled machine. Propping from foundations for delayed assembly of the rear launch tails and earlier dismantling of the front noses requires additional equipment and increases the cost of assembly of the machine.

■ Means and methods considered by the bridge contractor in the bidding phase, delivery routes of materials and precast components, permissible erection loads on the deck, and special operations such as jacking, skidding, pier cap stabilisation gear, hammer stabilisation with the gantry and the like. In the presence of adjacent bridges, special operations may include reversed erection of the second deck, simultaneous erection of the two decks, or reverse launching and repositioning of special equipment. Information on means and methods typically does not include the contractor's bid estimates.

■ Surveying of the equipment assembly area and of the landing area at the end abutment. If temporary erection towers are needed, geotechnical information is provided for the design of foundations.

■ Surveying of aerial and underground utilities and power lines that can affect the assembly and operations of special equipment.

- Access limitations to the equipment assembly area and environmental requirements in the areas adjacent to the bridge.

Construction method, length and weight of the typical span, and span modularity influence the design of bridge erection equipment. The total length of the bridge influences the optimum level of production mechanisation, and the number of decks to erect, the distance between adjacent decks, and the possibility of reverse launching on the completed bridge are additional parameters that influence the manufacturer's proposal.

11.2.2 Performance requirements and technical specifications

Performance requirements are identified for special bridge construction equipment in the bidding phase to estimate the construction cost of the bridge and to determine the time schedule of the project. The technical specifications for special equipment are tightly related to the product acceptance provisions of the prime contract, to the contractor's quality management policy, and to the type of machine and subsystems.

The general description of the project and the PR&TS for the erection equipment are the bases on which the manufacturer prepares the proposal. The manufacturer can hardly be held responsible for aspects of the project not included in the final version of the PR&TS included in the contract. The content of this section of the RFP may change slightly if the machine is returned to the manufacturer at the end of bridge construction, and it typically includes the following:

- Required productivity of the machine. Productivity is typically expressed in terms of erection rate (number of units produced or placed daily or weekly under normal conditions) or cycle time (number of hours or days to perform a complete erection cycle). In specific cases, performance requirements may also be expressed in terms of the expected labour demand for certain operations (Rosignoli, 2011a).
- General features of the machine: general description, expected length and weight, sequence of operations and level of automation in performing those operations, feeding and handling of construction materials, number and hoist capacity of carried cranes, means of concrete distribution and vibration, use of ground cranes for repositioning of supports.
- Load capacity of the completed bridge: loads allowed on the deck (stationary load during erection of segments or spans, travelling load during repositioning or transit of equipment); allowed vertical, longitudinal and transverse loads on the bridge piers; expected load ranges and eccentricity for the machine (Rosignoli, 2011b).
- Load capacity of the machine. Extra load capacity increases the initial cost but may facilitate future reuses. The bridge contractor rarely requires extra load capacity because heavier machines are more expensive and more difficult to assemble and operate. The general practice of returning special equipment to the manufacturer at the end of construction also discourages the bridge contractor from requiring extra load capacity.
- Modularity of assembly. Structural elements designed for a given bridge geometry can be reused in the future with new assembly configurations. Reuse is easier if the structural members can be dismantled into parts to be shipped in closed containers. The bridge contractor rarely requires modularity in a machine that will ultimately be returned to the manufacturer. Enhanced modularity also has the drawback of higher cost and crane demand for site assembly and final decommissioning. In most cases, modular design is an advantage only for the manufacturer, who may share the profits of easier depreciation of the investment with the client (Rosignoli, 2010).

- Access to the machine, distribution and load capacity of walkways and working platforms, covering, insulation, lighting and production assistance systems. Owners, contractors, unions and safety enforcement agencies consider easy access and a safe work environment as fundamental requirements, but comfortable work conditions also increase productivity, facilitate inspection and maintenance of the machine, and reduce accidents and the turnover of personnel.

- Design criteria for the structural and mechanical components of the machine. The design criteria reflect design standards for steel structures, forming systems, heavy lifters and crane supporting runway beams, may calibrate load and resistance factors of different industry standards to a common reliability index if necessary, and may prescribe load limitations or more demanding resistance factors for the assessment of second-hand components. The design criteria may also provide instructions for redundant design of structural components and control systems, and for general robustness and reliability of design.

- Number and features of the numerical models to implement for the design of the machine and the analysis of structure–equipment interaction, analyses to perform, and results to provide to the erection engineer and the independent design checker for validation of design (Rosignoli, 2007).

- Design criteria for the hydraulic, pneumatic, electric, electronic and lighting systems of the machine. The design criteria reflect industry standards for systems and plants, and specify the number and power of PPUs, the design pressure of hydraulic systems, the type and distribution of electronic control systems, the expected speed and level of automation in performing critical operations that affect the cycle time, such as launching and form opening and closure, and the type and distribution of form-mounted concrete vibrators and lighting systems for walkways and working areas.

- Specifications for acceptance and processing of raw materials: cutting of plates and shapes, flattening, drilling of holes, welding, fabrication of welded sections, permanent connections, field splices, turned parts such as pins and conical spindles, high-strength bars and anchor nuts, surface treatments, painting and such. The specifications include the number and type of controls to perform during fabrication. When the machine is returned to the manufacturer at the end of bridge construction, the specifications for acceptance and processing of raw materials are generally adjusted to the quality management policy of the manufacturer.

- Specifications for internal articulations and movement driving systems: hydraulic flexural and torsional hinges, articulated saddles with launch bogies or PTFE skids, temporary lock systems, double-acting cylinders for lifting and lowering of the machine, hydraulic geometry adjustment systems, launch winches and capstans, double-acting launch cylinders acting within perforated rails, friction launchers, and robustness and redundancy of launch and restraint systems.

- Specifications for staged acceptance of the machine, including workshop preassembly and dimensional control during fabrication, functional testing of systems, dimensional tolerances of the assembled machine, and acceptance criteria for the results of load testing.

- Criteria to follow for supervision of fabrication, shipping, site assembly and commissioning of the machine as well as for load testing of the completed machine and of those components that cannot be adequately loaded during load testing.

- Lifting capacity of the cranes available for site assembly. The lifting capacity of tower cranes serving pier erection may influence the design and assembly sequences of form travellers and lifting frames to be lifted on the pier tables. Heavier machines are generally assembled using specific cranes.

■ Time-schedule for the design and fabrication of the machine, including milestones for the approval of plan submittals, the purchase of materials and components, fabrication, preassembly, painting, shipping, site assembly, commissioning and load testing.

In the proposal, the manufacturer illustrates how the proposed machine meets or exceeds the contractor's requirements as expressed in the PR&TS. The manufacturer can rarely be held responsible for aspects of the project not defined in the final version of PR&TS included in the contract; ensuring that the PR&TS tightly corresponds to the obligations of the prime contract is, therefore, in the best interests of the bridge contractor.

11.2.3 Requested content for the proposal

Numerous features of the bridge project affect the design and operations of the erection equipment. Special equipment may also influence bridge design, and it certainly influences the technical and financial performance of the bridge contractor in developing the construction contract. The description of bridge project and PR&TS convey technical and contractual information to the manufacturer, while the requested content section of the RFP lists the issues that the manufacturer has to address in the proposal.

The erection rate and cycle time are indicators of the potential productivity of the machine but not of its efficiency and overall cost-effectiveness. The bridge contractor estimates labour demand and other costs associated with bridge construction during project pursuit, but the estimates are not included in the PR&TS. Describing productivity and operating costs of the machine is, therefore, up to the manufacturer. Optimistic expectations may lead to losses for the bridge contractor and legal disputes related to failure of the machine to perform as expected. Some offers will, therefore, convey optimistic expectations under vague disclaimers such as 'for normally unionised workforce and average weather conditions'. Such offers are often followed by extenuating negotiations when the competitors are gone and the bridge contractor is under time pressure.

The need to meet the PR&TS must be considered with commercial honesty but also with flexibility and common sense. A second-hand machine may meet only a part of the PR&TS, but the cost savings may justify low productivity or complex operations in short bridges. A second-hand machine may also offer higher levels of performance, which may open new perspectives of project organisation in spite of the higher cost of the equipment.

Reusing a second-hand machine may be in the best interests of the bridge contractor due to prompt availability. A second-hand machine, however, may be obsolete, have been operated and maintained carelessly, inefficient, expensive to recondition and assemble, or simply too heavy for the bridge. If the machine is returned to the manufacturer at the end of bridge construction, the leasing cost of a second-hand machine for a given period of time will also be similar to the leasing cost of a new machine. Unless a manufacturer's proposal is immediately preferable to the others, the proposals are compared based on cost estimates projected to homogeneous levels of performance. Evaluating the impacts of productivity, structure–equipment interaction and missing information is up to the technology consultant or the procurement personnel of the bridge contractor.

Changes to bridge design may be necessary to minimise reconditioning of a second-hand machine. Value engineering involves additional costs for the bridge contractor, delays in the time schedule due to the need for new approvals, and higher liability profiles. If the manufacturer's proposal requires changes to the bridge design, the proposal should identify the changes and should include adequate support calculations. The changes may include relocation of construction joints, different sizes of cross-section members or pier diaphragms, entity and application sequences of prestressing, different delivery and placement methods, and adjustments to the position of bridge bearings to accommodate the support systems of the machine.

Even a custom designed machine involves some interaction with the bridge to be erected, and structure–equipment interaction typically has financial implications. A light custom machine may save prestressing throughout the bridge, which may or may not offset the higher cost of a brand-new machine in relation to the number of spans to be erected. Production mechanisation saves labour costs and also indirect costs because of the shorter project duration, but the machine is necessarily more expensive. In the light of so many variables, the bridge contractor specifies the required content of the proposal to acquire all the information necessary for the subsequent evaluation process. A proposal that does not require value engineering typically includes the following:

- Concrete strength and curing requirements resulting from the proposed erection method.
- Entity and location of the loads applied to the bridge, under both service and out-of-service conditions and during launching.
- Expected cost of shipping, site assembly and final decommissioning of the machine, with indications about the expected labour and crane demand.
- Layout and details of temporary works, with estimated quantities of materials: earthworks, concrete foundations, reinforcement, location and details of temporary supports and storage platforms, demolition at the end of site assembly or upon decommissioning of the machine.
- Details of anchor bars, recesses, holes, embedded items and the like required for anchoring the machine to the bridge and to handle precast components. The proposal must also specify what anchor items will be provided by the manufacturer with the machine and what items will be at bridge contractor's expense.
- Time schedule, including the design (new machine or reconditioning), delivery of raw materials, fabrication, shipping, site assembly, commissioning and load testing. Should the machine be an assembly of new and second-hand components, the plans will specify what members are new and what members are used.

Additional information is requested in relation to the specific types of machine. Proposals for beam launchers and self-launching gantries for the erection of precast segments describe hoists (strand jacks, winches with reeved ropes, pull cylinders, hoist bars), lifting capacity, lifting and haulage speed, driving systems of winch trolleys and hoist crabs, PPU and hydraulic systems, lifting beams (radio controls, geometry adjustment capability), spreader beams, hangers (geometry adjustment capability, lock and adjustment devices) and walkways and working platforms for access, maintenance and operations.

The proposals for form travellers, forming carriages and MSSs are more complex due to the greater number of operations performed by these machines. The proposal includes forming details, of box girders location of construction joints, and filling sequence for the casting cell. One-phase casting requires description of concrete discharge into the bottom slab, webs and top slab so as to maintain fresh areas of concrete for consolidation and to avoid cold joints.

The filling schemes for two-phase casting also show the location of the horizontal joints in webs and pier diaphragms. The proposal describes inner and of the casting cell outer shutters (materials, thickness, dimensions of form modules), covering and insulation, camber and geometry adjustment devices, splicing of form modules, connections between pier cap, bottom form table and adjacent form modules, fixed vibration systems, walkways and working platforms, location of concrete feeding points, concrete distribution arms, duration of concrete filling, raising rate of concrete in the webs, allowed difference in elevation of fluid concrete, expected form deflections, and number of ties and anti-floating systems. If the MSS is designed for full-span cage prefabrication, the proposal also describes the winch trolleys, cage carrier and cage insertion procedures.

The proposals for all types of self-launching machines also describe pier support systems (tower–crossbeam assemblies, W-frames on through girders, crossbeams with hydraulic support legs, pier brackets), self-repositioning capability of the pier supports versus the need for ground cranes, launch supports (articulated bogies, PTFE skids), launch and lock systems (capstans, double-acting cylinders acting into perforated rails, friction launchers), launch speed, and robustness and redundancy of launch, shift and restraint systems.

The proposals for portal carriers with underbridge and telescopic span launchers fed by tyre trolleys are further complicated by the custom nature of these machines, the need for compatibility with the productivity of the precasting lines, the load capacity of the precast spans at the planned days of curing, the bridge length cantilevering out from the leading crossover embankment, the presence of tunnels and steep gradients along the delivery route, and many other challenges of full-span precasting on large-scale HSR projects.

The proposals for modular assemblies of SPMT illustrate the number of units to be connected laterally or longitudinally for each support location, the super-works necessary to support the span on the SPMT platform (hydraulic jacks, grillages on modular towers, stabilising devices, strong-backs and cross-frames, shims, etc.), the sequence of operations to position the SPMT platforms under the span and to engage the load, and the allowable limits for loss of support by any pair of wheels or axle line during operations.

All proposals should explicitly list the items not included in the proposal. Missing items are unwelcome surprises during site assembly, and are often the cause of delay. Missing items such as embedded items, anchor bars, electrical wiring to components, installation of hydraulic and compressed air lines, dust collection equipment and safety equipment can also be costly.

The manufacturer is not paid to produce the proposal, and therefore the process cannot be unreasonably expensive. However, the bridge contractor should not consider proposals not supported by the requested documentation as they may delay the negotiation process and distort competition against the interests of the bridge contractor.

11.3. Design documents

The contract between the bridge contractor and equipment manufacturer includes a time schedule for the design, fabrication, shipping, site assembly, commissioning and load testing of the machine. The time schedule includes milestones for the manufacturer to provide sets of plans that identify the machine in progressively greater detail. The final set of design documents is signed and sealed by the manufacturer and equipment designer and typically includes the technical report, design report, as-built plans of the machine, technical specifications, assembly manual, operation manual and risk analysis.

11.3.1 Technical report

The technical report describes the main features of the machine and of the bridge it will be used to erect. The PR&TS included with the contract between the bridge contractor and equipment manufacturer is divided into two parts: the performance requirements are incorporated in the technical report and the TS, integrated in the instructions of the equipment designer and suppliers of subsystems and plants, become the technical specifications that rule fabrication, site assembly, commissioning and operations of the machine.

Separating the PR&TS into two documents acknowledges the different role that the technical report and technical specifications may have in the contractual relationships with the other parties involved in the construction project. The technical report is mostly addressed to the approval processes specified in the contracts between the bridge contractor, prime contractor, owner, EOR, GEC, erection engineer and insurance carrier. The technical specifications may also be distributed to all parties, but mostly affect the contractual relationships between the equipment manufacturer, equipment designer, and workshops and sub-manufacturers involved in the fabrication of the machine.

The technical report does not override the contractual PR&TS, and typically discloses only aspects of technical nature. With this document, the equipment manufacturer and equipment designer undertake ownership of the design of the machine and contractual liability for its performance. The technical report describes the scope of work, features of the machine, and codes and standards used for its design; it also includes contractual submittals (shop drawings, product specifications) and illustrates the QA/QC procedures. The document typically starts with an introduction that summarises basic information about the machine in relation to the following aspects.

- General description of the machine, accompanied by three-dimensional (3D) views with nomenclature and location of the primary components, weight and main dimensions, accesses and general layout of walkways, working platforms and coverings.
- Technological features: number, lifting height and capacity, and line pull and operating speed of winch trolleys, portal cranes and monorail hoist blocks; load capacity and launch kinematics of self-launching frames and casting cells; number, load capacity and technical features of launch saddles and support cylinders; operations and redundancy of launch and restraint systems; flexural and torsional hydraulic hinges and other internal articulations, steering angle, loaded and unloaded operating speed, drive control systems, number and power of PPUs and power supply systems.
- Casting phases, filling sequence and duration, deflections of the casting cell, residual support action after application of prestress, delivery of materials on the ground or on the deck, handling capability of materials and demand for external means, special features of materials (self-compacting concrete, retarding admixtures).
- Productivity: operations, level of automation, technological limitations (restraints on overpassed areas, use of cranes on the ground or on barges, need for partial dismantling during repositioning), erection cycle, cycle time, expected labour demand.
- Loads applied to the permanent structures during operations and repositioning, recesses in the deck and the piers, holes, embedded items, tie-downs and anchor systems, lock devices

for service and out-of-service conditions, and other such information needed for the bridge contractor, EOR, GEC and erection engineer to understand the interaction of the machine with the bridge to be erected.

- Positioning tolerances of support systems, tie-downs and anchor systems in relation to the geometry-adjustment capability of the machine.
- Safety and redundancy of structural and mechanical systems, electro-hydraulic actuators, logical control systems, PLC and control software, and electric systems.

The introduction should be as concise as possible. The subsequent chapters of the technical report contain complete information on the machine, and the introduction is only aimed at illustrating to a technical reader the operations performed by the machine and how the machine will interact with the bridge during construction. The information on the bridge is typically limited to the aspects that may generate interaction with the machine.

11.3.2   Design report

The design report is signed and sealed by the equipment designer (sometimes also by the equipment manufacturer as contracting party), and includes analyses and design checks for the machine in compliance with the PR&TS included in the contract. The design report is submitted for approval to many contracting parties, and its final version may therefore include supplementary information requested by the independent design checker, technology consultant, erection engineer, bridge contractor and insurance carrier. The EOR, GEC and owner rarely play an active role in the approval process.

The design report includes the following:

- Bases of structural design: design standards, calibration of specialty standards to the reliability index of the primary structural design standards, load and resistance factors, load combinations, nominal and characteristic values of actions.
- Design standards for mechanical equipment (capstans, bogies, articulated launch saddles, winches) and components (motors, reduction gears, brakes, ropes).
- Design and industry standards for hydraulic systems (typically ASME B-30.1 and ISO 10100 standards) and pneumatic, electrical and electronic systems, data links and communication systems, programmable subsystems, process controllers and PLC.
- Strength and elastic properties of structural materials for the machine and the bridge at the different stages of the erection process.
- Bridge geometry, density of fresh and dry concrete, load capacity of the machine, maximum load eccentricity.
- Description of the structural behaviour of the machine, including the load paths and assumptions made about load distribution.
- Cross-sectional properties, geometry and weight of structural members, dimensional tolerances.
- Permanent loads and masses applied to the machine. Distributed loads include a percentage allowance applied to the structural members for welded attachments such as stiffeners and field splices, inner and outer forming systems, concrete vibration systems, insulation and covering, walkways and working platforms, hydraulic distribution systems, and electric and lighting systems. Localised loads include the PPU, hydraulic pumps, fluid tanks, rollers and launch bogies, support cylinders, motors, hoist blocks, concrete distribution arms, stressing jacks and other such mechanical devices.

- Service loads and masses applied to the machine in operation. Distributed loads include reinforcement cage, weight and lateral pressure of fluid concrete, load eccentricities resulting from the casting sequence, and load redistribution at the application of prestress. Localised patch loads include wheel loads of carried cranes and support reactions applied by the launch bogies.
- Equivalent lateral loads or geometry distortions representing the design geometry tolerances of the machine.
- Design hoist load of carried cranes. This includes the payload, payload hangers, spreader beams, lifting beam, hook, lower block of sheaves, and 50% of the suspended hoisting rope.
- Lifting capacity of carried cranes, centre of mass of the crane bridge assembly (PPU, hydraulic systems and fluid tanks are typically applied on one side and the centre of gravity is offset from midspan of the crane bridge), weight and maximum lateral eccentricity of the hoist crab assembly, and load distribution on the wheels at maximum and minimum load eccentricity.
- Meteorological loads: wind during service and out-of-service conditions, thermal differences and gradients, snow and ice.
- Approach to modelling and analysis, including simplifications, discretisation, description of structural models and load cases, description of the types of analysis (static, dynamic, linear, nonlinear with P-delta or full material nonlinearity) (CNC, 2007).
- Structural software programme used for the analysis, and input and output data.
- Description of the load conditions analysed for the machine: normal operations (this typically corresponds to an envelope of load effects resulting from multiple load cases and geometry configurations), launching and repositioning, out-of-service wind, load testing, impacts and other exceptional loads.
- Analysis of site assembly and decommissioning with identification of the loads applied to temporary supports and stabilisation devices, location of the centre of gravity of subassemblies and partial assemblies, and safety margins from instability of equilibrium in every step of the sequence.
- Analysis of operations: casting processes and sequences, load imbalance during filling of the casting cell, load redistribution at the application of prestressing, dynamic loads from carried cranes, handling of precast elements, movements of the machine or parts of the machine in loaded and unloaded conditions.
- Step-by-step analysis of launching and repositioning to determine a load envelope representing the design demand on structural members and mechanical components and the critical D/C ratio for the stability of equilibrium.
- Analysis of out-of-service wind conditions with the anchored machine. Long trusses with a low first lateral frequency may require dynamic analysis of wind-induced resonant vibrations in order to check lateral bracing.
- Analysis of structure–equipment interaction. Flexibility of supports and anchor points, actions induced by movement, loads applied to the completed bridge, and loads applied to the machine during the application of prestress and release of the span, are load conditions that affect most types of bridge construction equipment. Additional information may be requested by the erection engineer in relation to dynamic wind loading on tall piers and other complex types of interaction that may affect the design of the permanent structure.
- Generation of the load combinations for SLS, ULS, FFLS and EELS design. This order is not casual. The design approach incorporated in some standards for permanent steel structures is based on designing for strength (ULS) and checking for the other limit states.

This is rarely the best approach with bridge construction equipment because design is often driven by the SLS. A machine with excessive deflections or vibrations will crack concrete, concern personnel and lead to inefficient operations whatever the D/C ratio for strength may be.

■ SLS, ULS (strength and stability of equilibrium), FFLS and EELS checks for every structural member or group of members of the machine. Structural checks are also performed on connections, field splices, mechanical components and anchorages to the bridge.

11.3.3 Plans

The plans of a bridge erection machine identify structural members, mechanical components, auxiliary plants and control systems with sufficient detail for fabrication, site assembly, commissioning and operations. Sophisticated equipment includes systems of different type and nature; the drawings of these machines are therefore more complex than those for conventional steel structures.

If the plans are not of a shop-drawing level, they include information for detailing by third engineers. Outsourcing of detailing is frequent for wooden forms, hydraulic and pneumatic systems, electric circuits, lighting plants, and electronic process control systems. The shop drawings are subjected to the equipment designer's approval. If modifications occur during fabrication, the shop drawings are replaced with as-built drawings.

Bridge construction equipment is typically designed and detailed using metric (SI) units. Even if the machine is initially operated in a country that adheres to the imperial system, design in SI units will facilitate future reuse of the machine and prompt the availability of spare parts. The plans must be understandable by personnel with limited familiarity of the SI system units, and therefore only the most common units should be used. Dimensions and dimensional tolerances are typically expressed in millimetres. Electrical, hydraulic and pneumatic systems are described using UNI or ISO symbols and additional annotations as necessary (CSA, 1975).

Reusing a machine on new projects often requires reconditioning. This suggests organising the set of drawings into blocks of plans of related elements, according to the following criteria (CNC, 2007).

■ The machine is divided schematically into blocks, every block is divided into subassemblies, and every subassembly is divided into members, connections and technological systems. The drawings for every member, connection and technological system are grouped at subassembly level, and the sets of plans for every subassembly are grouped at block level.

■ The plans are numbered according to a tree structure. A first code AA identifies the block, a second code BB identifies the subassembly, a third code CCC identifies the member, connection or technological system, a fourth code DD identifies the drawing within the set of plans for that member, connection or technological system, and an increasing number EE identifies the revision of the plan and future modifications. The AA-BB-CCC-DD-EE code used in the plans corresponds to AA-BB-CCC marks put on members and technological systems.

■ The plans for structural and mechanical members, connections and technological systems include lists of items indicating the marks, materials, dimensions, weight and quantity of every item in order to facilitate traceability and control of fabrication. At the end of fabrication, the design plans are replaced with as-built drawings to incorporate the modifications made during fabrication. The final plans define the dimensions, geometry tolerances, weight, location of the centre of gravity, surface treatments of structural members (painting) and mechanical parts (nickel or hard chrome plating, hard coat or baked enamel finish), and the location of the identification mark. The final plans also include details of welds and field splices, the type and number of bolts, stressing loads and tightening couples, approval dates, and the names of the technicians that participated in the design process.

■ The set of plans for every subassembly includes a list of plans, 3D views of the subassembly with mark location for members and technological systems, plans of structural and mechanical members, schemes of technological systems, weight of members and components, and 3D views of the subassembly indicating dimensions, geometry tolerances, cambers, geometry-adjustment devices, weight, support and lifting points, and the location of the centre of gravity.

■ The set of plans for every block includes a list of plans, sets of plans for the subassemblies included in the block, connection schemes for the technological systems of the subassemblies, and 3D views of the block indicating dimensions, geometry tolerances, cambers, geometry-adjustment devices, weight, support and lifting points, and location of the centre of gravity.

■ An initial set of plans and 3D views of the machine identifies the blocks, the tree structure of the plan book, and the codes used to mark members and components. This set of plans includes the general list of plans and a table containing the weight of the individual blocks. Three-dimensional views indicate dimensions, cambers, geometry-adjustment devices, total weight, support and lifting points of small machines that can be handled as units, and the location of the centre of gravity of the machine.

■ The plans of walkways, stairs, ladders and working platforms are typically collected together in a separated set. When modules of access systems are applied to subassemblies or blocks of the machine prior to lifting, their plans are referenced in the set of plans of the subassembly or block, the access systems are shown in the 3D views, and their weight is included in the lifting weight. General 3D views of the machine show the network of access systems with indications of load capacity and evacuation routes.

■ Schemes of hydraulic systems identify symbols used to identify the components on the plans, location, maximum operating pressure and flow rate of hydraulic pumps, location of oil heaters, heat exchangers, return line filters and tanks, diameter and wall thickness of cold-drawn steel tubing (3/8 in., 1/2 in. and 3/4 in. tubing is typically used for open-loop and distribution systems in high-pressure plumbing applications with 1/16 in., 3/32 in. and 1/8 in. minimum wall thickness, respectively), stainless-steel fittings, gauges, control valves (three-way directional control for single-acting systems, four-way directional control for double-acting systems, remote locking for positive load holding, needle, relief, check, sequence, manual versus solenoid for independent control of multiple cylinders or functions, auto-damper or snubber on gauges), manifolds, couplers, hoses (thermo-plastic versus heavy-duty rubber), connections between subassemblies and blocks of the machine, hydraulic cylinders (single-acting, double-acting, hollow plunger, pull cylinders, strand jacks, locknut for mechanical load holding, stop ring, pressure and stroke transducers, built-in relief valve to prevent overloading, return spring for fast retraction regardless of hose length and system losses, plunger wiper seals for reduced contamination, tilt saddle, tonnage, stroke, material, weight), and list of spare parts. Three-dimensional views of the machine show the location and nomenclature of the hydraulic control systems. The design of programable control systems for hydraulic operations (sensors, data links and communication systems, programmable subsystems, process controllers and microcontrollers, PLC controllers and actuators) is described in technical reports issued by specialist designers of electronic control systems.

- Schemes of pneumatic systems identify the symbols used to identify the components on the plans, the location, maximum operating pressure and flow rate of air compressors, the volume of air-holding tanks, air regulators/filters/lubricators, the diameter and wall thickness of steel tubing, fittings, gauges, valves, manifolds, couplers and hoses, the connections between subassemblies and blocks of the machine, air motor hydraulic pumps and form vibrators, hand needle vibrators, and a list of spare parts. Three-dimensional views of the machine show the location and nomenclature of the pneumatic systems.
- Electric schemes identify the symbols used to identify the components on the plans, the location and power of the PPU, the system voltage, the power of motors and auxiliary systems (monorail hoist blocks, electric vibrators, lighting systems, electronic control systems) and a list of spare parts. Three-dimensional views of the machine show the connections (flexible wires, retractable wires with suspenders, bus-bars with sliding shoes) and the location of switchboards, control boards, secondary boards, distribution lines, emergency circuit-breakers and end-of-stroke switches.
- A specific set of plans illustrates the systems used to anchor the machine to the bridge. The anchor systems include ducts for through tie-downs and anchor bars, recesses in the pier walls, pockets for hanger bars in the box wings, embedded items and the like. The plans specify the positioning tolerances for the anchor systems and the conditions that require drilling cores for new anchorages.
- A set of plans illustrates the sequence of operations for repositioning of the machine, and the loads applied to the bridge during erection and the different phases of launching. Maximum and minimum loads are specified for every phase to facilitate identification of the design-governing load conditions for the bridge.

The shop drawings are submitted to the bridge contractor for review and approval prior to fabrication or assembly of components. Approval of shop drawings within the specified deadline is essential for the equipment manufacturer and his subcontractors and suppliers to meet the delivery schedule. The text of the shop drawing review stamp is typically subjected to approval by the insurance carrier of the bridge contractor, in order to avoid transfer of liability.

11.3.4 Technical specifications
The technical specifications are issued by the equipment designer and govern the fabrication of the machine in compliance with the PR&TS included in the contract. They often include instructions related to specific aspects of the machine or tasks assigned to subcontractors, suppliers, weld inspectors and the like. These provisions are often too specialised to be part of the PR&TS and beyond the field of interest of the bridge contractor when the machine is returned to the manufacturer at the end of the contract.

The technical specifications rule all the technical and technological aspects for the machine to comply with the design assumptions and the contractual obligations on the manufacturer. They cover all the components of the machine and govern the materials used, fabrication, site assembly, commissioning, load testing, operations and final decommissioning.

Collecting all the specifications together in one document allows any party, from the painter to the site assembly crew, to know what to do and where to refer to for instructions in case of need. This document also facilitates reconditioning of the machine for new projects.

11.3.4.1 Materials

The technical specifications identify the physical properties of the materials to be used in the machine and the documents to be provided by their suppliers to certify the actual properties and ensure traceability. As far as the structural members are concerned, the basic properties used for qualification of steels are strength and resilience.

Materials and components are subjected to reception/acceptance by the QC department of the equipment manufacturer. If fabrication of components is subcontracted, the QC department of the subcontractor is responsible for reception/acceptance of the raw materials, and the QC department of the prime manufacturer is responsible for reception/acceptance of the semi-worked components from the subcontractor after checking their compliance with the delivery conditions of the subcontract. Materials and semi-worked components that cannot be received are separated and marked to avoid their accidental use in the machine.

Steel plates and shapes are accompanied by certificates issued by the QC department of the mill or by third-party laboratories that certify the mechanical, physical and chemical properties of the steel according to specified testing standards. If certificates are not available, specimens are extracted and tested.

The acceptance checks for raw plates and shapes include the control of certificates, corresponding cast numbers to certificates, plate dimensions, type of shape and dimensions, compliance with specified dimensional tolerances, and surface conditions. Ultrasounding is often required for plates thicker than 20 mm to check for the absence of metallurgical defects. Excessive out-of-flat tolerances may require cold straightening.

Semi-worked components are marked permanently and accompanied by the acceptance documents of raw materials. The acceptance checks for semi-worked components may require complex geometry controls. Mechanical, hydraulic, pneumatic, electrical and electronic components typically derive from serial productions that are certified by the supplier in compliance with industry standards.

11.3.4.2 Fabrication

Also, in this case some instructions in the technical specifications derive from the PR&TS, and others are added by the equipment designer. Fabrication of the machine is supervised by the QC department of the manufacturer, who cooperates with the equipment designer in identifying fabrication processes complying with the manufacturer's quality management policy. The bridge contractor may assign auditing of manufacturer's QC to an external QA institute, which cooperates with equipment designer and manufacturer in defining the QA plan and the QC checks to perform. At the end of this initial planning effort, the fabrication chapter of the technical specifications identifies the following:

- Equipment designer's instructions related to fabrication.
- Allowed geometry tolerances at the end of assembly to ensure proper alignment of the machine.

- Welding procedures. These include welding standards (ASME, AWS, EN), types of weld (hand welding, submerged arc, shielded arc), qualification of welders for the different types of weld, symbols used to identify the type of weld on the plans, chamfering and edge preparation, welding equipment and process parameters (preheat, post-heat, weld heat input), flux and consumables, number and application sequence of seams, and thermal treatments for strain relieving. The welding procedures for major complete-penetration welds are submitted to the QA institute for review and approval. Plates thicker than 30–40 mm may be ultrasounded to identify imperfections and delamination prior to welding.
- Welding control procedures. QC procedures are different for workshop welds and field welds, and the latter should preferably be avoided. QC procedures specify the percentage length of welds to be checked using radiography or ultrasounding. Critical welds and thick complete-penetration welds are often checked over their entire length. Fillet welds cannot be imaged using radiography or ultrasound and are checked using a magnetic particle or dye-penetration test.
- Members to be preassembled in the workshop, geometry tolerances of the subassemblies, alignment and angular tolerances of end plates and connections, and surface flatness of field splices. Workshop preassembly avoids systematic geometry errors, ensures consistent fabrication geometry, and facilitates site assembly. Workshop preassembly should start as early as possible during fabrication.
- Members to be load tested in the workshop. Load testing at the end of commissioning is performed using one of the many possible configurations of the machine. The members that cannot be fully loaded in that configuration can be tested in the workshop.
- Painting (surface preparation, number and thickness of coats, properties of coats, colours). Bridge erection machines are high-profile items of equipment, frequently in the eye of the public and the technical press, and painting has public relations value. In addition to the obvious importance of corrosion protection, rust stains on concrete are difficult to clean and affect the appearance of the final structure. Preparation and protection of the lap surfaces of slip-critical connections are also specified (Figure 11.1).
- Control criteria for components that cannot be validated using calculations to assess their capability to perform as intended. This may apply to hydraulic systems, electronic control systems, electric circuits and the like.
- Testing and final setting of the PPU, mechanical components, hydraulic, pneumatic and electric systems, and electronic control systems.

During the initial planning of fabrication, the QC department of the manufacturer prepares a control plan that includes identification of materials, activities to be controlled, frequency of inspections and controls, control staff and technical means, planning of controls in relation to the fabrication time schedule, identification of the responsibility within the QC process, and systems of documentation. The control plan is submitted to the QA institute for approval.

The machine is progressively fabricated as per the plans. Semi-worked components are handled as per the plans and kept far from the ground and gathering of water. Mechanical, hydraulic, pneumatic, electrical and electronic components are stored and handled as per the supplier's instructions. All components are re-inspected prior to being incorporated in subassemblies to ensure the absence of deterioration during storage. Deteriorated components are repaired or marked for replacement.

Figure 11.1. Painting and preparation of faying surfaces for a box underbridge

Fabrication within quality management systems follows traceability criteria. Traceability allows the origin, location and next fabrication steps for every product or lot of products, in every moment of fabrication, to be known. At reception/acceptance, materials and components are identified, codified, marked permanently by specialist personnel, and registered along with their QC certificates. The QC checks performed during fabrication include references to the same codes.

New codes may be used for completed subassemblies to simplify traceability during the next steps of fabrication, yet ensuring back reference to the initial components. The final codes are registered in the as-built plans along with indications on where the marks are physically located. If necessary, starting from the marks on the machine it is possible to follow the inverse sequence of operations up to the initial acceptance of materials.

The first steps of fabrication are tracing and cutting. Tracing may be computerised using optimisation algorithms that minimise waste material. Plasma or automatic oxy-cutting is typically used for cutting. Hand flame cutting of secondary elements requires edge checking for irregularities or excessive hardness. The edges are machined to remove the layer affected by cutting, unless subsequent welding ensures melting.

Cold or hot processes may be used for folding. Cold folding and calendering are ruled by EN 10025 or equivalent standards in relation to the type of steel and may require final thermal treatment. Hot folding occurs in the temperature range 950–1050°C and requires effective thermal-control procedures. Folded plates are frequently used for forms or channel ribs.

Oxy-cutting and punching are rarely used to drill holes in primary structural elements unless the holes are then widened mechanically by 3–5 mm to remove the affected layer. Lap plates are stacked onto the main plate and drilled in one operation to ensure hole alignment. Robotised drilling ensures minimal geometry tolerances at every angle and does not require stacking.

The best design of a welded connection includes the least possible quantity of welding. However, welds designed to achieve full capacity of connecting members enhance the flexibility of assembly of the machine, and ensure ductility and robustness in case of accidents. The welds are therefore oversized and may require several seams.

Welders are qualified for specific types of welding. The connecting members are immobilised prior to welding. Ceramic plates are fixed with magnets beneath the edges to support complete-penetration welds. Overhead welding is typically avoided. The seams are inspected and cleaned prior to the application of a new seam according to the specified welding sequence. Non-destructive controls are typically performed only on the completed weld.

Workshop preassembly consists of multiple tests on members and subassemblies. Checks include splice alignment, geometry and flatness of members and splices, need for forcing members into position, and location and alignment of the end splices to minimise cumulative geometry errors. Preassembly is recommended in modular machines and is indispensable when through pins are used for the field splices due to the impossibility of correcting splice geometry by shimming.

Preassembly begins as soon as possible to avoid systematic errors and to ensure consistent geometry of fabrication. Preassembly involves additional workshop costs and may be limited to critical subassemblies or may include the entire machine. When the available space is insufficient for full preassembly, the end module of the subassembly is used as starter for a new subassembly. If frontal splices require shimming, the shims are permanently welded to one of the end plates.

After painting, the structural members are completed with mechanical parts and hydraulic and pneumatic steel tubing. Hydraulic and pneumatic systems use flexible hoses at points subjected to movements. Most power-distribution systems and lighting systems use flexible cables that are supported by ring brackets and installed during site commissioning. When the machine has a fixed PPU, bus-bars with sliding shoes or retractable cables with multiple suspenders are used to supply travelling cranes. The non-structural components are tested and pre-set by specialist personnel as per the supplier's instructions.

11.3.5 Assembly manual

The operation manual for a simple machine often provides instructions for site assembly and commissioning, but when the machine is more complex these operations may require a specific assembly manual. The assembly crews are specialist personnel and the operations and controls they perform are more thorough than those performed by the operating crews during bridge construction.

The assembly manual describes step-by-step sequences of operations for site assembly and commissioning of the machine. If site assembly affects permanent structures, the manual is integrated with structural checks made by the erection engineer and submitted to the EOR or GEC for approval. The assembly manual includes instructions for reception/acceptance of members and subassemblies, site assembly, commissioning, load testing and final decommissioning of the machine.

11.3.5.1 Shipping and reception

Improper storage, handling, shipping, assembly, operation and decommissioning may cause structural damage to bridge construction equipment. This chapter of the assembly manual specifies the shipping and reception of items within the quality management policy of the manufacturer, defines whether damaged items can be accepted or must be brought back to the workshop for repairs, and is also used at the storage yard for reception of the machine at the end of bridge construction.

The chapter on shipping and reception identifies the following.

- Shipping documents. Every delivery is accompanied by a shipping list, which specifies the marking of members and subassemblies, the number of pieces, and the dimensions and weight. When multiple members are tied together for shipping, the total weight is marked on the bundle to facilitate handling.
- Support and lifting points. Large and heavy subassemblies must be supported and lifted at precise locations identified by the equipment designer to avoid damage. Lifting accessories such as spreader beams or slings with active control of load distribution may also be necessary.
- Maintenance conditions. Members of second-hand machines that were inaccessible during pre-delivery inspections may turn out to be damaged at delivery. Acceptance inspections include alignment of members and field splices, angular tolerances, oval-shaped holes in field splices and lap plates, and the like.
- Acceptance conditions. Handling and shipping may damage structural members and systems, even when they are shipped in closed containers. Instructions specify if a damaged item can be accepted for assembly or must be sent back to the workshop for repairs. The acceptance conditions for elements of a new machine typically coincide with the fabrication tolerances.
- Checklists.

Acceptance or rejection of damaged items falls under the responsibility of the assembly supervisor. When site assembly is subcontracted, acceptance of deliveries transfers liability from the carrier to the assembly team. The equipment designer is often consulted to assess the influence of damage on the strength and functionality of the machine, and the decisions are registered in the assembly manual. At the end of shipping, the assembly supervisor issues an acceptance certificate for the components of the machine.

11.3.5.2 Site assembly and decommissioning

The assembly of such complex machines involves precise sequences of operations. The MSSs are among the most complex machines to assemble because of the number of functions they perform. Assembly requires positioning of temporary supports, erection of support towers, application of pier brackets or W-frames on through girders, positioning of crossbeams and launch saddles, staged assembly and in-air splicing of truss modules, multiple launch operations for load balance, assembly and application of carried cranes, assembly of the outer frame of the casting cell, application of outer forms, walkways and working platforms, assembly of the inner form, and connection of hydraulic, pneumatic and electric systems.

The self-launching gantries are simpler to assemble due to the absence of a casting cell and the related technological systems. Assembly of form travellers and forming carriages is simpler and faster. Lifting frames, derrick platforms, straddle carriers and tyre trolleys are workshop assembled and recomposed on site in a few operations. Portal carriers with underbridge and telescopic span launchers are mostly assembled on site due to their dimensions and weight.

Adhering to the specified sequence of operations is a fundamental safety rule of site assembly and decommissioning. Performing the same operations but in a different order or without completing the previous operations may lead to instability and collapse of subassemblies or the entire machine. The sequence of operations is clearly described in the assembly manual, which also includes the following.

- Plan view of the assembly yard, with the location, geometry and loads of temporary foundations. Preparation of the assembly yard includes earth moving, soil compaction and drainage. Overhead gantries and MSSs are assembled behind the starter abutment or on the leading span of the completed bridge (Figure 11.2). Underslung machines are typically assembled at the abutment span.
- Assembly plans for the temporary towers and surveying instructions for alignment and verticality.

Figure 11.2. Assembly of a single-girder overhead MSS on an existing bridge

- Stacking of members and subassemblies: maximum number of stacked units, maximum height of stacks, location and features of the separation devices.
- Instructions for field splicing: alignment and angular tolerances, surface preparation, stressing of high-strength bars, bolt tightening (torque and sequence), use of direct tension indicators, load-testing procedures, and number of bars and bolts to be tested.
- Assembly and decommissioning plans. The plans include the sequence of operations, location and orientation of subassemblies prior to lifting, lock and stabilisation devices to avoid involuntary displacement or overturning, crane location and capacity for every major lift, support and lifting points of the subassemblies, lifting accessories, weight and location of the centre of gravity before lifting and after application, and maximum wind speed allowed during in-air operations.
- Instructions for completion, testing and setting of hydraulic, pneumatic and electric systems.
- Commissioning operations for the machine.
- Checklists.

Starting from the assembly sequence, the supervisor works out a time schedule that defines assembly crews, number and capacity of the cranes, auxiliary equipment, logistics of the assembly yard, and shipping sequence and delivery dates for the individual items. Just-in-time delivery minimises storage in the assembly yard and may avoid double handling, but may delay site assembly if members are damaged during shipping.

At the end of commissioning, the assembly supervisor issues a certificate of completed assembly. Reception/acceptance forms of deliveries, communications intervened with equipment designer and storage yard, modification or reconditioning of members, shimming of field splices, checklists, inspection and QC logs and the final certificate are included in the assembly manual.

11.3.5.3 Load testing

Bridge construction equipment should always be load tested upon completion of site assembly. New load tests and comparisons with previous tests should follow every major reassembly. Load testing provides feedback to the equipment designer, shows the functions of the machine to site personnel in the presence of the supplier's representatives, who may clarify doubts and uncertainties, increases confidence in the machine, and facilitates final setting for production.

Heavy lifters may be subjected to static and dynamic tests. According to FEM 1.001 (FEM, 1987), the static test load is 110–140% of the nominal hoist load. According to BS EN 1991-3: 2006 (BSI, 2006) the test load is not less than 125% of the nominal hoist load when it is applied without the use of crane drives. No dynamic allowance is used to amplify the test load for structural checks.

Load testing takes place in the absence of wind and consists of slowly lifting a progressively increasing load until reaching the full test load. The load is lifted at different locations to reach full design stresses in critical members of the machine, and the deflections are surveyed and compared with the theoretical values. The test load may also be lifted at locations not related to normal operations of the machine to achieve launch stresses or to check local components.

According to FEM 1.001 (FEM, 1987), the dynamic test load is 110–120% of the nominal hoist load. BS EN 1991-3: 2006 (BSI, 2006) specifies that the dynamic test load shall be not less than 110% of the nominal hoist load, and the crane supporting structure is checked for the load-test conditions by amplifying the hoist test load by a dynamic factor φ_G:

$$\varphi_6 = 0.5(1.0 + \varphi_2)$$

The test load is moved by the drives in the same way that the crane will be used. The movements are tested individually at increasing speed until reaching the maximum speed. Blackout tests may be performed to check the dynamic response of the machine at the intervention of emergency brakes. Blackout tests during load lowering are recommended for cable cranes and lifters with large load deflections.

MSSs, form travellers and forming carriages are subjected to static tests. The load is increased in steps and the deflections surveyed at each step. The full load is kept for a certain time, the equipment is unloaded and the residual deflections are surveyed. Testing an MSS during casting of the first span is not recommended because loading cannot be interrupted. The deflections of the casting cell could also differ from the expectations, which would result in unsatisfactory span geometry. The test load may be attained by applying a waterproof membrane to the outer form and by filling the casting cell with water. Additional load may be generated with the inner form and by suspending concrete blocks (Figure 11.3).

Additional tests are often performed during launching. Launching causes load reversal in several machines, and deflections close to the elastic values confirm no slippage of field splices. The tip deflections of launch noses and tails are easily measured along with the load in the realignment

Figure 11.3. Load testing of a twin-truss overhead MSS

cylinders. Hydraulic pressure at the support cylinders during lifting to span-casting elevation confirms the weight and load distribution of the MSS and gantries for span-by-span erection.

Testing loads and procedures are described in the load testing manual. The manual is prepared by the equipment designer and includes geometry-control procedures (measuring equipment and location of the reference points on the machine), location and data of permanent benchmarks and reference points at the site, load-testing procedure, structural analyses showing that load testing will not overload the machine, calculated deflections and reactions at the planned steps of loading and unloading and after removal of the test load, allowed deviations of measured deflections and reactions from the calculated values, results of previous load tests, and checklists.

After performing a general inspection of the machine, load testing is preceded by checks of the QA/QC documents of fabrication and assembly, spot checks of compliance with the shop drawings, inspection of anchorages and articulations, spot checks of bolt tightening and bar stressing, surveying of plan and vertical alignment of the machine, verticality checks in the tower–crossbeam assemblies, and placing and zero reading of topographic targets.

The expected results of load testing are:

1 Deflections that increase in linear proportion with the test load.
2 Maximum deflections not larger than the calculated value.
3 No slippage of field splices.
4 Absence of cracks, deformations and other effects that can endanger safety and durability of the machine.
5 Residual deflections upon unloading reasonably close to the calculated value.

Additional investigation is necessary when the expected results are not achieved. Large-scale tests are expensive, and accurate structural analysis prior to testing, installing good instrumentation, and gathering as much information as possible are a good investment to avoid arbitrary conclusions regarding the unsatisfactory results of load testing based on sparse evidence. Robust information may provide the basis for a persuasive argument to a concerned bridge contractor under time pressure.

11.3.6 Operation manual

The operation manual identifies the actions to be, and not to be, performed for the safe and efficient use of bridge construction equipment. The operation manual accompanies every machine certified for production. The first draft of the manual is often written by the equipment manufacturer to identify the actions and sequences of operations to be considered in the design of the machine. The equipment designer adds information, plans and instructions until the final version of the manual is produced.

The operation manual is signed and sealed by the equipment designer and manufacturer, checked by the technology consultant, countersigned by the erection engineer in relation to structure–equipment interaction, and submitted to the EOR or GEC for approval. At the end of the approval process, the operation manual is given to the equipment supervisor and CEI.

Besides specifying procedures for safe operations and prevention of accidents, the operation manual identifies possible problems of the machine and its interaction with other machines and the bridge being erected. Because of space limitations on most bridge construction equipment and the risks of working in-air, the operation manual often also includes recommendations for good practice, such as storing small objects in containers, no storage of materials along transit routes, collection of rubble as soon as it is produced, and evacuation routes to be kept clear of obstructions.

Ensuring respect of the procedures as stated in the operation manual is part of the quality management system for the project, for which the primary responsibility for implementation and coordination lies with the prime contractor. The bridge contractor and CEI incorporate parallel checks within their own QA procedures. QA/QC checks, field observations, shop drawings and the operation manual for the machine are part of the project records and are kept as appropriate for permanent record-keeping purposes as required by the owner.

The operation manual is often divided into three blocks: description of the machine, description of operations, and instructions on inspection and maintenance.

11.3.6.1 Description of the machine

The description of the machine includes the following:

- Nomenclature of the main components. The components are described using 3D schemes, rendering or photographs to facilitate immediate identification. Description includes access paths for safe operations, inspection and maintenance.
- Description of the mechanical parts and the hydraulic and pneumatic systems. Most bridge erection machines are complex systems in which thermal engines or electrical motors activate mechanical devices for the operations of the machine, either directly or by means of hydraulic or pneumatic systems. For every part of the machine the operation manual includes a description, working principles, the sequence of operations, schematic plans, a list of spare parts, schemes of electric circuits, and instructions for operations, maintenance, repair and prevention of accidents.
- Three-dimensional views of the machine showing the electrical wiring and the location of PPU, motors and auxiliary systems, switchboards, control boards, secondary boards, distribution lines and emergency circuit-breakers, along with lists of spare parts and instructions for operations, maintenance, repair and prevention of accidents.
- Accessible electronic control systems, such as linear displacement transducers, stroke transducers, end-of-stroke switches, inclinometers, anemometers and related wiring and data links. Setting, maintenance and repairs are performed by specialist personnel, and the description of these systems and their location is mainly aimed at avoiding involuntary damage.
- Technical features of the support and lock systems. These include load capacity and geometry adjustment of articulated saddles and support cylinders, working and peak pressures, hydraulic control systems for flexural and torsional hinges with synchronised operations of multiple cylinders, power and flow rate of hydraulic pumps, load-retention valves and similar safety systems, and setting of load and displacement sensors. Detailed information is provided for the lock systems in relation to the loads applied to the bridge and the load imbalance due to staggered launching or asymmetrical traversing. The path of such loads to the bridge foundations changes during operations, and these changes must be well understood by the operating crew.
- Technical features of the launch systems. Information on friction launchers and launch cylinders acting on perforated rails includes the tonnage and effective stroke of launch

cylinders and hoisting jacks, launch cycle automation, working and peak pressure of the launch systems, power and flow rate of hydraulic pumps, electronic synchronisation controls for multiple cylinders, and setting of load and displacement sensors.

Information on capstans includes haulage rope (breaking load, diameter, composition, service life), scheme of capstan (including reeving and the number and diameter of sheaves), line pull and drum diameter of the winch, location and operations of brakes and tensioning counterweights, and protection systems for the rope. All launch systems are identified in terms of launch forces, elements that react to the launch forces and transfer them to the anchor points, locks to be removed prior to launching and to be reapplied upon launch completion, and special anchor systems to resist out-of-service conditions.

■ Weight and capacity of carried cranes for every hoisting configuration (between the runways and from side overhangs), operating speed for hoisting and translation, hoist crab, hoisting rope (breaking load, diameter, composition, service life), design line pull of the hoisting rope and use of sheaves for reeving, capstans or motorised braked wheels to drive translation, articulated bogies of crane bridge and hoist crab, power supply (on-board PPU or use of retractable cables or bus-bars for external supply), operations of lifting beams and load adjustment systems.

11.3.6.2 Description of operations

Along with the description of the machine, the description of operations is a primary tool for the formation of equipment supervisor and operating crews. Most bridge erection machines are prototypes, and even experienced supervisors need to know and intimately understand the operations and control means of the machine. One of the biggest single causes of failure of bridge erection equipment is the lack of experienced site supervision.

The description of operations includes the following:

■ Operations and controls to be performed for bridge construction. Step-by-step analysis of bridge erection and task interaction is indispensable for safe operations of the machine, reduces interferences and makes them less critical, and leads to more efficient erection processes. The advantages of step-by-step analysis are particularly evident when multiple activities take place in few locations.

  – Erection of precast beams requires instructions for beam delivery (on the leading span of the bridge or at the abutment), use of straddle carriers to load the launcher, beam handling, plan alignment of the launcher during launching and traversing, number of pier crossbeams required for operations, beam placement and release, and interaction with bridge bearings.

  – Erection of precast segments requires instructions for segment delivery (on the ground or the deck, which may require identifying truck crossing areas), location and capacity of ground cranes and lifting frames, plan and vertical alignment of the gantry during span erection, handling of segments, application of epoxy and gluing bars, staged application of prestressing, stabilisation of the hammer with pier cap gear or with the gantry during balanced cantilever erection, span release during span-by-span erection, and interaction with sliding bearings and fixity points of continuous spans.

  – Erection of precast spans requires instructions for lifting, transportation and placement of the spans, including final adjustment of the support reactions. One operation manual is used for portal carriers with underbridge, while two manuals are often used for tyre trolleys and telescopic span launchers as the operations of the two machines are less interrelated and the machines are sometimes supplied by different manufacturers.

- In-place casting with an MSS requires instructions on plan and vertical alignment of the MSS, camber and crossfall of the casting cell, form operations, cage delivery and insertion, casting phases, number and location of concrete pumps, concrete feeding lines, peak hourly demand on batching plant and concrete delivery lines, filling sequences and directions, allowed load imbalance during filling, concrete vibration, staged application of prestressing, opening and closure of the casting cell, operations of the inner tunnel form, span release and interaction with sliding bearings and fixity points of continuous spans.

- In-place casting with form travellers and forming carriages requires instructions similar to those for an MSS, although they are typically simpler. Transverse load balance must be carefully controlled in these machines during filling due to their light weight.

- Span-by-span erection of macro-segments combines operations and controls for in-place casting with those for handling of heavy precast segments. Casting cells, on-deck casting operations and such heavy erection gantries apply significant loads to the leading span and the front piers of the bridge, which often restrains load geometry.

- Application of prestressing requires instructions on storage and handling of strand coils, cleaning of anchorages, fabrication of tendons, operations to perform on the sliding bearings prior to application of prestressing, handling of stressing jacks, stressing forces and stages, and final grouting of tendons.

- The operations of complex auxiliary machines, such as cranes on the ground or barges, cage carriers, segment loaders, straddle carriers and portal cranes, are typically defined in individual operation manuals.

- Parameters to control during operations and warning and safety ranges for those parameters. Values exceeding the warning range are still acceptable but must be reported to the equipment designer for analysis and interpretation. Values exceeding the safety range require interruption of operations and consulting the designer.

- Instructions for application and removal of front and lateral pier brackets, W-frames on through girders, pier towers, and tower crossbeam assemblies, complete of plan and elevation tolerances, verticality tolerances, surveying and stressing of anchor systems.

- Location of the structural, hydraulic, pneumatic and electric connections to be released prior to repositioning the machine, access to connections, list of the auxiliary means needed for connection release complete of weight and storage location, operations to perform, safety instructions, tightening couple of bolts for reconfiguration of connections.

- Checks to perform prior to repositioning the machine: wind speed, surveying of landing supports and next anchor points, verticality of pier towers and temporary piers, clearance required for operations, disconnection of systems, removal of hanger bars, proper functioning of launch devices and anchor systems, cleanliness and lubrication of launch bogies and PTFE skids, setting of sensors, functioning of alarms, presence of personnel and materials on the machine, segregation of the area beneath the machine, availability of shims for packing.

- Operations to perform when repositioning the machine. The launch procedure is illustrated as a sequence of operations that follow a chronological order of actuation. Safety of launching depends on adhering to the specified sequence of operations. Performing the same operations but in a different sequence or without completing the previous operations may lead to instability and collapse of the machine. The launch procedure typically includes release of locks, activation of launch supports and devices, launching, release of launch supports and devices, geometry adjustment, and activation of locks. Launching an MSS involves additional operations in relation to opening and closure of the casting cell and checking form detachment from the deck soffit at span release. Launching involves landing on new supports and releasing rear supports; the support systems may be repositioned during launching or in independent operations. The operations are numbered progressively and are recalled in the design report for the relevant structural checks. The launch procedure describes systems and components of the machine actively involved in the operations, support points and forces applied; it typically includes lists of obligatory operations, operations to be performed upon approval of the equipment designer, and prohibited operations. Special emphasis is given to the most delicate phases of launching, when the machine is not anchored to the bridge. When plan or vertical curvature, span length, or deck crossfall vary along the bridge, the launch procedure is described for every atypical span. When the machine includes two independent halves, the launch procedures may be different for the inner and outer girder in curved bridges; this is often the case for the underslung MSSs due to the clearance requirements of the open casting cell. Staggered launching of the two halves may be limited in order to control transverse bending in the piers. Launch plans identify the geometry requirements for the machine to follow the launch trajectory, the traversing operations, the allowed lateral displacements, and the allowed differential in lateral displacements. The launch plans include the first and last span of the bridge, and illustrate structures that may interfere with launching, such as abutment walls and adjacent bridges.

- Location of the centre of gravity of the machine and loads applied to the bridge in the different phases of launching. Launching and traversing involve higher risk profiles than anchored operations, and the launch instructions describe risk-mitigation measures to ensure stability and avoid unforeseen displacements of the machine or its parts. The instructions describe the position and geometry of the casting cell, location of carried cranes and equipment, allowed loads on walkways and working platforms, and weight and location of stored materials.

- Controls to perform prior to, during and after launching to ensure that the machine responds according to the design assumptions. Launch forces, support reactions and deflections provide immediate information on the structural response of the machine. Pressure gauges are installed at launch and support cylinders, and the load transferred by the cylinders is often expressed as oil pressure to simplify checks. Pressure valves can be pre-set to provide warning in case of overloading. Automatic pressure relief or stoppage of operations in response to overloading is not recommended, as it may prevent or slow down emergency operations.

- Fail-safe procedures to complete the launch with minimal disruption to the erection cycle: for example, by cross-linking of hydraulic controls at a reduced operation speed.

- Conditions that require stoppage of operations and anchoring the machine, with instructions on the anchoring procedures.

- Contingency plans in the event of a major breakdown or equipment failure: for example, means by which a machine can be rendered sufficiently safe to enable traffic to pass underneath until the source of the problem has been found and corrected.

- Checklists for the main operations. The checklists discourage complacency and deviations from the prescribed procedures, and ensure absence of obstacles to operations and adequate structural capacity of the bridge. The checklists require annotation of parameters in the chronological sequence in which they become available, which helps the equipment supervisor to follow the correct sequence. Annotating measured values near the expected range for those values facilitates checking of anomalies, increases confidence in the machine and provides feedback to the equipment designer.

11.3.6.3 Inspection and maintenance

Most bridge erection machines require periodic inspection and maintenance during operations. Connections, members or entire sections of a machine may become loose and fail due to multiple load reversals, vibrations, local buckling or deterioration of members.

The results of inspections and controls are registered in the operation manual to ensure compliance with the quality management policy of the project and to facilitate reuse of the machine in the future. Some machines include thermal engines and mechanical systems that also require frequent maintenance. The instructions on inspection and maintenance include the following:

- Service intervals for thermal engines and hydraulic systems. Service intervals and the operations to be performed are typically defined by the manufacturer. Daily checks are performed on oil and cooler levels of the PPU, while hydraulic systems require much longer inspection intervals. Tyre trolleys and portal carriers operate full load for most of their service life, and the mechanical components are heavily loaded and subjected to accelerated wear and tear.
- Maximum duration (worked hours) of mechanical components, seals and wipers of hydraulic cylinders, hoses, electronic sensors and the like prior to reconditioning or replacement. The duration of most mechanical devices depends on materials and use conditions. For example, aluminium cylinders are frequently used for hand operations on bridge construction equipment because they are light and easy to carry and position; however, aluminium has a lower fatigue life than steel and should not be used in high-cycle applications.
- Guidance on the reuse of components subjected to sustained axial tension, such as hoisting ropes and strands, hanger bars, tie-downs, splice hardware, bolts, etc. Bolts for slip-critical connections are often resold for less demanding applications at every decommissioning of the machine. Hanger bars, tie-downs and anchor nuts are replaced after 20–30 reuses at 50–60% of the tensile strength. The service life of ropes is typically defined by the manufacturer.
- Inspections and controls to be performed, with indication of the frequency and operations. CNR 10027 (CNR, 1985) specifies two levels of inspection on an MSS: one before every casting cycle, and one before reusing the machine on a new project. The first-level inspections are limited to the effects of impacts or unforeseen events and ensuring the presence of every component. The second-level inspections also include corrosion protection, field splices and the need for maintenance. Both inspections are performed by the equipment supervisor or qualified technicians.
- Register of incidents, with indications of the relevance of the incidents and the revisions and repairs made. As-built plans for repairs made to structural members are added to the set of plans of the machine.
- Checklists for inspection and maintenance.

Not every component of bridge construction equipment is designed to last as long as the entire machine. Some components are replaceable, others are to be maintained, and others are permanent. Determining the influence of structural members, mechanical components and technological systems on the safe operations of the machine helps in identifying an economically appropriate service life for every component, and ultimately the maintenance plan for the machine. Postponing maintenance or replacement may lead to premature failure of more critical components, and may thus amplify the maintenance cost.

11.4. Risk management

Special bridge construction equipment provides a safe and reliable work environment if designed, fabricated and operated properly. However, conceiving, designing, fabricating and operating special equipment safely requires complete understanding of the behaviour of the bridge and the erection equipment during construction, as well as of the structure–equipment interaction.

Permanent steel structures are assembled once, are used for long periods of time and are never moved. Bridge erection equipment is assembled and dismantled many times, is used for short periods of time, and is moved regularly during the use. Assembly, inspection and maintenance flaws are more likely to occur, and the high structural efficiency of these machines makes them more susceptible to defects and errors.

Complex and sophisticated machines that handle heavy loads under such demanding conditions are subjected to multiple risk scenarios. Innovation itself also brings risks. In addition to the typical risks of bridge construction, new risks arise as the new-generation machines become 'smarter'. The impacts of software flaws on operations controlled by a PLC are one example of the risk-mitigation challenges of some of these machines.

Accidents are often the result of a chain of events, errors and omissions that ultimately result in a failure of the system (HSE, 2011). Identifying the root cause of failure is often difficult; however, most accidents with bridge erection equipment occur during repositioning and have structural causes related to design. Risk analysis and mitigation should therefore start in the design phase, when the decisions are easier to implement and the cost of changes and correction of errors is lower (ISO, 2009a).

ISO 31000 defines risks as the effects of uncertainty in objectives, whether positive or negative. Risk management requires the identification, assessment and prioritisation of risks, followed by actions aimed at minimising the probability and impacts of unfortunate events (ISO, 2009b). Risks can result from uncertainty in financial markets, credit risk, legal liabilities, project failures, accidents and natural causes. Risk-management strategies include avoidance, reduction, transfer and retention of risks.

Risk analysis of mechanised bridge construction investigates operations performed by bridge erection equipment and structure–equipment interaction to mitigate risks that have caused problems in the past with similar equipment. Risk analysis also investigates aspects of the work environment that can affect the operations of special equipment. Knowledge risks result from situations where there has been insufficient preparation, relationship risks result from ineffective collaboration between contractual parties or site crews, and process risks result from ineffective operational procedures. These risks affect the productivity of personnel and may also decrease profitability, service, quality, reputation and brand value.

Risk analysis is a transdisciplinary process that requires the participation of the bridge contractor, technology consultant, erection engineer, equipment supplier and equipment designer. Risks are about events that cause problems, and risk identification can therefore start with the source of problems or with the problems themselves. The process involves the identification and characterisation of the risks of mechanised bridge construction with that specific type of equipment, assessment of the vulnerability of the bridge and equipment to specific risks, identification of ways to reduce those risks, and prioritisation of risk-reduction measures in relation to safety, operations and maintenance (Homsi, 2012).

The risks are identified based on project objectives (events that may endanger achieving objectives of the project team), scenarios (events that trigger undesired scenario alternatives), taxonomy (groups of possible risk sources are defined on the basis of shared characteristics and are ranked, and groups of a given rank are aggregated to form super-groups and create a hierarchical classification), and known risks of the specific type of special equipment.

Risk assessment may be qualitative or quantitative. Qualitative assessment is based on a risk matrix that is compiled at the end of the analysis to identify resources at risk, threats to those resources, factors that may increase or decrease the risk, and consequences it is wished to avoid.

Creating a risk matrix enables a variety of approaches. One can start from resources and consider the threats they are exposed, and the consequences of each threat. Alternatively, one can start from the threats and examine what resources would be affected, or one can start from the consequences and determine what combination of threats and resources would be involved. The qualitative influence may be classified into four levels.

1 Highly critical: failure would cause collapse and irremediable loss of the machine.
2 Critical: failure would require interruption of operations or load limitation and immediate repair.
3 Sub-critical: failure would allow continued full-load operation and could be repaired outside of normal operating hours.
4 Not critical: failure would not affect operations.

A major difficulty in quantitative risk assessment (probability of failure and expected consequences) is determining the probability of occurrence, as no statistical information is available on past incidents of bridge construction equipment. Best educated guesses are made to evaluate the potential severity of consequences and the probability of occurrence, and to properly prioritise the implementation of the risk-management plan.

The most widely accepted approach for risk quantification identifies the composite risk index (CRI) as the impact of a risk event multiplied by the probability of its occurrence. The impact of a risk event is assessed on a scale of 0 to 5, where 0 and 5 represent the minimum and maximum possible impact, respectively. The probability of occurrence is also assessed on a scale from 0 to 5, where 0 represents zero probability and 5 represents 100% probability. The CRI ranges from 0 to 25, and this range is usually divided into three sub-ranges. The risk assessment is 'low', 'medium' or 'high' based on CRI sub-ranges defined as 0–8, 9–16 and 17–25, respectively. As both the impact of a risk event and the probability of its occurrence can vary in relation to the measures taken for risk prevention, the CRI is re-evaluated periodically to intensify or relax mitigation measures as necessary.

The risks of mechanised bridge construction are managed using a prioritisation process (Scheer, 2010). Risk events with higher CRIs are handled first, and risk events with lower CRIs are handled in descending order. In practice, however, risk events with the same CRI may have a lower impact but a higher probability of occurrence, or a higher impact but a lower probability of occurrence, and prioritising the assignment of limited resources for risk mitigation is often difficult.

After identifying and assessing the risks associated with the use of special equipment, risk-mitigation measures are formulated according to one or more of the following strategies.

1   Avoidance: eliminate, withdraw from or not become exposed; for example, by avoiding a particular high-risk construction method. Avoidance, however, also means losing out on the potential gain that accepting (retaining) the risks may have allowed.
2   Reduction: optimise (e.g. design new bridge construction equipment with adequate built-in risk control and containment measures) and mitigate (periodically reassess risks that are accepted in ongoing processes as a normal feature of business operations and modify the mitigation measures). Risks can be positive or negative, and risk management requires finding a balance between negative risks and the benefits of the operation or activity, and between risk reduction and effort applied.
3   Sharing: transfer risks to third parties through insurance or outsourcing.
4   Retention: accept the loss, or benefit of gain, from a risk when it occurs. Risk retention is a viable strategy for small risks where the cost of insuring against the risk would be greater over time than the possible loss. Risks that are not avoided or transferred are, by default, retained: this includes risks that are so large or catastrophic that they either cannot be insured against or the premium would be unfeasible.

Some ways of managing risk fall into multiple categories. Outsourcing fabrication of components may share and reduce risk if the subcontractor is capable of more effective risk management, and the prime supplier can concentrate on business development and customer support. Risk-retention pools retain the risk for the group and spread the risk between the prime supplier and the subcontractors; no premium is exchanged within the group, rather potential losses are assessed for all members.

Implementing the risk-management plan includes purchasing policies for the risks to be transferred to the insurance carrier, avoiding all risks that can be avoided without sacrificing the project goals, reducing others, and retaining the rest. Ideal use of these strategies is not always possible, as some risks may involve unacceptable trade-offs.

11.5.   Safety
Preventive measures and safety plans are the last steps in the planning and implementation of a safe work environment. A bridge contractor can proactively enhance safety and mitigate the risks of mechanised construction long before the start of site operations (Homsi, 2012).

1   Select the proper team. Choose qualified and experienced bridge designers, erection engineers, QA/QC technicians and supervisors. Set up a task force to establish design criteria, PR&TS and safety standards to be incorporated in the design of special equipment. If team coordination seems inadequate or the project manager does not have recent familiarity with mechanised bridge construction, hire a technology consultant to coordinate the team.

2 Shortlist the equipment suppliers from firms with a high reputation and financial solidity. Insert performance bonds and indemnification provisions for one or both parties into the contract to transfer risks of losses and damages to the party that has the ability to insure for such risks or may be factually more culpable, or to distribute them between the parties.

3 If the equipment designer is selected by the manufacturer, hire an independent design checker with proved specific experience. Structural failures due to design flaws may have unforeseeable consequences and involve high collective risks that cannot be mitigated with prevention measures and safety plans. Pay attention not only to the design calculations, but also to the implementation of design assumptions, detailing, method statements and control procedures. Prevention of failure starts at the design stage.

4 Design and fabricate bridge construction equipment according to state-of-the-art standards for permanent steel structures. Use the same suite of standards for bridge and erection equipment to ensure calibration of load and resistance factors and coherent load assumptions. Allow for tolerances in the design to accommodate minor variations that will unavoidably arise during fabrication, assembly and operation, and communicate such tolerances to the equipment supervisor.

5 Ensure an adequate flow of communication during the design, fabrication and operation of equipment. Long subcontracting chains may lead to loss of communication, the problems not dealt with during planning and design must be solved on the site, the risks of incorrect operations are not always evident in such complex machines, and human error is the prime cause of accidents.

6 Duplicate the elements of the machine the failure of which may lead to collapse, or design them with redundancy to create alternative load paths in case of failure. Design field splices and anchor systems with active systems of forces, and design the members that receive those forces for the breaking load of connectors. Specify mechanical locknuts on all hydraulic cylinders having long-term action. Design the support systems to follow the deck gradient and crossfall with minimal transfer of horizontal loads to the anchor systems.

7 Design proactively for safety by incorporating elements that would anticipate and mitigate the risks of failure due to human error. Apply locks to prevent excessive screw extension. Apply mechanical blocks to multiple articulations to prevent generation of mechanisms. Apply end-of-stroke buffers to crane runways and design every member between the buffers for the full design load of the crane. When an action cannot be prevented, apply visible markings to warn the operator. Keep in mind that the equipment will be used by workers, who can push it beyond its limits, either on purpose or inadvertently.

8 Provide safe and effective access for personnel to operate, inspect and maintain the equipment. Inspection and maintenance schedules for components that cannot be reached will likely be ignored. Provide proper lighting that minimises shadows. In many cases the erection equipment is the principal means of accessing the work areas, and proper access reduces the risks and increases productivity. Keep in mind that access will also be used by inspectors and visitors not used to working in-air.

9 When considering a second-hand machine, know the entire history of equipment (modifications, accidents and repairs in particular) and review the operation manual in relation to inspections and maintenance. Age of equipment, number of projects, storage conditions, operating environment, maintenance, modifications and accidents are all factors that may require modifications or redesign prior to certifying, re-rating or developing the operating parameters of second-hand equipment. Verify traceability and quality of materials, systems, mechanical components, fabrication, handling, modifications and repairs.

10 Prevent human error (Reason, 1990) with QA/QC in the design, detailing, fabrication, site assembly and operation of special equipment. Ensure QA/QC also in the design and construction of the permanent structure to avoid substandard construction practices such as missing reinforcing steel or embedded items, inadequate concrete, improperly stressed tendons, improper curing and not using the latest shop drawing revisions. Ensure QA/QC also in the design and erection of temporary structures used in association with special equipment such as scaffolding, shoring towers, working platforms, props and temporary foundations. Ensure proper understanding of the magnitude of the loads applied to the bridge and the risks of cutting corners. Load test special equipment at the end of site commissioning.

11 Assign supervision of site assembly, operation and decommissioning of special equipment to a competent supervisor who has adequate technical preparation and is constantly present on the site. The supervisor is also responsible for inspecting and maintaining the equipment and for raising the flag in case of need. Centralise site decisions on the equipment supervisor, and ensure strict adherence to the assembly manual and the operation manual. Both manuals are based on experience to avoid risk situations already encountered in the past. If multiple crews operate the machine, assign a deputy supervisor to every shift.

12 Train a dedicated crew to the tasks described in the operation manual. Perform regular inspections and technical audits to maintain high standards of safety and quality throughout the project duration. Use checklists for all major operations to avoid the risks of short cuts, complacency and normalisation of deviance.

REFERENCES

BSI (2006) BS EN 1991-3: 2006. Eurocode 1. Actions on structures – Part 3: Actions induced by cranes and machinery. BSI, London.

CNC (Confederación Nacional de la Construcción) (2007) *Manual de Cimbras Autolanzables.* CNC, Madrid, Spain.

CNR (Consiglio Nazionale delle Ricerche) (1985) CNR 10027. Strutture di acciaio per opere provvisionali. istruzioni per il calcolo, l'esecuzione, il collaudo e la manutenzione. CNR, Rome, Italy.

CSA (Canadian Standards Association) (1975) *Falsework for Construction Purposes. Design.* CSA, Toronto, Canada.

FEM (Fédération Européenne de la Manutention/European Federation of Materials Handling) (1987) FEM 1.001. Règles pour le calcul des appareils de levage. FEM, Brussels, Belgium.

Homsi EH (2012) Management of specialized erection equipment: safety. *Structural Engineering International* 22: 148–153.

HSE (Health and Safety Executive) (2001) *Reducing Risks, Protecting People. HSE's Decision-making Process.* HSE, Bootle, UK.

HSE (2011) *Preventing Catastrophic Events in Construction.* HSE, Bootle, UK.

ISO (International Organization for Standardization) (2009a) *Risk Management – Principles and Guidelines.* ISO, Geneva, Switzerland.

ISO (2009b) *Risk Management – Risk Assessment Techniques.* ISO, Geneva, Switzerland.

Reason J (1990) *Human Error.* Cambridge University Press, Cambridge, UK.

Rosignoli M (2007) Robustness and stability of launching gantries and movable shuttering systems – lessons learned. *Structural Engineering International* 17(2): 133–140.

Rosignoli M (2010) Self-launching erection machines for precast concrete bridges. *PCI Journal* 2010(Winter): 36–57.

Rosignoli M (2011a) Bridge erection machines. In *Encyclopedia of Life Support Systems*, Chapter 6.37.40, 1–56. United Nations Educational, Scientific and Cultural Organization (UNESCO), Paris, France.

Rosignoli M (2011b) Industrialized construction of large scale HSR projects: the Modena bridges in Italy. *Structural Engineering International* **21(4)**: 392–398.

Scheer J (2010) *Failed Bridges: Case Studies, Causes and Consequences*. Wiley-VCH, Weinheim, Germany.

Bridge Construction Equipment
ISBN 978-0-7277-5808-8

ICE Publishing: All rights reserved
http://dx.doi.org/10.1680/bce.58088.415

Institution of Civil Engineers

publishing

Chapter 12
Forensics

12.1. Introduction

Accidents and major failures of bridge construction equipment are often followed by legal claims and disputes. Industrialised work environments involve high investments, and disruption of production causes economic losses and financial exposure (Rosignoli, 2011a). Chains of liabilities often force one party to claim for reimbursement and, when confronted with the claim, the counterparty will likely file similar claims against other parties.

Bridge construction projects involve multiple levels of contractual relationships. Responsibilities and liabilities are complex and often criss-crossed, and additional obligations arise between the parties as a matter of law. Basically, all parties on a bridge construction project have obligations and responsibilities to one another and to a wide range of third parties, and can be held liable for injuries or damages resulting from breaching standard of care or duty to perform (Ratay, 2009).

Each of the contracts between the owner, designers, subconsultants, professionals, contractors, subcontractors, manufacturers and suppliers may be used as the basis of a claim by one party against another within the liability limitations and the statutes of limitation/repose identified by the contract. In response to a major accident, therefore, the parties with significant interest in the failure assemble failure response teams as part of their legal strategy for dealing with the failure.

The failure response teams typically include a senior-level manager of the affected party, a forensic engineer and a lawyer responsible for the legal strategy; an insurance professional is sometimes added in the most complex cases. The forensic engineer may provide advice, testimony or both. Some engineers are chosen as undisclosed consultants to obtain the best technical solution, and others are chosen as testifying experts for their ability to convey expert opinion on the causes of and responsibility for a failure.

Technical, business and legal analyses may differ, but the ultimate goals of a failure response team are identical: to implement immediate steps to avoid further harm and to mitigate losses and financial exposure, identify the causes of failure, and determine legal accountability. Property and commercial general liability insurance policies typically cover business interruption, personal injuries and damages to third parties; however, the remediation costs are often recovered only in the presence of a builder's risk policy, as special equipment fails during bridge construction. The support of the insurance carrier should always be sought when determining the remedial plan.

Investigating the failure of special equipment requires long-term practice of structural engineering, technical and business practice with mechanised bridge construction, and the ability to

415

conduct an investigation. As most failures originate claims, the engineer must also be familiar with legal proceedings and working with lawyers. Bridge construction equipment is often fabricated abroad, and the international arena adds further complexity.

At the end of the investigation, the engineer's conclusions will be used to evaluate the strength of a claim and as an evidentiary tool during resolution of a dispute. In other instances the engineer's primary role is education: if a judge or an arbitrator appoints the engineer as a confidential consultant, the engineer will educate judge, jury or arbitrator on technical issues and industry standards, and will prepare them to make decisions on the dispute.

The engineer brings technical expertise, business practice and personal credibility into the forensic arena. When confronted, the engineer must demonstrate thorough knowledge of special equipment and mechanised bridge construction, and must possess the confidence and ability to deliver professional opinions that are ethically and technically correct (Ratay, 2009).

12.1.1   Incidents and failures

Failure of bridge construction equipment is often thought of as catastrophic, but in most cases it just involves an unacceptable difference between expected and actual performance (Rosignoli, 2007). Failure may imply distress, excessive deformations, vibrations and other forms of incapacity to perform the intended or required function. Some failures are related to strength, while others are functional deficiencies that do not affect load capacity. Functional issues that take days or weeks to solve may turn out to be more harmful than the rupture of a component that is easy to replace.

Failure may also be related to contractual or operational aspects that have no relationship to the structural and functional performance of a machine. Delays in delivery, shipping and assembly costs, productivity, labour demand, service life of forming systems, need for working platforms or ground cranes, and poor finishing of concrete surfaces are often behind these types of claims (Homsi, 2012a).

Structural failure of special equipment may be due to increased loads, reduced structural capacity or a combination of both. Loads may increase due to incorrect operations, functional failure of control systems and other reasons not considered in the design. Structural capacity may diminish due to inadequate design or fabrication, assembly errors, inadequate maintenance, and corrosion, deterioration, damage or fatigue (Rosignoli, 2010). The ultimate purpose of failure investigation is to determine the contribution of all these factors.

Failures are investigated to determine the causes and understand the mechanisms in order to avoid similar events in the future, to identify remediation methods to repair the machine and return it to service, to assure the contractor that the machine has been repaired properly and can be used again, and within legal claims (Rosignoli, 2011b). Investigation often discovers additional distress situations, distress investigation often discovers latent defects, and defect investigation often discovers deviations from design standards that may range from harmless to unacceptable breaches of the standard of care.

12.1.2   Standards of care and liability

Resolving disputes related to the failure of special equipment often requires assessing the performance of engineering professionals. Designers, erection engineers, independent design checkers and inspectors may be accused of professional negligence, which in many countries means an engineer's failure to exercise the skill of a normally competent practitioner.

The fact that an engineer makes an error and that error causes damage is not sufficient to lead to professional liability. Professional liability derives from proving that the engineer's services were professionally negligent and breached the standard of care of the profession.

Professional negligence is different from the suboptimisation resulting from collaborative design. Special bridge construction equipment is often a prototype, design involves multiple parties, and those parties may have different priorities, perceptions and goals. Collaborative design leads to solutions that satisfy most of the participants and dissatisfy the fewest, and is rarely an optimisation process (Ratay, 2009).

Liability against construction teams rarely follows principles of standard of care. Contractors and manufacturers are bound by express and implied warranties of the contract, regardless of the industry standards, lower levels of error are tolerated, and negligence is often measured in terms of ordinary liability.

Ordinary liability results from failing to exercise the care that every reasonable person would use. Other possible sources of liability for engineering professionals include express or implied warranties and negligence *per se*, when the engineer violates a law or a regulation, the violation causes the type of damage that the law intended to prevent, and the damaged party is a member of the category of persons that the law intended to protect.

More stringent forms of liability may apply for equipment manufacturers. The law often imposes strict liability on the manufacturer for harm caused by the product, whether or not failure is related to negligence. On the other hand, bridge construction equipment is rarely operated by the manufacturer, and strict liability may apply only when it can be proved that failure was caused by original defects that could not be detected and were not exacerbated by substandard operations.

Several theories of liability may be used in legal disputes related to the failure of special equipment. Contract-based liability, tort-based liability, strict liability and indemnity may or may not apply. Because of the absence of industry standards and the multiplicity of roles and responsibilities in a bridge construction project, decisions on disputes other than negligence or strict liability are mostly based on the contractual obligations between the parties, local laws and regulations, and sometimes the specific circumstances (Ratay, 2009). The success of litigation is uncertain and the costs are often huge; most legal disputes are therefore settled out of a court.

The insurance carrier plays a major role in the settlement of a dispute. Although many policies enforce the duty to defend the insured from the damages associated with the failure, some carriers will undertake such a duty only after reserving their rights to contest payment for the claim a second time. Defining the role of the insured in the selection of the counsel and in the defence and settlement of the case, the criteria for resolving disputes between the carrier and the insured, and the extent of the reservation of rights is therefore in the best interests of the insured.

12.2. Emergency response to failure

The first steps following failure of bridge construction equipment are critical. They may prevent further damage, may guide the research of causes of failure and contributing factors, and may influence the success of subsequent forensic investigations. Most of the evidence related to failure is of perishable nature and must be documented and preserved.

The engineer that leads the emergency response to failure is often the most qualified person on the site, may recommend actions and persuade the site management, and with such decisions may affect the abundance or scarcity of evidence. Robust evidence avoids arbitrary conclusions and provides the basis for a persuasive argument.

A structural engineer may also be asked to investigate bridge construction equipment that has not collapsed but shows signs of distress or functional failure. The emergency response must in this case determine whether the machine is stable or collapse is imminent, and, in the case of partial collapse, if the machine has reached a stable equilibrium condition or is in a transient phase of progressive collapse.

After failure of bridge construction equipment, site activities are disrupted or suspended and personnel are available. Site engineers, supervisors and surveyors are the best candidates for an emergency response team. They have some knowledge of the equipment and often some first-aid training. They may assist the rescue teams and may help in identifying wreckage, applying marks, and taking photographs.

12.2.1 Safety

The engineer that leads the emergency response to failure is often requested to assess the safety and stability of the machine, for a variety of reasons.

1 To identify the safest routes through the debris and the zones that must be avoided until stabilised (Figure 12.1). This is often necessary for the rescue teams to reach victims and for site personnel to stabilise the structure.
2 To identify components which are in immediate danger of further failure and to evaluate methods to stabilise them. Whether or not time and resources are invested in temporary remediation depends on what other harm may occur if failure is not stabilised promptly.
3 To evaluate demolition or dismantling techniques. The load paths are rarely evident after a major failure, and stressed components can release load in an uncontrolled way. The load paths become evident as the demolition proceeds, and the process must therefore be monitored and adjusted in real time.

Forensic investigation must wait until the structure has been stabilised. The risks associated with working in the proximity of failed equipment may be difficult to assess, but must be addressed (Figure 12.2). Risks that may be appropriate during rescue and stabilisation operations are rarely justified during a forensic investigation.

12.2.2 Documentation of conditions

After a failure, the conditions of the structure become evidence (Ratay, 2009). This evidence plays a fundamental role in determining the origin, mechanisms, root causes and contributing factors of failure, in formulating and denying failure theories, and in ensuring that the prevailing theories are consistent with the physical evidence. Part of this evidence is of perishable nature or will soon be altered, and must be preserved or documented as quickly as possible.

The conditions of the machine are documented in field notes, photographs and videos. Field notes include sketches, observations, references to photographs, measurements, inventory of materials

Figure 12.1. Rescue teams reaching victims

and equipment stored on the machine, notes on deterioration, description of fracture surfaces, records of samples removed, procedures and results of field tests, notes of persons met and information gained, and a record of the instructions given. The notes should be legible, accurate and well organised. The pages should be numbered and dated. The notes should remain factual in nature and should refrain from stating conclusions, as they might have to be disclosed to third parties in the future.

Photographs document the observations, assist in confirming or excluding failure theories, and are persuasive in reports and presentations. Hand-held cameras may be sufficient for general information, while professional equipment is necessary if photogrammetric techniques will be used. Three-dimensional models can be generated with photogrammetric conversion of digital photographs, although these models do not have the precision of laser scanning.

LiDAR is an optical laser scanning technology that measures the distance to a target by illuminating it with narrow laser beams. Micropulse LiDAR can map a wide range of surfaces via backscattering reflection, with very high resolution. Data acquired from different locations are converted into a full 3D scan. Some LiDAR scans are photorealistic and can be used as a means of remote inspection.

Figure 12.2. Working in proximity of a collapsed twin-truss gantry

A nomenclature is established to label the members of the machine prior to taking photographs and removing them from the site. Using marks as per the design plans is preferable but, if the plans are not available or the machine includes interchangeable members, it may be necessary to develop a different nomenclature. When rescue operations are underway or access is hazardous, the members are labelled as they are removed.

If the identity of a member is unknown, the member is marked arbitrarily to distinguish it from other unknown members. If the orientation of a member is known from its context within the debris, the orientation is marked on the member. When members are cut or disconnected, match marks will facilitate later reconstruction. When multiple persons assign the marks, their initials are incorporated in the marks to avoid duplication.

In the case of a major failure it is often necessary to return to the site as material is removed and new evidence is uncovered (Figure 12.3). Document review and the first structural analyses will also indicate additional areas of investigation. Load tests may be necessary at the end of demolition to determine the structural capacity of the remaining structures and foundations.

12.2.3 Preservation of evidence

Components that are suspected of being associated with the initiation or propagation of failure are obvious candidates for preservation. Members that failed due to consequential damage can rarely provide much information. In the initial stages of a forensic investigation it is not always

Figure 12.3. Reaching the pier on the ground requires removal of gantry and span

possible to distinguish consequential damage, and retaining all damaged elements is therefore advisable when practically possible.

Unfailed components can also be useful. They can be used for load testing of members and connections, to study quality of fabrication, adherence to design and differences with failed components, to explain the aspect and functioning of an unfailed component, and to demonstrate that a certain component did not fail.

Examination of fracture surfaces may help in determining load magnitude and failure suddenness, whether fracture originated at a pre-existing defect and fatigue played a role, the order of magnitude of the load cycles experienced by the machine, and other useful information. Such information may clarify whether failure was due to improper design or fabrication, or to improper assembly or use of the machine, and may be instrumental in assigning liability.

Part of this information exists at a microscopic level and can be obscured or destroyed by corrosion. Exposed steel surfaces corrode rapidly, especially in a humid environment. Fracture surfaces should be readily protected with a pigmented acrylic lacquer that keeps the surface intact for future evaluations and can be easily removed with a solvent.

Bridge construction equipment may also fail because of substandard concrete strength. Curing specimens should be tested immediately to determine the strength of concrete at the moment of failure. Tested specimens may be retained for additional analyses. Cores may be extracted from the failure area or from previously built parts of the bridge to obtain additional information on concrete strength.

12.2.4   Gathering information

In the initial stages of a forensic investigation, some areas of interest may be common to all parties. These provide opportunities to pool resources, avoid duplication of costs and establish a common knowledge basis. Potential areas of cooperation include gathering construction documentation, surveying, 3D laser scanning, identifying members, documenting dismantling and demolition, and laboratory testing.

Gathering construction documentation is a top priority. It includes documents associated with the machine (operation manual, design report, as-built plans, field reports, checklists) and general information registered by the contractor (shop drawings and technical specifications for the bridge, change orders, field reports, inspection reports, material testing reports, climatological data and project correspondence). Although the right to disclosure of evidence varies from country to country, the QA/QC documents of the project are the owner's property, and no party can deny disclosure upon the owner's notice.

The photographic archive of the project is another source of information. With the spread of digital cameras, the quantity of photographs taken in the field is often huge. One of these photographs, even if taken for different reasons, may show the root cause of failure. Other photographs will indicate the overall quality of construction: the order and cleanliness of the site, storage of materials and equipment on the deck, number and activation of tie-downs, and the like.

Information provided by eyewitnesses and persons with project knowledge can assist in formulating or denying theories and in guiding the investigation. The interviews are conducted as soon as possible in order to capture the recollections while they are fresh in the memory, to minimise the influence of other persons, and to identify other possible witnesses.

Although the information sought from the interviewees depends on the type of machine and failure, the lines of questioning are more or less common. The interviewer studies the plans and nomenclature of the machine to facilitate communication with the interviewees, but long-term familiarity with the specific type of machine and bridge construction method is indispensable for effective discussion with the operating crews and to guide the interviews in the most promising direction. The author's line of questioning is typically as follows.

1   Operations performed by other erection crews at the moment of failure. Most interviewees will have no clue about the tasks performed by other crews in a large-scale project, and a long list of checkmarks on 'don't know' at the top of the interview form will make them feel concerned that they are appearing reticent. Detailed information about crews operating far from the interviewee will pose questions of credibility of the account. Vague initial questions related to others will help in understanding the body language of interviewees. Talking about crews unlikely to be related to the failure will understate the interviewer's preparation, which often relaxes the interviewees.

2   Operations performed by the machine at the moment of failure. Several operations are possible with the same configuration of the machine. For example, a precast segment may be accidentally released during lifting, translation, rotation, lowering or load transfer to the hanger bars, and in the meantime the gantry can be traversed and the hoist crab can be shifted along the crane bridge.

3   Opinions about the cause and the sequence of failure. Vibrations, oscillations, loudness and type of noise, the growth of rumble, clouds of dust, movements of the machine or its shadow, the suddenness of the events, the reactions of other workers, and the escape direction of crew members, may all provide unconscious information to a person who has spent weeks of work on that machine. Previous problems (wind-induced vibrations, collision with buffers or fixed obstacles, impacts when picking up segments from barges on a rough sea, diagonal lifting, imbrication of rope systems, loosening of connections, buckling, etc.) may also be connected to failure. Knowing where failure initiated or which member failed first may help in quickly focusing the investigation.

4   Opinions about possible triggering events, such as loads on the deck or the erection equipment. If an operation was prohibited and the interviewee says 'We did it many times without any issues', don't remind him or her that the operation was prohibited – just say 'OK, but what went wrong this time?'

5   Operations performed by the interviewee at the moment of failure. In the author's experience, this question is the last one to ask. Human error (Reason, 1990) is the prime cause of accidents of bridge construction equipment, the interviewee could have caused the accident or could believe he or she has caused it, and concerns about being held liable can break the line of confidence created with the previous questions. Nevertheless, this question is necessary. An experienced engineer should be able to soon regain the confidence of the interviewee, and to relieve him or her of the stress of hiding the truth, especially if failure caused wounding or fatalities of members of the erection crews.

If time allows, different lines of questioning may be developed for different interviewees. Interviews should start with personnel not directly involved with the failure in order to better focus the line of questioning for the operating crews. The operators of other shifts should also be interviewed. Eyewitness accounts are rarely fully reliable, as most observers are not trained in structural engineering and witnessing a traumatic event can leave distorted memories. Nevertheless, eyewitness accounts can be useful to obtain clues or evidence.

12.2.5   Preliminary evaluations

The evidence is typically analysed as it is gathered. Bridge construction equipment may fail due to technical errors during design and fabrication, operational or procedural errors, or events beyond the reasonable control of the parties. Categorising the possible causes into one of these areas may help in guiding the legal strategy and the next steps of the investigation.

The initial information is evaluated to identify as many failure scenarios and possible causes as possible. The emphasis is on formulating theories rather than denying them; later forensic investigation will exclude unlikely possibilities. Structural analysis may help in assessing the viability of different theories, in identifying additional failure scenarios, and in understanding the sensitivity of the structural system to different loads and operations (Ratay, 2009).

The preliminary evaluations may show the need for additional expertise in specific areas. Additional resources should be identified rapidly for them to join the team and guide the investigation. The preliminary evaluations may also identify missing documents to acquire, additional field investigations or samples needed, additional persons to interview, and additional questions for the persons already interviewed.

12.3.   Forensic investigation

Forensic investigation of civil engineering structures is based on the use of science and technology in the investigation and establishment of facts or evidence to be used in civil and criminal justice or other proceedings. In the case of failure of bridge construction equipment, forensic investigation often starts during the emergency response to failure.

Functional failure is often easy to investigate, as the machine can be inspected and tested, while the investigation of major structural failures involving partial or total collapse is more complex. Structural failure may be due to overload, understrength or a combination of both; the investigation is therefore focused on why the demand exceeded the structural capacity, or why the capacity did not meet the demand.

Prompt identification of causes and contributing factors helps in guiding the investigation toward a rapid conclusion and diminishes its cost. When the machine can be repaired, the investigation is performed under time pressure due to the accruing costs of missed production and financial exposure.

At the beginning of the investigation, the team creates an investigative plan. The investigations do not follow standard protocols because every failure is different and involves different profiles of responsibility and liability. The huge costs of bridge collapses involving fatalities and wounding often justify large investments in forensic advice in preparation for court proceedings and defence against criminal implications. Minor failures just require understanding the causes in order to strengthen the machine, restart production and negotiate a settlement.

12.3.1   Investigative plan

An investigative plan typically includes the following steps.

1 Accepting or declining the request for professional services. General contractors, subcontractors, equipment manufacturers, attorneys, insurance carriers, arbitrators and judges are different types of client. The client's reputation, the risk of the client becoming bankrupt as a result of failure, and payments subjected to others' payments or to a positive result of the claim should be considered with care.
2 Definition of the objectives of the investigation. This includes identification of scope of work and deliverables, and negotiation of budget and payment conditions. Some clients will prefer a 'cost plus' assignment to keep the scope of work flexible, while others will prefer a lump-sum assignment to transfer part of the financial risks to the investigative team.
3 Review of the documentation gathered during the emergency response to failure.
4 Formation of the investigative team and formulation of the investigative plan.
5 Collection and review of additional documents.
6 Laboratory analyses to determine the basic properties of materials.
7 Theoretical analyses. Failure analysis is more complex than analysis for design purposes, and is often integrated with other types of scientific evaluations. Analysis of the response degradation of cyclic shear plasticity (Figure 12.4), or thermal flow analysis in materials

Figure 12.4. Analysis of the response degradation of cyclic shear plasticity with custom-written material model

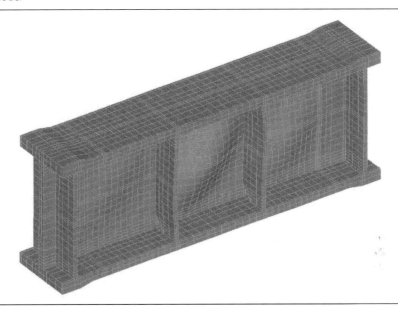

developing time-dependent heat of hydration often requires custom-written material models compiled as a dynamic link library (DLL) from Fortran subroutines. The costs, learning curve and intricacies of high-end software programmes for finite-element analysis suggest subcontracting these tasks to specialist professionals.

8 Fabrication and load testing of specimens. Fabrication of subassemblies is rare due to the cost of large sections of bridge construction equipment. Fabrication of specimens of connections is more viable and provides information on the feasibility of welding, the load behaviour of the connection and its capability of forcing plasticity in the connected members. Solid 3D finite-element analysis is used to determine the expected response of the connection during load testing, the best location and direction for the strain gauges, and the expected readings.

9 Analysis of data, synthesis of information and development of failure theories and scenarios.

10 Preparation and delivery of reports and presentations.

The investigative plan is revised continuously to account for new information, evolving theories and client's instructions within the agreed budget and scope of work.

12.3.2 Parties and communications

The communications between the contracting parties are often scrutinised during a forensic investigation. Communications include contractual documents and formal systems for handling the interactions between the parties.

Poor communication between the bridge contractor, EOR, erection engineer, and equipment designer is often the root cause of failure of bridge construction equipment. Even in those few cases where design plans and PR&TS provide an exhaustive picture of scope of work and contractual requirements to the equipment manufacturer, that information dates back to the RFP and the contract negotiation phase. Several design changes may be made during final design, and these changes may affect loads, dimensions, clearances, location of anchor bars and tie-downs and many other aspects of the machine and its interaction with the bridge.

The bridge contractor's project manager is in charge of the communications with the owner and EOR, and of distributing information to the construction team. Part of this information is kept confidential, and the rest is distributed to the parties. A strong technical background and recent familiarity with mechanised bridge construction are necessary for the project manager to understand all the possible implications of instructions received or decisions made and to enforce proper communications with the erection engineer and the designer of special equipment.

12.3.3 Design errors

Design errors often lead to failure. Design errors may be due to professional negligence, insufficient understanding of the structural system and the applied loads, limitations of structural models and analysis methods, and the dramatic effects that small details may have on the behaviour of such complex and sophisticated machines (Harridge, 2011).

Several engineering professionals may be held liable for design errors related to bridge construction equipment: the equipment designer in relation to the performance of the machine, the erection engineer in relation to structure–equipment interaction, the independent design checker, and the bridge contractor itself in relation to the distribution of information between the parties.

Most structural failures are due to multiple design flaws. Single errors rarely lead to failure because of the conservativeness of the design standards and the robustness and redundancy of most of these machines. This complicates a clear definition of the most common design-related causes of failure of bridge construction. Collapses and major failures provide an overview of the most critical design aspects, and lead to the following recommendations.

1   Communication between the parties is essential in bridge construction projects. Competent engineering practice includes timely distribution of information.
2   Design the machine proactively for safety (Homsi, 2012b).
3   Take great care in the implementation of the design assumptions.
4   Use repetitive design details and take great care over those details: the smallest detail can cause a major problem.
5   Assign the review of shop drawings to an experienced practitioner: this is often the last opportunity to correct errors or misinterpretation of contractual documents.
6   Perform independent design checking on every component of special equipment. When foundations and temporary piers are designed in the field, have their design checked as well. Choose one competent practitioner for the entire project in order to mitigate the risk of losing sight of the interaction between different components. The cost of independent design checking is a small percentage of the cost of special equipment, which is often a minor part of the overall financial exposure. Failure of special equipment may lead to long-term stoppage of an entire production line and to major economic losses.
7   Facilitate and control the interaction between the equipment designer and the erection engineer. Decisions between higher costs of special equipment and higher costs of the permanent structure are very frequent. If the project manager does not have recent

familiarity with mechanised bridge construction, seek the advice of a competent technology consultant.

The performance of engineering professionals is assessed on the grounds of standard of care and the sources of liability admitted in the jurisdiction ruling the contractual relationships. Bridge construction equipment is often designed abroad, and a professional engineer registered in the jurisdiction where the equipment is going to be used must therefore certify its compliance with the design and safety standards enforced in that jurisdiction. In many jurisdictions such certification implies assumption of professional responsibility for the design and the engineer is seen as a successor engineer adopting the work of another engineer. As a consequence, statutory rules or ethics of professional conduct often require that the engineer perform a full independent design check prior to signing and sealing the plans.

Further complexity derives from the distinction between ethical and legal actions. Many actions of engineering professionals may be grounds for disciplinary proceedings related to violations of codes of ethical or professional behaviour, in spite of being perfectly legal. This distinction may weaken the protection offered by professional liability, as opposed to other theories of liability, due to the broader definition of professional negligence and breach of the standard of care.

Expert testimony on the work of engineering professionals requires review of design plans and calculations, shop drawings, as-built plans, and documents showing subsequent modifications. Quality, clarity, accuracy and thoroughness of calculations indicate the level of care and diligence exerted by the engineer; however, the calculations only substantiate the engineer's judgement and, if forensic investigation confirms the correctness of design, poor documentation is rarely considered a breach of the standard of care.

Design and fabrication standards are important points of reference in a forensic investigation, even though adherence to a standard does not necessarily ensure satisfactory performance of a machine. Design assumptions and load and resistance factors reflect the intended performance of the machine, not necessarily the actual performance. Investigating failure requires understanding why the machine failed rather than checking whether or not its design complied with a design standard. Bridge construction equipment is therefore analysed under the conditions and circumstances that existed at the time of failure. Properties of materials and connections and functioning of mechanical and control systems are modelled as they existed rather than as they were intended to be in the design. If analysis does not show distress, the next steps will focus on consequential damage from previous operations.

12.3.4 Defects of fabrication and assembly

Steel is manufactured under controlled conditions, and material defects are not frequent. Most defects of bridge construction equipment are due to improper design and detailing, errors during fabrication, storage, shipping and assembly, improper operations, and deterioration over time due to inadequate maintenance and repairs.

1   Misaligned permanent connections may be due to poor geometry control during fabrication or inadequate immobilisation of members prior to welding.
2   Damage of field splices may be due to improper fabrication, handling, shipping, assembly, use and decommissioning of the machine. Rust and loss of section are due to gathering of

moisture at the faying surfaces and accelerated corrosion processes due to electrochemical cells; irregular faying surfaces may eventually cause local distress and fatigue cracking.

3 Misaligned holes are due to improper fabrication. Welding plugs and drilling new holes is the typical remedy for slip-critical connections with lap plates. Missing connectors are often critical in frontal splices because load eccentricity and prying action can cause tensile failure of the connection on load reversal.

4 Tolerance in verticality of long-stroke hydraulic cylinders is a common cause of unanticipated load eccentricity and structural distress. Bending in the plunger overloads the cylinder and shortens the service life of seals and wipers.

Material testing and field measurements assist in evaluating the quality of fabrication. Daily workshop reports and field reports assist in establishing whether roles and responsibilities were being performed correctly during the fabrication and site assembly.

12.3.5 Improper operations

Most structural failures of bridge construction equipment, and a lesser extent of functional failures, are due to human error. Improper operations include disregarding safety regulations and established procedures, carelessness, lack of training, poor engineering oversight, shortcuts taken in haste to meet the time schedule, complacency, and normalisation of deviance. Training sessions, checklists, inspections and frequent technical audits are primary tools for avoiding human error.

Failing to follow established procedures is among the most common causes of accidents. The desire to simplify and accelerate the operations with a 'better', 'faster' or 'less expensive' way of doing things is a positive thing, and many innovations do originate in the field; construction personnel, however, rarely possess the engineering skills to evaluate the potential problems thoroughly, and often do not understand when engineering assistance must be sought (Homsi, 2012b). New-generation bridge construction equipment may be complex to understand, even for experienced structural engineers. Any modification to the established procedures must therefore go through a proper approval process.

Another major cause of failing to follow procedures is complacency. The operations of special equipment are slow and repetitive. The crews perform the same tasks for long periods of time and become desensitised to the risks involved in those operations. Complacency is particularly dangerous when crews accustomed to a certain routine do not recognise a special case and fail to apply appropriate procedures.

Inexperience and lack of training are additional contributors to field accidents. Although operating special equipment is relatively simple, the complexity and sophistication of these machines require skill and attention. Supervisors and operating crews must be trained to fully understand the criticality of operations. Experienced supervisors are able to train the workforce and to command respect and adherence to procedures. 'Play of the day' discussions at the beginning of the shift are an effective way to discuss the upcoming activities and raise crew awareness, to identify special cases and procedures, and to recognise hazards and safety concerns (Homsi, 2012b).

In relation to the explosion of the space shuttle Challenger, managers and engineers at NASA have been described as succumbing to 'normalisation of deviance', the progressive acceptance of sequential minor errors and failures that accumulate and culminate in a major catastrophe (Vaughan, 1996). The operation manual of bridge construction equipment identifies risks already encountered in the past with that particular type of equipment, and multiple minor failures to follow the established procedures can have cumulative effects.

Operations are typically investigated by interviewing the operating crews. Missing checklists and field reports or poor quality of these documents over time may indicate carelessness and complacency of the equipment supervisor and site management. Lack of inspections and technical audits turn the attention onto the contractor's project management.

12.3.6 Defects due to deterioration, maintenance and repairs

Deterioration over time is normal for structural and mechanical components and for hydraulic, pneumatic, electric and electronic control systems. Deterioration may be accelerated by defects that originated during the design, fabrication, assembly and operation of special equipment. Inspections, maintenance, repairs and replacements lengthen the service life of a machine and delay structural and functional failures due to deterioration.

The load history of a machine may be relevant to forensic investigation. Damaged members may weaken the machine and cause failure during subsequent operations. Fatigue cracks may weaken the connections, and a single overload or impact may cause distortion or local buckling. Even if local distress does not lead to immediate failure, it may trigger a later failure that would not have occurred otherwise.

Defects may initiate failure of materials and systems, may influence the rate of deterioration, and may be involved in multiple cause-and-effect relationships. Identifying defects and causes of deterioration is important both for the forensic investigation and to extend the service life of equipment that shows remediable levels of distress.

12.3.7 Defects in the bridge

Prestressed concrete and composite bridges are susceptible to several design- and construction-related pathologies. If the problem is not noticed immediately, it may be difficult to distinguish poor workmanship of bridge construction from poor performance of erection equipment.

1 Cracking of concrete is one of the most frequent complaints. Many elements of a prestressed concrete bridge are not prestressed, and thus cracking should not be a surprise; however, cracking may also be due to improper casting procedures, the formation of cold joints (Rosignoli, 2002), and underestimated deflections and residual support action exerted by an MSS.
2 Steps in the soffit at the construction joints are due to inadequate anchoring of the casting cell to the previous segment. Excessive geometry tolerances, inaccurate elevations and undulations in the deck surface are often due to poor workmanship; incorrect elevations may also be due to the time-dependent behaviour of concrete and inaccurate evaluation of the weight of special equipment.
3 The weight of forming carriage, thickness imperfections in the concrete deck, and the different age of deck segments complicate camber analysis in continuous steel girders with cast in-place concrete slab. Transverse slab cracks radiating from the top flanges are often due to thermal shrinkage.

4   Joint distress in the top slab of precast segmental box girders may be due to poor workmanship, insufficient prestressing, excessive joint shimming during gluing or distortion or thermal warping of segments. Diagonal cores may be extracted from the epoxy joints to evaluate the adhesion of epoxy by slant shear test carried out as per ASTM C882 (ASTM, 2012).

Forensic investigation of construction defects may be particularly complex. Compliance with plans and contractual specifications, however, is relatively easy to check.

12.3.8   Structural analysis

Failure analysis is different from analysis for design purposes. Design analysis is based on loads specified by the design standard and on conservative capacities to develop structural systems with a consistent reliability index. Analysis of failure is based on the actual loads applied to the structure and the actual capacity of the structural members. Neither of these is well represented by the design standards.

Actual loads vary and the structural capacity is predicted inaccurately by design standards conceived to achieve uniform levels of structural safety rather than to understand why a structure failed. The load capacity of special equipment may be lower than expected due to defects in materials, design, fabrication or assembly, and to improper use, damage and deterioration, or may be higher due to post-critical domain, enhanced properties of materials, ductility and redundancy.

Forensic investigation often requires refined numerical analysis. Physics prevails on code equations, and non linear effects and post-damage capacity must often be evaluated. Identifying the design standard provisions that the machine does not meet is not necessarily a key factor in failure analysis. Deviations from the design standard may raise issues of standard of care but become defects only when they may lead to failure.

When there is no evidence of diffused plasticity, linear analysis often provides reasonable estimates of the failure load. Nonlinear analysis is necessary when plasticity and large deformations have occurred. Most structural analysis programmes allow starting the analysis from the deformed geometry of a previous load case, many programmes handle plasticity, large displacements and contact problems, and only a few high-end programmes interface effectively with custom-written material models.

The first steps in a forensic structural analysis are modelling the machine in the configuration that led to failure and applying the loads that were certainly present at the time of failure. Loads include self-weight and horizontal forces due to the inclination of supports.

■   The hydraulic pressure at the support cylinders as registered in the checklists identifies the total weight of the machine. The weight of construction materials and equipment stored on the machine may be checked by examining the debris, but the original position may be difficult to determine after a collapse, especially when rescue operations have been necessary. Interviews with the operating crews are often the most reliable source of information.
■   Inadequate lubrication and wear and tear of the launch bogies increase the longitudinal loads applied to the supports. The total frictional resistance is easy to determine from the launch pressure registered in the checklists.

- The weight of concrete elements is affected by tolerances in density and geometry, and by the weight of reinforcement and prestressing. Precast elements often break into large parts when released, and geometry and density are easy to measure. Load distribution may be difficult to establish when an MSS fails during filling; comparing the quantity of concrete delivered by the batching plant for the previous spans with the theoretical quantity provides an overall picture of the geometry controls implemented on the MSS.
- Determination of wind loading is more complex. Damage to trees and structures near the site may help in assessing the wind speed registered by the anemometers located on the machine using the Beaufort scale. The application of advertising banners and protection nets is easy to check during inspections. Previous cases of wind-induced vibrations are investigated by interviewing the operating crews.
- Thermal loads may increase due to internal releases of degrees of freedom that do not work properly. Accurate calculation of the effects of thermal differences and gradients is almost impossible after failure.
- Human error is always investigated. The checklists identify the configuration of the machine at the time of failure and the last operations performed. The application of anchorages and tie-downs are easily checked during inspections.

Failure analysis reveals facts concerning both demand and capacity. An approach based on load and resistance factors is therefore recommended for failure assessment, even when the machine was originally designed with allowable stresses.

12.3.9 Laboratory analysis

Laboratory analysis is used to determine the actual properties of materials for accurate modelling and comparison with the values specified in the design documents.

Specimens extracted from failed members may be tested to determine the yield and tensile strength of the steel, elongation, elastic modulus, fracture toughness, nil-ductility transition temperature and hardness. Chemical composition and grain analysis are rarely necessary.

The welds may be examined by means of conventional techniques or by metallography. When brittle fracture is a concern, the fracture toughness of base metal, weld metal and coarse-grained heat-affected zone may be determined using Charpy V-notch tests or equivalent procedures. Plates with fusion lines on the surface may be cut into flat heat-affected specimens for testing of fracture toughness. The fracture surfaces may be examined using an optical or scanning electron microscope.

The metallurgical structure of steel varies throughout the plate thickness. These variations are not revealed by standard tests, and it may be necessary to perform non-standard tests at locations and orientations corresponding to failure. Deviations from standard test results determined with non-standard test procedures are not necessarily a sign of material non-compliance with standard specifications (Ratay, 2009).

When concrete components are involved in a failure, laboratory analysis may include compressive and tensile strength, elastic modulus and density. Adhesion of epoxy may be determined by means of slant shear tests on cores extracted diagonally from the joint. Laboratory and *in situ* testing of soils may be necessary if failure may be related to the foundations.

12.3.10 Reporting

Investigation of structural failure identifies at what load level the members failed and what sequence of member failures led to the final configuration of the machine. Investigation of functional failure is simpler, and rarely requires analysis of structural deterioration. Both investigations may detect multiple failure scenarios in relation to material properties, fabrication, operations, maintenance and the like. Each scenario is compared to the design documents, field observations, structural analysis and laboratory tests to weight its likelihood and to converge on a small set of likely causes of failure. Alternative theories and strong and weak points of the conclusions are conveyed to the client's lawyer to address potential challenges of the opposing party.

The final report of a forensic investigation describes the investigation, summarises the findings, and proposes conclusions based on supporting documents. The report must be comprehensible to a non-technical audience and must support a clear and concise argument in a convincing manner (Ratay, 2009).

The conclusions of the investigation are used to assess the strength of the claim and as an evidentiary tool during the resolution of the dispute. A party that perceives his or her case as weak will often avoid incurring additional costs and the business disruption of a dispute, and will consider settling the issue.

Although most disputes related to failure of bridge construction equipment are settled by private negotiation or arbitration, and very few of them make their way to a court, the conclusions of the final report should be stated so as to ensure relevance and admissibility in court proceedings should the engineer be called to offer sworn testimony. Unwarranted, misleading or confused conclusions may cause inadmissibility of evidence in a trial. Sometimes, evidence that logically should be admitted is excluded on technical grounds, and the draft final report should therefore be submitted to the client's lawyer for scrutiny in relation to the rules of procedure and evidence of the competent jurisdiction.

12.4. Case studies

Most major failures of bridge construction equipment lead to disputes, most disputes lead to claims and forensic investigations, and most forensic investigations are conducted to assign liability to the opposing party.

The parties rarely reach consensus on the root cause of failure because, when the liability exposure becomes evident, the succumbing party often prefers to interrupt accruing of costs and to negotiate a settlement. Even in those few cases that make it to court, the complexity of the topic and the multiplicity of possible causes often lead to decisions by the judge or jury that are not accompanied by technical consensus between the parties.

The case studies below derive from the forensic engineering activity of the author. Comments and conclusions represent the author's point of view, and not necessarily the point of view of the counterparty.

12.4.1 Inconsistent load path

Stability of equilibrium requires a continuous and consistent load path from the point of application of the load to the bridge foundations. Although this concept seems evident, load-path

Figure 12.5. Three-hinge load path in a pendular W-frame

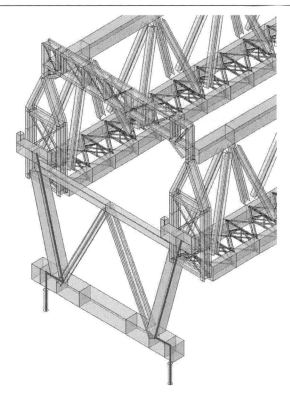

issues do arise when the processes of analysis and detailing lose sight of the overall structural concept and become inconsistent.

The bottom crossbeam of the rear pendular W-frame of a twin-truss overhead gantry had a box section throughout the width, and the support cylinders were fixed at the ends. Traversing the gantry in a curved bridge required shifting the cylinders laterally to place them over the deck webs, and the crossbeam was windowed at the ends to allow this operation. As a result, the constant of torsion of the two 3.6 m windowed end portions became less than 1% of the original box-section constant.

When the cylinders are at the ends of the crossbeam, the vertical load path becomes a three-aligned-hinge scheme where the central cylindrical hinge has minimal rotational stiffness because of the poor constant of torsion of the windowed section (Figure 12.5). The buckling factor of the W-frame at segment lifting was as low as $\gamma_{\mathrm{LC1}} = 1.15$.

Interestingly, this structural deficiency was highlighted by local buckling modes of the pendular leg preceding the sequence of out-of-plane buckling modes of the trusses. This further highlights the need for accurate numerical modelling of such complex machines. Hollow triangular stiffeners were welded to the crossbeam to increase the constant of torsion of the end sections and to stabilise the frame.

12.4.2 Inconsistent execution of field splices

Load testing of a twin-box-girder overhead MSS detected large inelastic deflections. Inspection upon unloading showed that plastic rotations had occurred at multiple lap splices in the main girders. A box girder was launched to shift a splice section to a self-weight counterflexure point in order to relieve bending, and the flange lap plates were removed. Here are the findings.

1   The faying surfaces of the slip-critical connection had been painted with normal paint instead of paint as per RCSC specifications.
2   The bolts had been recycled from previous bearing splices of thinner plates, and the length of the non-threaded portion of the shanks was only 70% of the total thickness of the plate sandwich. As the bolt heads were all located at the top lap plate, the shanks bore only at the top lap plate and flange plate.
3   The low slip coefficient of painted faying surfaces and insufficient tightening of the bolts caused splice slippage and bearing transfer of axial load at the flanges. Because of the gaps between thread and hole at the bottom lap plates, flange axial load was transferred only through the top lap plates, which were designed for 50% of that load.
4   Yielding of the top lap plates at both flanges caused inelastic rotation of the web splices. Relative displacements up to 45 mm were measured between the two flanges, and notches of bolt threads were found in the holes of the bottom lap plates (Figure 12.6).
5   During load testing, the equipment manufacturer shimmed the solutions of continuity at the splices of the top compression flange to diminish rotations. This raised the centre of rotation of the splices to the top flange level, and caused yielding at the bottom holes in the webs.

Figure 12.6. Thread notches in the bottom lap plate

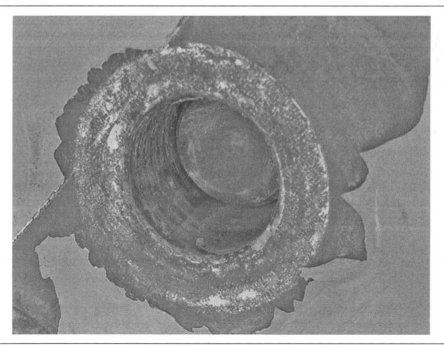

Such situations are incompatible with safe operations of an MSS. The box girders were propped from foundations and the field splices were dismantled, sandblasted and reassembled using friction bolts to get the expected slip-critical connections. This caused a 2-month delay in the contractor's time schedule.

12.4.3 Inadequate communications

In an 820 ton, 103 m twin-box-girder overhead MSS, the left girder (on the right-hand side in Figure 12.7) progressively shifted leftwards during launching due to excessive clearance at the lateral guides. When the front cantilever was 48 m long, the equipment supervisor decided to realign the girder to land on the launch saddles at the new pier with the proper alignment.

A crossbeam was installed close to the main support cylinders at the root of the cantilever to support two flat jacks on PTFE plates. The plan was to lift the box girder beneath the web-flange nodes with the flat jacks, and to pull jacks and girder rightward along the crossbeam on the low-friction contact. The procedure was misunderstood and, after lifting the girder with the flat jacks, the PTFE plates were installed between the girder and the main support cylinders, and the flat jacks were released.

Figure 12.7. Flange buckling

Figure 12.8. Recovery of the box girder

In the initial stages of realignment the bottom flange resisted transverse bending generated by the increasing lateral eccentricity in the support reactions. Eventually, both the left outer flange and the central flange panel yielded and buckled upwards. This generated two low-friction inclined planes that gave rise to uncontrolled rightward sliding of the box girder. The end block of the crossbeam halted the girder when the left support cylinder was almost disengaged (see Figure 12.7). Overturning of the girder would have caused collapse of the MSS. The girder was realigned using the original sequence of operations (Figure 12.8), repaired and reused.

12.4.4 Excessive extraction of geometry adjustment devices

Hydraulic cylinders and screw jacks are used for most vertical-geometry-adjustment systems of special equipment. When the screws are extracted excessively (Figure 12.9), local buckling can cause collapse. Long-stroke cylinders are also susceptible to lateral bending, which overloads the plunger, seals and wipers. In both cases the effects of tolerances of verticality increase with the extraction length.

The allowed extraction of cylinders and screw jacks should be specified using signs or coloured marks. Excessive extraction can be prevented with blocks welded at the bottom of the screw.

Figure 12.9. Excessive extraction of screw jacks

12.4.5   Inadequate bracing
Figure 12.10 shows the lateral bracing of a 19 m tall pier tower supporting an MSS. While the upper portion of the tower (from the pier cap upwards) is adequately braced longitudinally, the lower portion is definitely underbraced. Forward buckling is controlled with light tubes, and rearward buckling with iron wire.

Lateral support governs the load capacity of slender compression members, and both the limit slenderness ratio and the notional design load of lateral braces were grossly violated in this pier tower. The ironworkers operating in the casting cell were evacuated, and pier bracing was stiffened.

12.4.6   Instability of temporary piers
Collapse of a twin-truss overhead gantry for span-by-span macro-segmental construction caused the death of one worker and serious injury to 14 others. Two spans of the bridge fell to the ground and two bridge piers also collapsed. The toll of human life was low because the two crews were within the box cell at the moment of collapse and the top slab protected them from the gantry falling onto the deck on the ground.

The gantry was not being used at the moment of collapse and fell due to loss of support. Out-of-plane buckling in the temporary piers supporting the macro-segments during stitching was

Figure 12.10. Inadequate lateral bracing of a pier tower

recognised as the most probable root cause of collapse. The numerical model used for the forensic investigation is shown in Figure 12.11.

Investigation detected multiple design flaws in the temporary pier. Load eccentricity of macro-segmental construction and the P-delta effect of thermal deck displacements (the fixity point was at the starting abutment) had been disregarded, and the effective length of the braced tower columns was also underestimated.

Investigation also detected multiple assembly errors, such as missing diaphragms and lateral braces. The vertical adjustment screws were extracted excessively and diagonal turnbuckles were not applied to the screws to resist lateral forces. Lateral ties between modular legs were applied far from the panel nodes and at different locations from leg to leg, which led to asymmetrical buckling modes. Such mistakes occur frequently in modular towers due to the large number of holes available for assembly.

In spite of so many inaccuracies, the buckling factor for the segment weight was $\gamma_{LC1} = 1.39$. Tolerances in verticality, thermal stresses in the temporary pier, and thermal deck displacements probably triggered buckling. Out-of-plane buckling should always be checked with prudent load factors in modular structures of this type.

Figure 12.11. First buckling mode of the right temporary pier

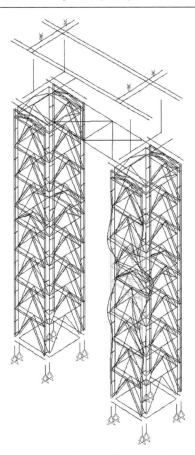

12.4.7 Distress of midspan closure joints

Balanced cantilever erection of precast segmental bridges involves a few repetitive operations: match-casting, handling and gluing of segments, application of prestressing, and midspan closure. When deck pathologies are diffused and the segments have been cast and handled properly, structure–equipment interaction is often seen as the only possible cause of distress.

A full two-span gantry was suspected to have caused top slab distress at most midspan closures of a balanced cantilever bridge with 100 m spans. Step-by-step analysis of staged construction using 3D solid models that included the gluing bars and 3D tendons showed the absence of permanent stress anomalies and accurate control of shear-lag in the wide deck slab.

Sophisticated thermal analyses of the short-line match-casting were necessary to identify thermal warping of segments as the most probable root cause of the distress. Bow-shaped segments are a particular problem during balanced cantilever erection. The size of the gap increases progressively as each joint is closed, and this increases the axial compression in the deck wings (Figure 12.12). Time-dependent shortening of overloaded deck wings eventually pulls match-cast surfaces apart at the midspan closures.

Figure 12.12. Side wing overloading due to thermal warping of segments during match-casting

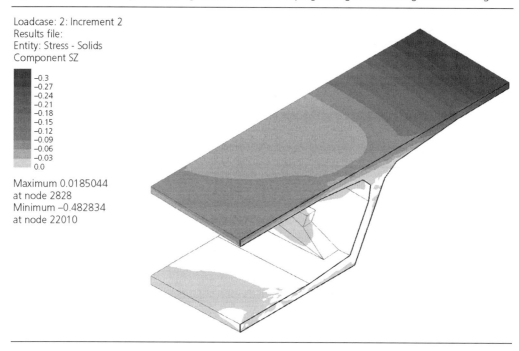

Loadcase: 2: Increment 2
Results file:
Entity: Stress - Solids
Component SZ

-0.3
-0.27
-0.24
-0.21
-0.18
-0.15
-0.12
-0.09
-0.06
-0.03
0.0

Maximum 0.0185044
at node 2828
Minimum -0.482834
at node 22010

Figure 12.13. 3D solid model for time-dependent analysis of staged erection of gapped segments

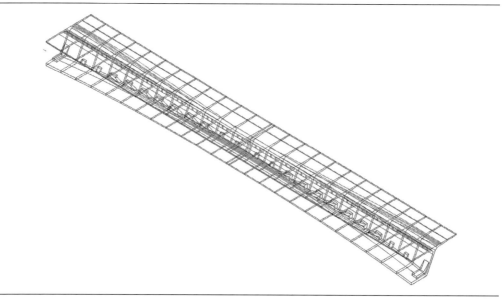

Analysis of the time-dependent stress redistribution within 3D solid models with gapped contact surfaces at the epoxy joints was necessary to investigate this complex functional bridge failure. Figure 12.13 shows the geometric complexity of the model and the 3D layout of cantilever and continuity tendons (transverse top slab prestressing is not shown for the sake of simplicity).

REFERENCES

ASTM (2012) ASTM C882 Standard test method for bond strength of epoxy-resin systems used with concrete by slant shear. ASTM International, West Conshohocken, PA, USA.

Harridge S (2011) Launching gantries for building precast segmental balanced cantilever bridges. *Structural Engineering International* **21(4)**: 406–412.

Homsi EH (2012a) Management of specialized erection equipment: selection and organization. *Structural Engineering International* **22(3)**: 349–358.

Homsi EH (2012b) Management of specialized erection equipment: safety. *Structural Engineering International* **22**: 148–153.

Ratay R (2009) *Forensic Structural Engineering Handbook*, 2nd edn. McGraw Hill Professional, New York, NY, USA.

Reason J (1990) *Human Error*. Cambridge University Press, Cambridge, UK.

Rosignoli M (2002) *Bridge Launching*. Thomas Telford, London.

Rosignoli M (2007) Robustness and stability of launching gantries and movable shuttering systems – lessons learned. *Structural Engineering International* **17(2)**: 133–140.

Rosignoli M (2010) Self-launching erection machines for precast concrete bridges. *PCI Journal* **2010(Winter)**: 36–57.

Rosignoli M (2011a) Industrialized construction of large scale HSR projects: the Modena bridges in Italy. *Structural Engineering International* **21(4)**: 392–398.

Rosignoli M (2011b) Bridge erection machines. In *Encyclopedia of Life Support Systems*, Chapter 6.37.40, 1–56. United Nations Educational, Scientific and Cultural Organization (UNESCO), Paris, France.

Vaughan D (1996) *The Challenger Launch Decision: Risk Technology, Culture, and Deviance at NASA*. University of Chicago Press, Chicago, IL, USA.

Bridge Construction Equipment
ISBN 978-0-7277-5808-8

ICE Publishing: All rights reserved
http://dx.doi.org/10.1680/bce.58088.443

Glossary

Alignment wedge: Truss extension with inclined bottom chords for progressive recovery of the elastic deflection (front end) and progressive release of the support reaction (rear end) during launching.

Axle line: Row of pairs of wheels that are positioned along a line across the narrowest plan dimension of an SPMT, tyre trolleys and portal carriers for precast spans.

Balanced cantilever casting: Construction method where deck segments are cast symmetrically on either side of the pier with form travellers. After completion of the hammer, a midspan closure pour is cast with the previous hammer to make the deck continuous.

Balanced cantilever erection: Erection method where ground or floating cranes, lifting frames or overhead self-launching gantries apply precast segments symmetrically to the hammer. After completion of the hammer, in-place casting of a midspan closure pour with the previous hammer makes the deck continuous.

Beam launcher: Light self-launching gantry for erection of precast beams.

Bottom crossbeam: Crossbeam that supports the bottom form table of the casting cell and the rib frames for the outer forms of webs and side wings.

Cantilever tendons: Longitudinal post-tensioning installed in the top slab of balanced cantilever bridges to resist negative bending due to cantilever erection and the weight of erection equipment.

Capstan: Cable system for bidirectional drive of a trolley with a haulage winch.

Casting cell: Assembly of outer form, inner form and front bulkhead for casting of segments or spans.

Casting cycle: Sequence of operations for casting of segments or spans.

Casting run: Segments for a span or a cantilever that are cast on a long-line bed prior to transportation to storage.

Casting span: Distance between the supports of an MSS during span casting.

CEI: Construction Engineers and Inspectors.

C-frame: Adjustable frame that supports an overhead girder or suspends two underslung girders by rolling on the new segment or span during launching. Vertical cylinders are used to adjust frame geometry to crossfall, to lift the machine for span erection, and to lower the machine back on the launch bogies after application of prestress.

Continuity tendons: Longitudinal post-tensioning installed in the bottom slab of balanced cantilever bridges to resist positive bending in the midspan region.

Convertible gantry: Full two-span gantry for balanced cantilever erection of 100–120 m precast segmental spans that is also compatible with span-by-span erection of 40–50 m approach spans and abutment closures.

Crane bridge: Part of the carrying frame of a travelling crane that spans the crane runways and supports the hoist crab.

CRI: Composite risk index.

Crossbeam: Support girder for overhead gantries and MSSs equipped with adjustable support legs and articulated launch saddles.

Cylinder: Long-stroke steel or aluminium hydraulic jack. The plunger may be equipped with a locknut for mechanical safety when the cylinder is not operated and with a tilt saddle to lift sloped surfaces. Double-acting cylinders provide bidirectional hydraulic control and are operated with four-way valves. Single-acting cylinders provide hydraulic control only for extension, are operated with three-way valves, and may lodge return springs for fast retraction. Hollow-plunger cylinders use a central high-alloy steel bar for pulling. Pull cylinders are inverted systems the prime action of which is pulling instead of pushing. Built-in or external stroke transducers may be applied to the cylinder to provide feedback on plunger position to electronic control systems.

D/C ratio: Ratio of factored demand to factored capacity. The D/C ratio is the main parameter for assessing the stability of equilibrium and the structural capacity of members and connections in LRFD design.

DTI: Direct tension indicator. DTIs are single-use mechanical load cells used to indicate when the required tension has been achieved in structural fastener assemblies. DTIs are hardened washers incorporating small arch-like protrusions on the bearing surface that are designed to deform when subjected to compressive load. DTIs may be installed under either a bolt head or a hardened washer. As a fastener assembly is tightened, the arch-like protrusions are compressed and the change in distance between the base of the protrusions and the support surface correlates to the tensile force induced in the fastener; this can be measured by insertion of a tapered feeler gage. Some DTIs contain coloured material in the cavity of each protrusion that is expelled when the protrusions collapse upon tightening, indicating bolt tension visually.

Dynamic allowance: Ratio of the dynamic response to the static response.

Equalising beam: Articulated beam that balances the load in two rolls placed at the ends of the beam by rotating about a central pivot. Two equalising beams may be applied to the ends of a longer equalising beam for a four-roll assembly. Eight-roll assemblies with seven equalising beams are a practical limit for long bogies for heavy loads.

Extraction saddle: Articulated saddle at the rear end of a tyre trolley that supports the precast span during transportation and rolls or slides along the trolley during span extraction.

EELS: Extreme event limit state. Limit state relating to events such as earthquakes, ice loads and impacts of vehicles or vessels.

Factored load: Nominal load multiplied by the load factor specified for the load combination and the limit state under consideration.

Factored resistance: Nominal resistance multiplied by the resistance factor specified for the structural element, the type of stress and the limit state under consideration.

Falsework: Temporary construction used to support the permanent structure until it becomes self-supporting. Falsework includes steel or timber beams, girders, columns or modular shoring systems, piles and foundations.

Fast split: Assembly technique for overhead form travellers on short pier tables.

FFLS: Fatigue and fracture limit states. The fatigue limit states express restrictions on the stress range related to the expected number of stress range cycles. The fracture limit state is taken as a set of material toughness requirements.

Forced vibration: Vibration of a system where the response is imposed by the excitation.

Form brackets: Brackets overhanging from the main girder of an overhead MSS to suspend form hangers with adjustable connections.

Form hangers: Suspension frames for the outer form of an overhead MSS.

Form table: Rolling form for top slab strips.

Form traveller: Short self-launching MSS for in-place segmental casting of balanced cantilever bridges. A form traveller includes a self-launching frame, a casting cell and two levels of working platforms. The self-launching frame is used to advance the casting cell from segment to segment, to support the leading edge of the casting cell, and to cast the midspan closures at the end of hammer erection.

Forming carriage: Long overhead MSS that shuttles back and forth along the steel girder of a composite bridge for in-place segmental casting of the concrete slab. The sequence of segmental casting is studied to minimise permanent tensile stresses in the concrete slab in the negative bending regions of the continuous beam. The carriage includes an overhead space frame that suspends form tables for the side wings. The central strip of concrete slab between the webs is cast with a self-supporting form table.

Formwork: Temporary structure or mould used to retain fluid or plastic concrete in its designated shape until it hardens.

Free vibration: Vibration of a system that occurs in the absence of forced vibration.

Friction launcher: Support and launch device equipped with hydraulic systems that generate longitudinal movements of the supported member by friction.

Front crossbeam: Overhead crossbeam at the front end of a form traveller that suspends spreader beams for support and vertical adjustment of form tables and the front bottom crossbeam of the casting cell.

GEC: General engineering consultant.

Gluing bar: Post-tensioning bar used for pressing epoxy joints between precast segments.

Guidance means: System used to keep a crane aligned on a runway, or a truss aligned on a launch bogie, through horizontal reactions.

Hammer: Portion of a balanced cantilever deck comprising pier table and cantilever segments on either side. Upon completion, the hammer is connected to the adjacent hammer with a midspan closure pour and continuity tendons to make the deck continuous.

Hammerhead segment: Starter segment of a precast segmental balanced cantilever bridge.

Hanger: Bar or rope suspending a precast segment or a section of the casting cell from an overhead frame.

HDPE: High-density poly-ethylene.

Heavy lifter: Self-propelled machine equipped with hoist devices. Cranes, straddle carriers, derrick platforms, cable cranes, beam launchers, span launchers, lifting frames and portal carriers are designed with standards for heavy lifters.

HiFIT: High frequency impact treatment. Weld hammering device to increase the service life of dynamically loaded welded steel structures.

Hoist: Machine for lifting loads.

Hoist block: Underslung trolley that incorporates a hoist and travels on the bottom flange of a runway beam.

Hoist crab: Part of a travelling crane that incorporates a hoist and travels laterally on rails on the top of the crane bridge.

Hoist load: Non-factored design load for a hoist system. The hoist load includes payload, lifting accessories and a portion of the suspended hoisting rope.

HPC: High performance concrete.

HSR: High-speed railway.

Hydraulic hinge: Open loop system of hydraulic cylinders that allows uncontrolled differential movements at the supported points with uniform hydraulic pressure.

I/O: Input and output.

Launch bogie: Articulated assembly of cast-iron rolls on one or more equalising beams.

Launch nose: Front extension applied to the main girder to control negative bending and overturning during launching.

Launch saddle: Articulated assembly of launch bogies that supports the main girder and allows rotations in the longitudinal and horizontal planes and longitudinal and transverse movements. The bottom launch bogies may be replaced with PTFE skids for traversing.

Launch tail: Rear extension applied to the main girder to control negative bending and overturning during launching.

LiDAR: Light detection and ranging. LiDAR is an optical remote sensing technology that measures the distance to a target by illuminating the target with narrow laser beams.

Lifting arm: Hydraulic system to recover the deflection of the launch nose at landing at the new pier and to release the support reaction at the rear tail during launching.

Lifting frame: Self-launching beam-and-winch assembly for lifting and handling of precast segments. Lifting frames are used for balanced cantilever erection, for progressive erection and to load precast segments onto an underslung gantry during span-by-span erection.

Limit state: Condition beyond which a structure ceases to satisfy the provisions for which it was designed.

Load factor: Multiplier applied to force effects to account for the variability of loads, the lack of accuracy in analysis, and the probability of simultaneous occurrence of different loads.

Load multiplier: Multiplier applied to force effects to account for ductility and redundancy of the structural system.

Load test: Operation of load testing a system or component in order to establish its rated capacity.

Long-line casting: Method of match-casting precast segments on a long casting bed that supports a complete cantilever or span of segments.

LRFD: Load and resistance factor design. Reliability-based design methodology in which the effects of factored loads are not permitted to exceed the factored resistance of members and connections.

LRT: Light rapid transit.

Macro-segmental construction: Span-by-span and balanced cantilever construction techniques involving the use of precast segments the weight and dimensions of which prevent ground transportation.

Main frame: Longitudinal lozenge truss of an overhead form traveller comprising a rear triangular truss to the tie-down bogie and a front triangular cantilever that supports the front crossbeam.

Match casting: Method of casting precast segments where a segment is cast against an existing segment to create a matching joint. The segments are separated and reassembled in the structure.

Monorail hoist block: Hoist block that is suspended from a fixed runway.

Motorised saddle: Self-propelled trolley that lifts and supports the front tractor of a portal carrier for placement of precast spans. The saddle rolls along the underbridge in the first phase of launching and pushes the underbridge forward in the second phase of launching to clear the area under the span for lowering.

MSS: Movable scaffolding system for in-place casting of prestressed concrete bridges. A deck-supported MSS for balanced cantilever bridges is called a 'form traveller'. A girder-supported MSS for the concrete slab of composite bridges is called a 'forming carriage'.

Nominal resistance: Resistance of a member or connection determined based on the dimensions specified in the plans and the permissible stresses or deformations or the specified strength of materials.

One-phase casting: Casting technique for solid or voided slabs, ribbed slabs and box girders where the entire cross-section is cast at once.

Overhead machine: Bridge construction equipment where precast girders, segments or spans or a casting cell are suspended from a self-launching frame.

Overhead travelling crane: Machine for lifting and handling loads that moves on wheels along overhead runway beams. It incorporates one or more hoists mounted on crabs or underslung trolleys.

Payload: Net weight of a precast girder, segment or span.

PC: Prestressed concrete.

Pendular leg: Auxiliary support leg pinned at the top and bottom that carries no moment in the longitudinal plane.

Pier bracket: Support bracket for underslung gantries and MSSs that is anchored laterally to the piers. The pier brackets are used in pairs that are connected with external stressed bars to resist transverse bending. Shear keys in pier recesses are typically used to transfer vertical loads.

Pier table: Starter segment or assembly of segments of a balanced cantilever bridge.

PLC: Programmable logic controller.

PLR: Programmable logic relay.

Portal carrier: Multi-wheel carrier for on-deck transportation and placement of precast spans in combination with a full two-span underbridge at the leading end of the erection line. It includes two motorised tractors connected by a main girder equipped with hoist winches.

PPU: Power-pack unit.

PPWS: Prefabricated parallel-wire strand.

PR&TS: Performance requirements and technical specifications.

Progressive cantilever erection: Directional erection method where precast segments are erected in full-span cantilevers from one pier to the next using temporary props from foundations or stay cables to support the cantilever.

PTFE: Poly-tetrafluoroethylene (commercial name is Teflon).

P/W: Payload-to-weight ratio.

QA: Quality assurance.

QC: Quality control.

Rectangular truss: Modular truss with rectangular cross-section comprising two braced bottom chords and two braced top chords connected by framed vertical and diagonals. The depth of the truss is typically 2–4 times the width.

Redundancy: Capability of a structural system to carry load after damage or failure of one or more members.

Resistance factor: Multiplier applied to the nominal resistance of members and connections to account for variability of material properties, structural dimensions and workmanship, and uncertainty in the prediction of resistance.

Resonance: A system in forced vibration is in a resonance condition when any change in the frequency of excitation causes a decrease in the response of the system.

Reverse launching: Backward transfer of bridge construction equipment on the completed deck to erect a second parallel deck without decommissioning.

Rib frame: Outer framing system of the casting cell that supports the form panels according to the geometry of the deck section.

Runway: Rectangular rail welded to the top flange of a girder to support travelling cranes.

Scaffolding: Elevated working platform used to support personnel, materials and equipment but not intended to support the structure being constructed.

Segment cart: Support saddle for precast segments that rolls along an underslung gantry and provides lateral and ve'rtical geometry adjustment.

Self-launching gantry: Self-launching bridge erection machine that is used to lift, move and place precast segments. After completion of a hammer or span, the gantry repositions itself to erect the next hammer or span.

Short-line casting: Method of match-casting precast segments using a fixed bulkhead at one end of the casting cell and a previously cast segment at the other end. The form is only one segment long, from which the term 'short line' derives.

Simple span: A statically determined simply supported span.

SLS: Serviceability limit state. Limit states related to stress, deformations, vibrations and cracking under regular operating conditions.

Span-by-span casting: Construction method where a span is cast in place with a casting cell supported on ground falsework or an underslung self-launching frame (underslung MSS), or suspended from an overhead self-launching frame (overhead MSS). After application of pre-stressing, the construction equipment is repositioned on the next span.

Span-by-span erection: Erection method where the precast segments for the span are supported on ground falsework or an underslung self-launching gantry or are suspended from an overhead gantry. The segments are aligned, jointed and post-tensioned longitudinally in one operation to make a complete span.

Span launcher: Heavy telescopic self-launching gantry for placement of precast spans delivered on the deck by tyre trolleys. It includes a rear main frame and a front underbridge.

SPMT: Self-propelled modular transporter.

SPMT assembly: Group of SPMT units connected together into an integrated platform that operates under a unified computerised control system for steering and levelling.

SPMT unit: Top load-carrying platform supported by axle lines with independently steered pairs of wheels and hydraulic jacking systems enabling the platform to be raised, lowered and driven in any direction under computerised control conditions.

Spreader beam: Short beam suspending multiple hangers.

Square truss: Modular truss with rectangular cross-section comprising two or three braced bottom chords and two or three braced top chords connected by diagonals and diaphragms. The depth of the truss is typically 1.5 times the width.

Staggered launching: Launch procedure for twin-girder gantries and MSSs where the advance of launching is different in the two girders, thus creating a net transverse couple at the pier supports due to different support reactions.

Straddle carrier: Wheeled portal crane for handling of precast girders and segments.

Stressing platform: Working area for fabrication and tensioning of prestressing tendons.

Suspension-girder MSS: MSS for balanced cantilever in-place casting where two long casting cells shift along one or two self-launching overhead girders.

Telescopic gantry: Self-launching gantry comprising a rear main girder or twin-girder frame and a front underbridge connected by a hydraulic turntable.

Temporary pier (also called *shoring tower*): Modular structure on a temporary foundation that supports a load.

Tie-down bogie: Adjustable rear bogie that prevents uplift and overturning of a form traveller during launching.

Tower-crossbeam assembly: Support system for overhead gantries and MSSs comprising a support tower and crossbeams that sustain the main girders.

Traversing: Lateral shifting of a gantry or MSS along pier brackets or crossbeams for launching along curves.

Triangular truss: Modular truss with triangular cross-section comprising two braced bottom chords, one top chord and inclined diagonals.

Two-phase casting: Casting technique for box girders where the bottom slab and webs are cast first and the top slab is cast in a second time.

Tunnel form: Collapsible inner form for one-phase casting of box girders.

Tyre trolley: Self-propelled multi-wheel trolley used for on-deck transportation of precast spans. The spans are loaded onto the trolley in the precasting facility and extracted by a span launcher at the leading end of the erection line.

UHPC: Ultra-high performance concrete.

ULS: Ultimate limit state. Limit states related to strength and stability during the design life.

Underbridge: Front support box girder for telescopic gantries, span launchers and portal carriers.

Underslung crane: Overhead travelling crane that is supported on the bottom flanges of the runway beams.

Underslung machine: Bridge construction equipment where precast segments or a casting cell are supported onto a self-launching frame.

W/C: Water-to-cement ratio.

W-frame: Truss assembly of crossbeams and diagonals supported on through girders crossing the pier longitudinally to support underslung MSSs.

Winch trolley: Overhead travelling crane comprising a flat rectangular frame, one or two hoist crabs and longitudinal translation systems.

Bridge Construction Equipment
ISBN 978-0-7277-5808-8

ICE Publishing: All rights reserved
http://dx.doi.org/10.1680/bce.58088.453

Further reading

BSI (2009) BS EN 13670 : 2009. Execution of concrete structures. BSI, London.

International Federation for Structural Concrete (FIB) (1990) CEB-FIP Model Code 1990. FIB, Lausanne, Switzerland.

Standards Association of Australia (SAA) (2007) AS 5100. Bridge Design. SAA, Sydney, Australia.

Bridge Construction Equipment
ISBN 978-0-7277-5808-8

ICE Publishing: All rights reserved
http://dx.doi.org/10.1680/bce.58088.455

Index

Page references in italics refer to figures and tables.